Optimal Lightweight Construction Principles

Federico Maria Ballo · Massimiliano Gobbi ·
Giampiero Mastinu · Giorgio Previati

Optimal Lightweight Construction Principles

Springer

Federico Maria Ballo
Politecnico di Milano
Milan, Italy

Massimiliano Gobbi
Politecnico di Milano
Milan, Italy

Giampiero Mastinu
Politecnico di Milano
Milan, Italy

Giorgio Previati
Politecnico di Milano
Milan, Italy

ISBN 978-3-030-60834-7 ISBN 978-3-030-60835-4 (eBook)
https://doi.org/10.1007/978-3-030-60835-4

This Springer imprint is published by the registered company Springer Nature Switzerland AG
The registered company address is: Gewerbestrasse 11, 6330 Cham, Switzerland

To my love Giorgia and to my family
Federico

To Marta and Riccardo
Massimiliano

To Leda and Elisabetta
Giampiero

To Barbara, Davide and Fabio
Giorgio

Foreword

In the long time of working with students and young practising mechanical engineering designers, I have always marvelled at the mastery our young engineers have for using theories and computation methods to analyse quite complex structural problems—provided someone would first define for them free-body diagrams, basic topologies and cross-sectional geometries. There are many excellent texts on structural analysis and optimisation. This book is unique in its focus on how to gain early design insights grounded on sound theory so that you can make those crucial early design decisions on lightweight structures wisely and elegantly.

The book offers a deep understanding of how to approach competing design decisions in structural system design using Pareto optimality. The book starts with a review of multicriteria optimisation methods and the derivation of Pareto-optimal sets, then proceeds with specific important structures and associated loadings and a presentation of topology optimisation for single- and multi-body structures. A final chapter on automotive vehicle component structural design exemplifies the attribute that characterises the entire book: a great wealth of case studies and examples for every single topic discussed, drawn from the authors' vast experience working closely with industry. The book's examples and case studies alone make it a precious addition to the literature.

In the present era of computational dominance in all matters, we occasionally lose track of the knowledge behind the computations and, often most importantly, of the assumptions made in order for the computational models to be valid. This book is unique also in that the vast majority of the examples can be worked out "by hand", meaning they can be treated with explicit analytic expressions allowing you to follow closely the interactions dictating your design decisions. Once you have gained insights from these 'simple' examples, you can explore more complex problems with greater confidence.

I believe this book will serve well both novice and experienced structural designers, as well as offer ideas on teaching structural design in an academic setting.

Finally, let us recall that lightweight structural design is a core element in any efficient, sustainable mechanical design solution and a major driver in the evolution of a large number of artefacts from automobiles and aeroplanes to medical implements and user-oriented products. This book is a good guide for practising such responsible design thinking.

Ann Arbor, Michigan, USA
July 2020

Panos Y. Papalambros

Preface

The idea of writing a book on lightweight construction was born after the book *Optimal design of complex mechanical systems* was completed in 2006. Actually, the existing books on structural optimisation did not address the apparently simple problem of cross-shape optimisation of a beam. In 1992, Ashby produced a book with an outstanding and fundamental contribution on the topic, we would like to extend that contribution to Multi-Objective Programming, i.e. Optimization Theory.

In our book, we aim to provide to designers some practical hints to start a drawing, an apparently simple action that requires a profound knowledge of mechanics. When a designer drafts the preliminary sketch of a structure, he/she has in mind the basic knowledge of de Saint-Venant equations, defining the cross-shape of beams. Such basic available knowledge is taken from university courses on mechanical engineering design, which provide just information on stress, compliance and stability (buckling). No information is given on how choosing the dimensions of the cross section of a beam in order to obtain the best compromise on mass and compliance, given the admissible stress and buckling as constraints. In practice, university professors of mechanical engineering design provide to students just the basic information to do the job. The proper combination of such information, the only action that satisfies the technical requirements, is missing. In other words, it is like as the ingredients to cook a cake were provided, without the recipe. Our aim is to combine the said basic knowledge available from classical courses of mechanical engineering design with optimal design paradigms, based on Pareto theory.

Pareto theory (Multi-Objective Programming) allows to combine design variables (cross-sectional geometry, material) to find a desired compromise between relevant performance indices, namely mass and compliance. Obviously, Pareto theory takes into account relevant constraints on admissible stress and local stability (buckling). Our aim is to provide *once for all* the sound hints for the design of the cross section of a beam.

Taking into account the mentioned performance indices and constraints, we suggest newly derived paradigms to designers. In other words, in the book, we suggest simple design paradigms that are derived from an in-depth and theoretically sound analysis based on Pareto theory. Classical mechanical engineering design provides the competence to find *a* solution, we propose how to find *the* optimal solution. In the book, a number of examples are considered, e.g. the tube under bending, under torsion or inflated. Other examples are taken from the classic set of basic mechanical engineering design schemes.

The greatest majority of the examples produced in the book are dealt with in an analytical form. This provides general and theoretically sound conclusions.

We anticipate here to the readers that the output of the whole book is very simple. The new, and according to our knowledge, unreferenced, hint for designers is just using all of the room available when drawing the cross section of a beam. A thin-walled structure is to be preferred as well.

Following these suggestions allows designers to obtain structures which are stiff, lightweight and safe. Together with the well-known paradigm of thin-walled structures, we are proposing in the book an additional paradigm: *extended structures*.

Designers are encouraged to exploit all, or much, of the room available for the structure to be placed in the space. Despite the indication is simple, it is not trivial. It has been derived after the comprehensive studies presented in the book and based on Pareto theory. Someone could think that the mentioned suggestion is unimportant, minor or insignificant. This was the case of some scholars, possibly unused to design structures or machines, that were acting as peer reviewers of some of our papers from which the book is composed.

We have bright evidence that the newly derived paradigm is effective. By trial and error, engineers are slowly approaching what we address in a straightforward way in the book.

Comparing the old and new bicycle frames (see Fig. 1), we see that, to increase stiffness and reduce mass, the cross-sectional dimensions of beams have been steadily growing.

Fig. 1 Evolution of the frame of road bicycles. Left: 1978. Right: 2017

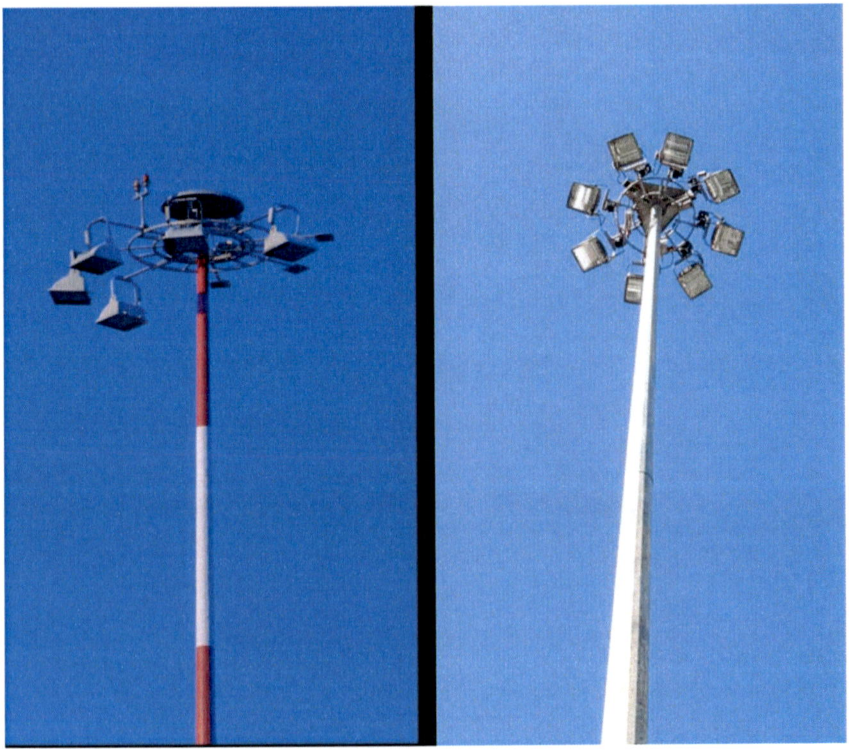

Fig. 2 Evolution of the structure of airport lighting poles. Left: older (*source* "yasin.ylmzz/Shutterstock.com"). Right: newer (*source* "PhotoSmileBeautiful/Shutterstock.com")

Airport lighting poles have undergone in the last decades a considerable expansion of the cross section of the pole structure (see Fig. 2).

A clear trend for truss structures to occupy all of the available room is shown in Fig. 3. The evolution of a car frame is shown, in which some of the beams have been magnified to increase stiffness and reduce mass.

The book addresses not only simple beams but also trusses and topological structural optimisation of continuous structures. This is done just to highlight that the simple paradigms developed for elementary structures can be combined for more complex applications.

Some examples presented in the book are based on the exploitation of Artificial Intelligence/Machine Learning (Global Approximation). Also in these cases, the addressed paradigms of thin-walled and extended structures can be inferred.

To perform the design of a structure, new technologies like Additive Manufacturing require Topological Optimisation. Often, optimised structures take the form of trusses. The paradigms addressed in the book could be used to attempt an interpretation of the involved results coming from Topological Optimisation.

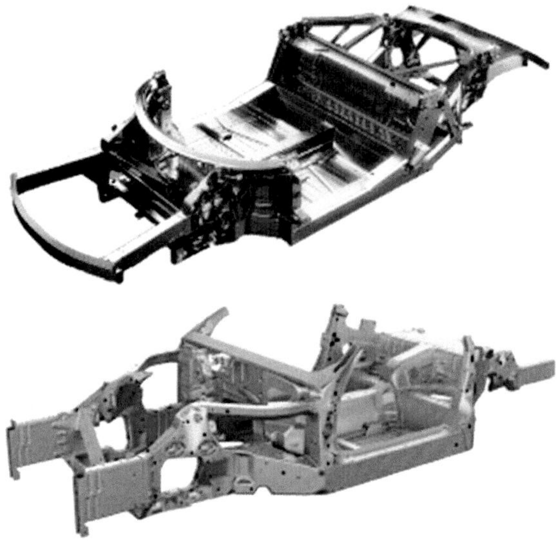

Fig. 3 Evolution of luxury sports car frame (courtesy of Ferrari S.p.A.)

The book does not focus on an important paradigm of lightweight design like Life Cycle Assessment. Nonetheless, the design methods addressed in the book could be applied to the case in which life cycle assessment is considered.

The book does not focus on bio-inspired structures, but it could suggest paradigms and methods to understand how nature has performed an apparent optimal design.

The authors have equally contributed to the drafting of the book. We do hope that our work will enable designers to reach a preliminary optimal dimensioning of structures in a natural way. We are confident that the book provides some lasting contribution for the sake of mechanical engineering design.

Milan, Italy
May 2020

Federico Maria Ballo
Massimiliano Gobbi
Giampiero Mastinu
Giorgio Previati

Acknowledgements The Italian Ministry of Education, University and Research is acknowledged for the support provided through the Project "Department of Excellence LIS4.0—Lightweight and Smart Structures for Industry 4.0".

Contents

Chapter 1
Engineering Design and Optimal Design of Complex Mechanical Systems: Definitions

The following chapter provides the lexicon used throughout the whole book. In the field of optimisation, a number of expressions (locutions) are used that refer to parameters, minimisation, maximisation, etc. The precise meaning of such expressions (locutions) has been derived from [143] and adapted to this book.

1.1 Engineering Design

The objective of engineering design, a major part of research and development (R&D) activity, is to produce drawings, specifications and other relevant information needed to manufacture products that meet customer requirements.

The main task of engineers is to apply their scientific and engineering knowledge to find one (or more) solutions to technical problems, and possibly to optimise those solutions within the requirements and constraints set by material, technological, economical, legal, environmental and human-related considerations. Design problems[1] become concrete tasks after the clarification and definition of the criteria which engineers have to adopt and to apply in order to create new technical products.

1.1.1 Stages of the Design Process

According to [161], there are roughly four stages of the design process (see Fig. 1.1)

1. *Conceptual design* determines the principle of a solution. Conceptual design is that part of the design process in which, by the identification of the essential

[1] A *design problem* (in engineering) may be defined as a set of requirements to solve a technical problem.

F. M. Ballo et al., *Optimal Lightweight Construction Principles*,
https://doi.org/10.1007/978-3-030-60835-4_1

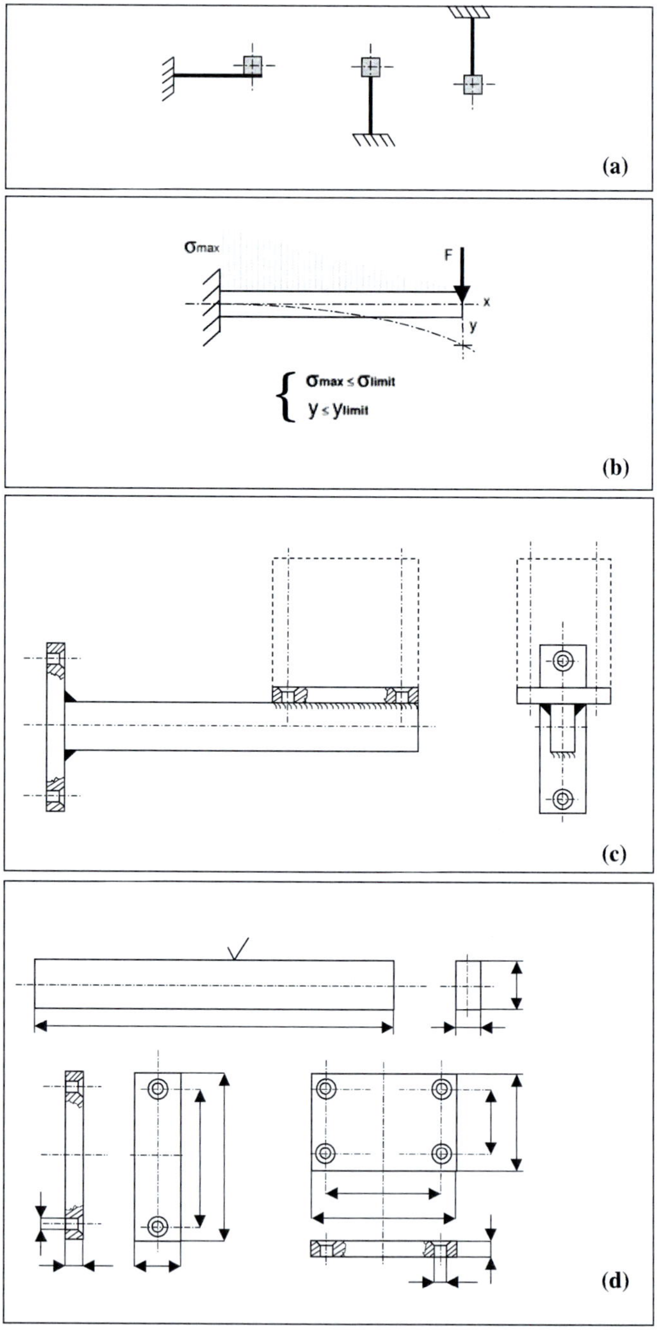

Fig. 1.1 Example of an engineering design process which develops in four stages. Adapted from [143]. First stage **a**: Three different *concept* design solutions to carry a body. Second stage **b**: *Feasibility* study. Third stage **c**: *Embodiment* of the design solution. Fourth stage **d**: *Details* of the design solution

problems, by the establishment of function and by the search for appropriate working principles and their combination, the basic solution is obtained through the elaboration of a solution principle.

2. *Preliminary design*. This stage can be considered as a part of conceptual design. The preliminary layout is obtained by refining the conceptual designs and ranking them according to the design specifications, and choosing the best as the preliminary design.
3. *Embodiment design* is that part of the design process in which, starting from the working structure or concept of a technical product, the design is developed, in accordance with technical and economic criteria and in the light of further information, to the point where subsequent detail design can lead directly to production [161].
4. *Detail design* is that part of the design process which completes the embodiment of technical products with final instructions about the layout, forms, dimensions and surface properties of all individual components, the definitive selection of materials, operating procedures and costs [161].

The above classification of engineering design in four phases is not the unique one,[2] anyway, it is rather general, at least for what concerns complex mechanical systems.

On the more general level, design consists of a loop: product design $\leftrightarrow$ manufacturing $\leftrightarrow$ marketing improvement $\leftrightarrow$ product design [143].

1.1.2 Creativity

Creativity is a very important component of some design phases. According to the level of creativity involved, engineering design can be classified into the following four categories [59] (see also Sect. 1.6):

Creative Design: A priori plan for the solution of the problem does not exist. Design is an abstract decomposition of the problem into a set of levels that represents choices for the components of the problem. The key element in this design type is the transformation from the subconscious to conscious.

Innovative Design: The decomposition of the problem is known, but the alternatives for each of its subparts do not exist and must be synthesised. Design might be an original or unique combination of existing components. It can be argued that a certain amount of creativity comes into play in the innovative design process.

Redesign: An existing design is modified to meet required changes in the original functional requirements.

Routine Design: A priori plan of the solution exists. The subparts and alternatives are known in advance, perhaps as a result of either a creative or innovative

[2] Actually it depends on the product developed.

design process. Routine design involves finding the appropriate alternatives for each subpart that satisfy the given constraints.

At the creative stage, the design is very fuzzy. As it moves to routine design, it gets precise and predetermined.

1.2 Optimal Design of Complex Mechanical Systems

1.2.1 Fundamental Hypothesis

Referring to the classifications introduced in the previous Sects. 1.1.1 and 1.1.2, let us assume that, given a design problem, a *conceptual* design solution has been found by a *creative* design activity.

At this stage of the design process, the *optimal design of complex (Mechanical) systems* can take place.

1.2.2 Single- and Multi-criteria Optimisation

When one thinks of optimisation in general, or about well-known optimisation problems, one usually thinks of problems minimising or maximising a single quantity or objective. Indeed, with many optimisation problems, one begins with the implicit assumption that all candidate solutions can be ranked unambiguously according to their cost or utility.

The goal of the optimisation process is then well-defined: one must find the highest ranked solution(s) possible.

But in real-world applications, problems with a single, well-defined objective to be optimised tend to be the exception rather than the rule. In engineering, (but also in finance, operational research, medicine, design, planning, scheduling, timetabling and many other domains), it is common to have problems with multiple requirements on system performances. Often, problems with a single objective may express what is most important or fundamental about a task in these domains, and they are mathematically simple to be treated, but they are not a faithful model of the real world. Unfortunately, in a problem with multiple objectives, it is generally impossible to obtain a total-ordering (a ranking) of all of the alternative solutions, without invoking further rules or assumptions. This means that "pure" optimisation, in which an unambiguously best solution is sought, may not be possible. This problem of ranking solutions arises whenever we must compare two solutions that offer a different compromise of the different performances. In this situation, the decision of the better solution may become somewhat subjective, or must rely on additional information, such as the "importance" of each performance.

1.2.3 Multi-criteria Optimisation (MCO)

Multi-Criteria Optimisation (MCO) is the discipline that deals with the optimisation of (engineering design) problems in which many conflicting criteria have to be accounted for. Criteria can be expressed either mathematically or not. In any case, the solutions cannot be ranked alone from their evaluation. Thus, in a broad sense, MCO really entails two different tasks, namely, search of a set of solutions and decision-making.

- Search is needed to find solutions,
- Decision-making is needed for ranking them.

This last task or activity of MCO is often called *Multi-Criteria Decision-Making (MCDM)*. MCDM refers to the methods for making choices between solutions that offer a different compromise of criteria. It is a scientific and mathematical discipline in itself, separate from search. MCDM essentially entails methods for scalarising the vector of objective functions (for the definition of "objective function" see Sect. 1.5), so that a total ordering of solutions can be obtained, from which the "best" can be chosen. Scalarising methods in turn involve techniques for equalising the ranges of different criteria, and for mathematically modelling the "preferences" that (human) expert Decision-Makers (DMs) have, considering compromise choices between solutions. For a concise but an extensive overview of methods for performing MCDM, see [152].

Some more problems of conceptual design are described in [41] where MCDM is considered in a framework called Multiple Criteria Decision Aid (MCDA). Managers (i.e. designers acting as DMs) are often not satisfied with the optimal solution found in a bare MCDM process, and they require something more, an effective *aid*. Actually

- Preferences are formed in a learning process;
- Often there is a set of designers with different opinions and preferences;
- An optimal solution is created, not found;
- Imprecision, interaction, flexibility . . . are needed;
- Support for ambiguity handling and uncertainty;
- There is a need to move from a rational to proactive approach;

1.2.4 Multi-objective Optimisation (MOO)

Multi-Objective Optimisation (MOO) is the discipline that deals with the optimisation of (engineering design) problems in which many criteria are expressed mathematically.

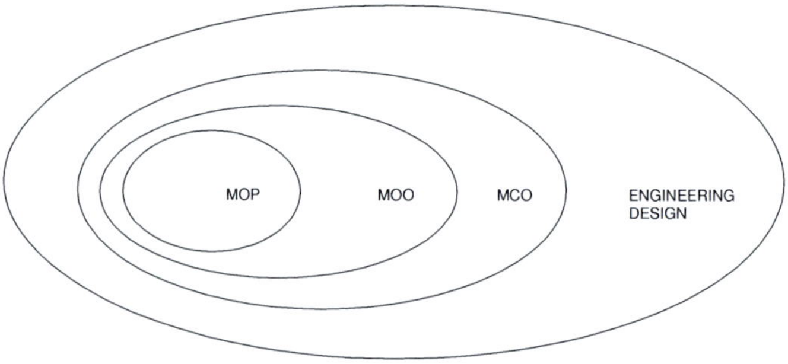

Fig. 1.2 Multi-objective programming (MOP) is a part of multi-objective optimisation (MOO) which is a part of MCO, which can be exploited in the engineering design process. Adapted from [143]

1.2.5 Multi-objective Programming (MOP)

Multi-Objective Programming (MOP) is a theory belonging to MOO (Fig. 1.2). By means of MOP, an engineering design problem (in which the many criteria are expressed mathematically) can be formulated and solved. MOP will be introduced in Sect. 1.10, after the following definitions relating to complex systems, system models, objective functions, design variables, (...) will be given.

1.3 Complex Systems

Scientists and engineers have not yet accepted a common definition of what is meant by complexity. According to the definition given in [143], a *complex system* is taken to be one whose properties are not fully explained by an understanding of its component parts. In a complex system, multiple interactions between many different components exist.

The complexity of a system has also been related to the complexity of the process to define it. For instance, how much effort would be taken to solve a problem. According to designers' point of view, complexity should be defined relative to what they are trying to achieve [143]. Simply, when a designer has to define the values of four (or more) design variables and a preferred compromise has to be found among four (or more) conflicting system performances, then we can say that the designer has to deal with a complex system.

Complexity can be also defined as a measure of uncertainty. Uncertainty arises because of many factors: lack of knowledge about a system, the interaction among its multiple components.

Complex systems are often mathematical structures and processes that involve non-linearity. Complex systems are frequently composed by both many interacting objects and non-linear dynamical sub-systems. The studies on complex systems include a wide variety of topics, such as

- Cellular Automata,
- Chaos,
- Evolutionary Computation,
- Fractals,
- Genetic Algorithms,
- Artificial Neural Networks,
- Parallel Computing.

1.4 System Models

Given a *physical system* (an actual or virtual set of physical objects), we will assume that we will be always able to derive a *mathematical system model* (or, briefly, a *mathematical model*) capable to describe the behaviour of the original physical system.[3]

If the mathematical system model will be derived by applying physical laws, it will be called shortly *physical model*. The mathematical system model can be composed by a set of algebraic or differential equations or a combination of them.

Often, given an actual mechanical system, it is convenient, in order to derive efficiently the corresponding physical model, to construct a *mechanical system model*. In this case, we would have (see Fig. 1.3) the actual mechanical system, the mechanical system model, the mathematical system model (i.e. a physical model of the original actual system).

In a Global Approximation approach, the physical model is substituted by another mathematical model capable to compute the system physical behaviour very quickly [143].

The mathematical system model can be stochastic if some parameters and/or variables in it are defined by one or more stochastic process.

1.5 System Performances, Criteria, Objective Functions

By definition, the physical behaviour of an actual physical system is described by a number of *system performances*.

System performances can be judged by a set of standards or principles strictly related to the design problem formulation. This set is called the set of *criteria*. Given a design problem, all the pertinent aspects must be represented by a complete set of

---

[3]We will consider mostly physical systems, however, all of the theoretical topics dealt with in the book may refer also to other systems (economic, social,...) that can be described by a mathematical model.

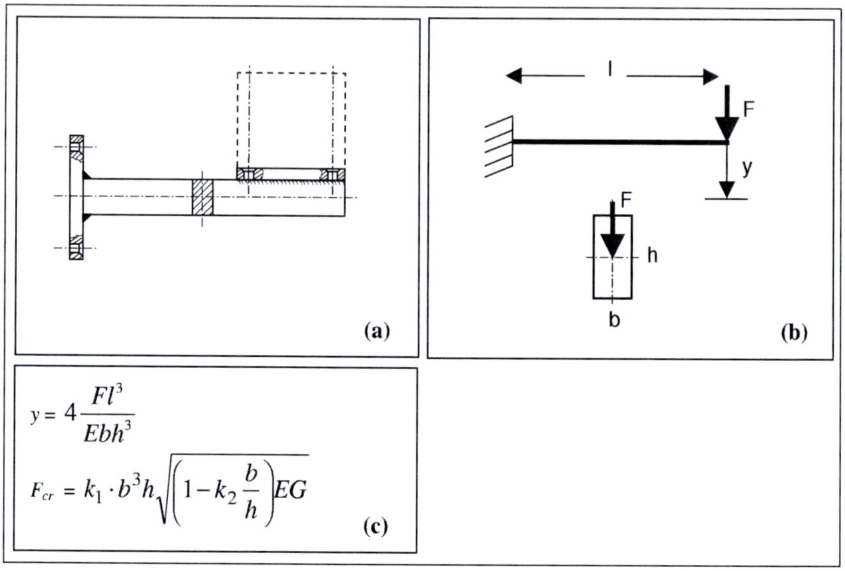

Fig. 1.3 Different models of an actual physical system. Adapted from [143]. **a**: Actual system. **b**: Mechanical system model. **c**: Mathematical system model

criteria. The set of criteria must be minimal, there is no smaller set of criteria capable of representing all the aspects of the problem.

If the system performances can be quantified and expressed in mathematical form, then the mathematical expressions are called *objective functions*. Objective functions are expressed as computable functions of a set of parameters called design variables (see Sect. 1.6). Objective functions are the quantities that the designer wishes to optimise.

Objective functions will be identified as

$$\mathbf{f} = \{f_1(\mathbf{x}), \ldots, f_i(\mathbf{x}), \ldots, f_{n_{of}}(\mathbf{x})\}^T$$
$$\mathbf{x} = \{x_1, \ldots, x_j, \ldots, x_{n_{dv}}\}^T$$

where $\mathbf{x}$ is a vector of design variables (design variables are defined in Sect. 1.6). Objective functions measure the goodness of the system being designed. In almost all applications, it is advisable to re-scale the values of the objective functions, that is, normalise the objective functions

$$f_{i\,norm} = \frac{f_i - f_{i\ \min}}{f_{i\ \max} - f_{i\ \min}} \quad i = 1, \ldots, n_{of} \tag{1.1}$$

With this transformation the range of each new objective function is [0, 1]. The normalised objective functions used in calculations need to be restored to the original scale for analysing and displaying.

The need to take into account of subjective as well as objective aspects of the design is a complicating issue. For example, the handling of a road vehicle is easily described in terms of linguistic descriptors rather than numerical values.

A common method to deal with subjective data is to convert it to objective values using a numerical scale.

Hierarchical decomposition can help [149, 150]. Subjective evaluations can be handled by decomposing a complex quantitative attribute into lower level sub-attributes to make assessments more meaningful and manageable for the designer.

1.6 System Parameters, Design Variables

In general, a mathematical system model embodies a set of parameters. Usually, designers are interested in finding one (or more) set of parameters that define one (or more) preferred configurations of the mathematical system. During the optimisation procedure, the said set of parameters is changed to find their preferred values. So, during the optimisation process, the mathematical system model parameters do vary. As parameters, by definition, do not vary, it is advisable to define these 'variable parameters' as *design variables*. Design variables are the variables that pertain to the mathematical system model during the optimisation process. Design variables can be grouped into a vector and will be identified as

$$\mathbf{x} = \{x_1, \ldots, x_i, \ldots, x_{n_{dv}}\}^T \tag{1.2}$$

Sometimes, the designer can choose to hold some design variable fixed in order to simplify the optimisation problem. These fixed quantities will be re-classified (obviously) as parameters.

The selection of the design variables is not unique, but it is very important to select these variables to be independent of each other.

Depending on how design variables are involved in the design process, the design can be reputed to be more or less creative. Referring to Sect. 1.1.2 the following classification of creativity in design is given [82]

- if the number of the design variables and the ranges of values they can take remain fixed during design processing, then the process is routine design;
- if the number of the design variables remains fixed but their ranges change, then it is innovative design;
- if the number of design variables changes too, then it is creative design.

1.7 Constraints

Constraints are conditions which must occur for the design, in order to function as intended. When constraints can be expressed in mathematical form, they take the form of inequalities and/or equalities. Constraints will be identified as

$$\mathbf{g} = \{g_1(\mathbf{x}), \ldots, g_i(\mathbf{x}), \ldots, g_{n_c}(\mathbf{x})\}^T$$
$$g_i(\mathbf{x}) \leq 0 \quad (i = 1, \ldots, n_c)$$

where $\mathbf{x}$ is the vector of design variables.[4] Constraints on design variables are generally direct limitations on the variables themselves (side constraints, called lower bounds and upper bounds)

$$x_{i_{low}} \leq x_i \leq x_{i_{up}} \quad (i = 1, \ldots, n_{dv}) \tag{1.3}$$

Constraints on system model behaviour are limits on the system output (objective function values). Sometimes these limits are given by the physical laws governing the system.

1.8 Space of Design Variables, Space of Objective Functions

We have already defined the design variables vector $\mathbf{x}$ and the objective functions vector $\mathbf{f}(\mathbf{x})$, we have to consider two spaces: the n_{dv}-dimensional space of the design variables ($\mathcal{X}$) and the n_{of}-dimensional space of the objective functions (C).

1.9 Feasible Design Variables Domain, Design Solution

The whole set of constraints define the boundaries of the *feasible design variables domain*. Obviously, the feasible design variables domain is contained in the space of design variables

$$\mathcal{F} \subset \mathcal{X} \tag{1.4}$$

Often it is not easy to find $\mathcal{F}$. A *design solution* is represented by a point in the feasible design variable domain.

[4]This is a general definition of constraints. If a constraint should be expressed by $g_i(\mathbf{x}) \geq 0$ setting $g_i'(\mathbf{x}) = -g_i(\mathbf{x}) \leq 0$ would restore the inequality $\leq$. Also, an equality constraint $g_i = 0$ can be expressed by using two additional inequality constraints $g_i'(\mathbf{x}) = g(\mathbf{x}) \leq 0$ and $g_i''(\mathbf{x}) = -g(\mathbf{x}) \leq 0$.

1.9.1 Conflict

Given a design problem, a number of criteria are formulated. As a rule in MCO *conflicting criteria* are managed. A "Criterium" is a standard or a principle that is used to judge something. Criteria are considered conflicting when the satisfaction of one criterium does not imply the satisfaction of the others.

1.10 Multi-objective Programming (MOP)

1.10.1 Non-linear Programming (NLP) and Constrained Minimisation

Before treating MOP problems some relevant definitions for scalar optimisation will be introduced.

The aim of a scalar optimisation is to select the values of a vector of design variables $\mathbf{x}$ subject to the effect of some constraints $\mathbf{g}(\mathbf{x})$[5] in such a way that one single-objective function $f(\mathbf{x})$ is minimised. In mathematical form, a scalar optimisation can be written as

$$\begin{aligned}
&\min_{\mathbf{x}\in\Re^{n_{dv}}} \; f(\mathbf{x}) \\
&\text{subject to :} \\
&g_i(\mathbf{x}) \leq 0, \quad i = 1, \ldots, n_c \\
&\mathbf{x}_{low} \leq \mathbf{x} \leq \mathbf{x}_{up} \quad (\mathbf{x} \in \mathcal{X}[\mathbf{x}_{low}, \mathbf{x}_{up}])
\end{aligned} \tag{1.5}$$

The subset $\mathcal{X} \subset \Re^{n_{dv}}$ is the design variable space of definition.

Let us notice that if an objective function $f(x)$ should be maximised (instead of minimised), the dual function $f'(x) = -f(x) \leq 0$ could be introduced and used.

Some important basic definitions, useful for the following explanation, are

Definition 1.1 (*Global minimum*) A point $\mathbf{x}^*$ (Fig. 1.4) is a global minimum if $f(\mathbf{x}^*) \leq f(\mathbf{x})\ \forall \mathbf{x} \in \mathcal{X}$.

Definition 1.2 (*Local minimum*) A point $\bar{\mathbf{x}}$ (Fig. 1.4) is called local minimum if a $\delta > 0$ exists such that $f(\bar{\mathbf{x}}) \leq f(\mathbf{x})\ \forall \mathbf{x} \in \mathcal{B}(\delta, \bar{\mathbf{x}}) \cap \mathcal{X}$, where $\mathcal{B}(\delta, \bar{\mathbf{x}}) = \{\mathbf{x} \in \mathcal{X} \mid \|\bar{\mathbf{x}} - \mathbf{x}\| < \delta\}$.

Definition 1.3 (*Convexity*)

[5] Considering an equality constraint $g_i(\mathbf{x}) = 0$ is equivalent to consider two inequality constraints $g_i(\mathbf{x}) \leq 0$ and $-g_i(\mathbf{x}) \leq 0$, so considering problems having only inequality constraints, equality constraints are implicitly considered.

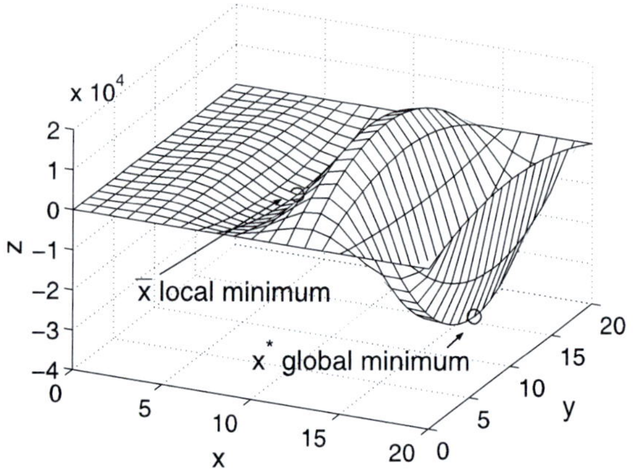

Fig. 1.4 Local and global minimum of the function: $z = 10(x^2 \sin(x)(-(y - 10)^2 + 100))$, $0 < [x; y] < 20$. Adapted from [143]

(i) A subset $\mathcal{X} \in \Re^{n_{dv}}$ is convex if

$$\alpha \mathbf{x}_1 + (1 - \alpha)\mathbf{x}_2 \in \mathcal{X} \quad 0 \leq \alpha \leq 1 \quad \forall \mathbf{x}_1, \mathbf{x}_2 \in \mathcal{X}$$

(ii) A real-valued function $f(\mathbf{x})$ is convex on the convex subset $\mathcal{X}$ if

$$f(\alpha \mathbf{x}_1 + (1 - \alpha)\mathbf{x}_2) \leq \alpha f(\mathbf{x}_1) + (1 - \alpha) f(\mathbf{x}_2) \quad 0 \leq \alpha \leq 1 \quad \forall \mathbf{x}_1, \mathbf{x}_2 \in \mathcal{X}$$

The *optimality conditions* are the conditions that can help in finding the minimum of a function $f(\mathbf{x})$. The optimality conditions are important because they can be used to develop numerical methods (see [152]) and to check the optimality of a given point (i.e. to check if that point is a minimum for the given function $f(\mathbf{x})$). Referring to the constrained problem (1.5) the optimality conditions can be given by the *Karush–Kuhn–Tucker* (KKT) *conditions*. (The optimality conditions can be expressed in several equivalent ways [152]).

A basic assumption for deriving the KKT conditions is that the minimum point is a regular point of the feasible set (see Sect. 1.9).

Definition 1.4 (*Regular point*) A point $\mathbf{x}$ is called a *regular point* if the gradients of all the active[6] constraints at $\mathbf{x}$ are linearly independent.

It is useful, before introducing the KKT conditions, to define the Lagrangian function, which for the problem (1.5) is defined as

$$L(\mathbf{x}, \boldsymbol{\eta}) = f(\mathbf{x}) + (\boldsymbol{\eta}^T \mathbf{g}(\mathbf{x})) \tag{1.6}$$

The Karush–Kuhn–Tucker (KKT) necessary conditions (first-order necessary conditions) are

[6] An inequality constraint $g_i(\mathbf{x}) \leq 0$ is active in $\mathbf{x}$ if $g_i(\mathbf{x}) = 0$.

Theorem 1.1 (Karush–Kuhn–Tucker (KKT) first-order necessary condition) *Let the objective function $f(\mathbf{x})$ and the constraint $g(\mathbf{x})$ functions of problem (1.5) be continuously differentiable at a vector $\mathbf{x}^* \in \mathcal{X}$. A necessary condition for $\mathbf{x}^*$ to be a local or a global minimum of $f(\mathbf{x})$ is*

$$\begin{aligned} &\nabla f(\mathbf{x}^*) + \sum_{i=1}^{n_c} \eta_i^* \nabla g_i(\mathbf{x}^*) = 0 \\ &\eta_i^* g_i(\mathbf{x}^*) = 0 \quad i = 1, \ldots, n_c \\ &\eta_i^* \geq 0 \quad i = 1, \ldots, n_c \end{aligned} \tag{1.7}$$

where η_i $(i = 1, \ldots, n_c)$ are the Lagrange multipliers (1.6).

The proof of the theorem can be found in [152]. The foregoing conditions are optimality necessary conditions for a constrained problem having regular functions.

If the objective function $f(\mathbf{x})$ and the constraint $g(\mathbf{x})$ functions are convex, then a local minimum is also a global minimum and the KKT first-order conditions are necessary as well as sufficient [152].

The second-order necessary conditions follow.

Theorem 1.2 (Second-order necessary condition) *Let $\mathbf{x}^*$ satisfy the first-order KKT conditions (1.7) for the problem (1.5). Let $H(\mathbf{x}^*)$ be the Hessian of the Lagrangian (1.6). Let $\mathbf{d}$ be a small non-zero feasible[7] change from $\mathbf{x}^*$. For $j = 1, 2, \ldots, n_c$ and for all j with $g_j = 0$, we have*

$$[\nabla g_j(\mathbf{x}^*)^T \mathbf{d}] = 0 \tag{1.8}$$

Defining $Q = [\mathbf{d}^T H(\mathbf{x}^*)\mathbf{d}]$, if $\mathbf{x}^*$ is a local minimum point, it must be true that

$$Q \geq 0 \tag{1.9}$$

for all $\mathbf{d}$ satisfying Eq. (1.8).

The proof of the theorem can be found in [152]. The inequality (1.9) is a necessary condition, a point that violates it cannot be a minimum point.

If in Eq. (1.9) only the inequality is valid, then $\mathbf{x}^*$ is an isolated local minimum.

It must be stressed the all the gradient-based optimisation methods [152] give only local minima unless some additional requirements are fulfilled such as convexity.

If the objective function $f(\mathbf{x})$ and the constraint $g(\mathbf{x})$ functions are convex, then the second-order conditions are necessary as well as sufficient [152].

[7] A small change $\mathbf{d}$ from $\mathbf{x}^*$ is called feasible if for any small $\mathbf{d}$, all the active constraints remain active.

1.10.2 Multi-objective Programming: Definition

The concepts introduced in the previous Sect. 1.10.1 for scalar constrained optimisation will be extended here to vector (constrained) optimisation problems. A vector constrained optimisation problem can be formulated as a *Multi-Objective Programming* (MOP) problem as follows. Find a vector of design variables (1.6) which satisfy a vector of constraints (1.7) and minimise a vector of objective functions (1.5). In mathematical form, given a design variable vector $\mathbf{x} \in \mathcal{X}$ which satisfies all the constraints $\mathbf{g(x)}$ and minimises the n_{of} components of the objective function vector $\mathbf{f(x)}$, the MOP problem can be written as[8]

$$\min_{\mathbf{x}\in\Re^{n_{dv}}} \mathbf{f}(\mathbf{x}) \quad \text{subject to}$$

$$\begin{aligned} &g_i(\mathbf{x}) \leq 0, \quad i = 1, \ldots, n_c \\ &\mathbf{x} \in \mathcal{X} \end{aligned} \tag{1.10}$$

where $\mathbf{f}(\mathbf{x}) = (f_1(\mathbf{x}), \ldots, f_i(\mathbf{x}), \ldots . f_k(\mathbf{x}))$ is the objective function vector, $\mathbf{x}$ the design variable vector, $\mathcal{X}$ the domain of definition of the variables (generally given as lower bound $\mathbf{x}_{low}$ and upper bound $\mathbf{x}_{up}$, $\mathbf{x}_{low} \leq \mathbf{x} \leq \mathbf{x}_{up}$), $\mathbf{g(x)}$ a vector of constraints. "Minimise" means that we want to minimise all the objective functions *concurrently*. Obviously, we do not consider the trivial case when there is no conflict between objective functions (see Sect. 1.9.1). This means that the selected objective functions must be at least partially conflicting. In general, in MOP, it is not possible to find a single solution that is optimal for all the objective functions simultaneously.

The main target of MOP is to provide a rational approach to find optimal design solutions in the presence of conflicting objective functions.

1.10.3 Pareto-Optimal Solutions and Pareto-Optimal Set

In a MOP problem, the concept of optimal solution (obvious for the scalar formulation) in not straightforward, it requires the special definition of *Pareto-optimal* solution.[9]

---

[8]Again, we will assume the vector function $\mathbf{f}(\mathbf{x})$ to be *minimised*. If $\mathbf{f}(\mathbf{x})$ were to be maximised, then one could retain the formulation presented in the following and consider $-\mathbf{f}(\mathbf{x})$.

[9]The correct name Edgeworth–Pareto-optimal is not practically used in literature [214]. Francis Ysidro Edgeworth and Vilfredo Federico Damaso Pareto were the two economists who developed the theory of MCO. Edgeworth [64], first in 1881, gave the definition of 'optimal solution' for a problem with two objective functions. Some years later, in 1906 Vilfredo Pareto [165] developed the definitions of Pareto-optimal solution for the n-dimensional case.

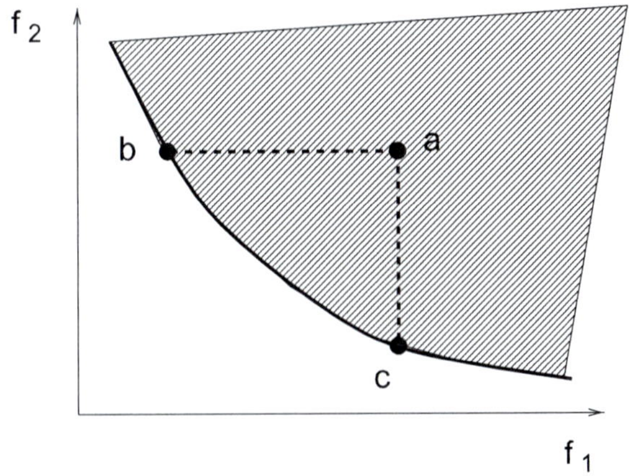

Fig. 1.5 Pareto-optimal set into the criterion space f_1, f_2. Adapted from [143]

Definition 1.5 (*Pareto-optimal solution*) Given a MOP problem with n_{dv} design variables and n_{of} objective functions, the Pareto-optimal (i.e. Non-dominated, Efficient or Non-inferior) (vector) solution $\mathbf{x}_i$ satisfies the following conditions $\nexists \mathbf{x}_j$:

$$\begin{cases} f_n(\mathbf{x}_j) \leq f_n(\mathbf{x}_i) \quad n = 1, 2, \ldots, n_{of} \\ \exists l : \; f_l(\mathbf{x}_j) < f_l(\mathbf{x}_i) \end{cases} \tag{1.11}$$

If the above conditions hold, Pareto-optimal solutions will be denoted by $\mathbf{x}_i = \mathbf{x_i}^*$. Notice that the number of Pareto-optimal solutions approaches infinity. The whole set of Pareto-optimal solutions constitute the *Pareto-optimal set*. Given a solution $\mathbf{x}_i^*$, if one tries to change it to improve one objective function, at least another objective function is worsened. For all non-Pareto-optimal solutions, the value of at least one objective function f_l can be reduced without increasing the values of the other components. A typical characteristic of a MOP is the absence of a unique point that would optimise all objective functions simultaneously.

Let us assume that a MOP problem with two objective functions and n_{dv} design variables has been formulated and solved, i.e. the Pareto-optimal set has been computed. Figure 1.5 refers to the two objective function space. The dashed area is the mapping of the design variable space $\mathcal{X}$ into the objective function space (see Sect. 1.8), the Pareto-optimal solutions lie on the curve $b - c$. The solution a is not obviously Pareto-optimal, all the solutions inside the area abc can reduce both the objective functions f_1 and f_2.

Any point in the Pareto-optimal set can be selected as an optimum solution. There is no generally accepted standard to judge (i.e to rank) the solutions.

The designer has to choose a preferred solution among those (and only those) belonging to the Pareto-optimal set. It is advantageous to find a number of Pareto-optimal solutions. The designer acts as a decision-maker (DM) and expresses preferences between two (or more) different Pareto-optimal solutions. The optimisation procedure generates information (Pareto-optimal solutions) for the designer who has

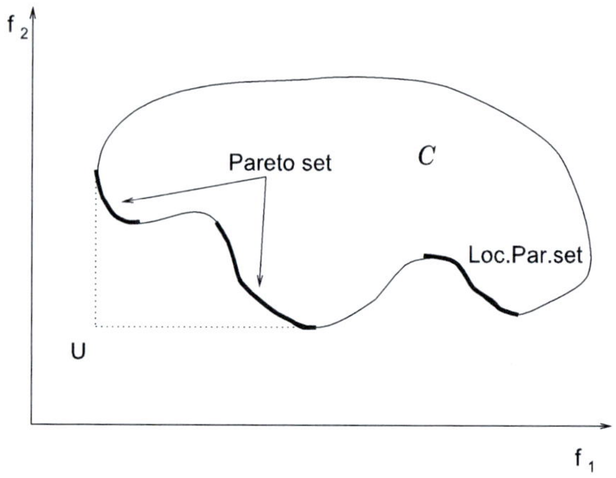

Fig. 1.6 Local and (global) Pareto-optimal sets in a two design variable space f_1, f_2. Adapted from [143]

to select a final solution. The designer will be interested in Pareto-optimal solutions only and the remaining solutions will be excluded. However, this cannot be the case if the problem has not been well formulated.

According to the Definition 1.5, selecting Pareto-optimal solutions requires a compromise or a trade-off. In Principle, it is not always necessary to resort to a trade-off to improve the design process.

The definition 1.5 introduces *global Pareto-optimality*. The concept of local Pareto-optimality is shown in Fig. 1.6. A solution $\mathbf{x}_i$ is locally Pareto-optimal if there exists $\delta > 0$ such that $\mathbf{x}_i$ is Pareto-optimal in the circle centred in $\mathbf{x}_i$ with radius δ. Any global Pareto-optimal solution is locally Pareto-optimal. The converse is valid for convex multi-objective problems.

The Pareto-optimal set always belongs to the boundary of the mapping of the design space into the space of objective functions ($C \in \Re^{n_{of}}$). There is an infinity of Pareto-optimal solutions. It is easy to guarantee their existence [152] since all that is required is that the mapping of the design variable space into the objective function space ($C \in \Re^{n_{of}}$) is lower bounded and closed. The Pareto-optimal set may be connected or have gaps (see Fig. 1.6 referring to a two-dimensional criterion space). We can guarantee that the Pareto-optimal set is connected when C is lower convex and closed [31, 66, 241].

The necessary conditions for determining Pareto-optimal solutions are summarised in the following theorem (necessary conditions)

Theorem 1.3 (Pareto-optimal necessary conditions) *Let us consider the problem (1.10) if* $\mathbf{x}^* \in \mathcal{X}$ *is a regular local Pareto-optimal solution for the vector function* $\mathbf{f}(\mathbf{x})$, *then there exist multipliers* $\boldsymbol{\lambda} = (\lambda_1 \lambda_2 \cdots \lambda_{n_{of}})^T$ *and* $\boldsymbol{\eta} = (\eta_1 \eta_2 \cdots \eta_{n_c})^T$ *such that*

$$\sum_{k=1}^{n_{dv}} \lambda_k \nabla f_k(\mathbf{x}^*) + \sum_{i=1}^{n_c} \eta_i^* \nabla g_i(\mathbf{x}^*) = 0$$
$$\eta_i^* g_i(\mathbf{x}^*) = 0 \quad i = 1, \ldots, n_c$$
$$\eta_i^* \geq 0 \quad i = n_{ec} + 1, \ldots, n_c$$
$$\boldsymbol{\lambda} \geq 0 \tag{1.12}$$

The proof of the theorem can be found in [152]. The Lagrangian function is in this case

$$L(\mathbf{x}, \boldsymbol{\lambda}, \boldsymbol{\eta}) = \boldsymbol{\lambda}^T\, \mathbf{f}(\mathbf{x}) + \boldsymbol{\eta}^T\, \mathbf{g}(\mathbf{x}) \tag{1.13}$$

If we consider $\boldsymbol{\lambda}^T\, \mathbf{f}(\mathbf{x})$ where $\boldsymbol{\lambda} \geq \mathbf{0}$ instead of the single-objective function $f(\mathbf{x})$, the Pareto-optimal necessary conditions are the same as the KKT necessary conditions for the general non-linear programming problems (1.12).

The idea of minimising one objective function subject to constraints on the other objective functions has been formulated to produce a necessary and sufficient condition for Pareto-optimal solutions by Schmitendorf [197].

Theorem 1.4 (Pareto-optimal Theorem) *A point $\mathbf{y}^* \in C$ is Pareto-optimal if and only if for each $j \in \{1, \ldots, n_{of}\}$*

$$y_j^* \leq y_j \;\forall \mathbf{y} \in C_j \tag{1.14}$$

where $C_j = \{\mathbf{y} \in C \,:\, y_i \leq y_i^*\; i = 1, \ldots, n_{of};\; i \neq j\}$.

The proof of the theorem can be found in [152]. The theorem is illustrated in Fig. 1.7. The point with components y_1^* and y_2^* satisfies $y_1^* \leq y_1 \in C_1$ and $y_2^* \leq y_2 \in C_2$.

As well as in the case of scalar optimisation (Sect. 1.10.1) second-order Pareto-optimality conditions can be expressed

Theorem 1.5 (Second-order necessary condition) *Let the objective functions of the problem (1.10) be twice continuously differentiable at a design variable vector $\mathbf{x}^* \in X$. Let be in $\mathbf{x}^*$, the active constraints linearly independent. A necessary condition for $\mathbf{x}^*$ to be Pareto-optimal is that vectors exist such that $0 \leq \boldsymbol{\lambda} \in \Re^{n_{of}}$ ($\boldsymbol{\lambda} \neq 0$) and $\boldsymbol{\eta} \leq\in \Re^{n_c}$ such that the first-order Pareto-optimal necessary condition in $\mathbf{x}^*$ is valid and, considering $H(\mathbf{x}^*)$ the hessian of the Lagrange function (1.13), the following inequality is also valid*

$$\mathbf{d}^T H(\mathbf{x}^*)\mathbf{d} \geq 0$$

for all $\mathbf{d} \in \{\mathbf{0} \neq \mathbf{d} \in \Re^{n_{dv}} : \nabla f_i(\mathbf{x}^*)^T \leq 0$ for all $i = 1, \ldots, n_{of}$, $\nabla g_i(\mathbf{x}^*)^T \mathbf{d} = 0$ for all i where g_i is an active constraint $\}$.

The proof of the theorem can be found in [152].

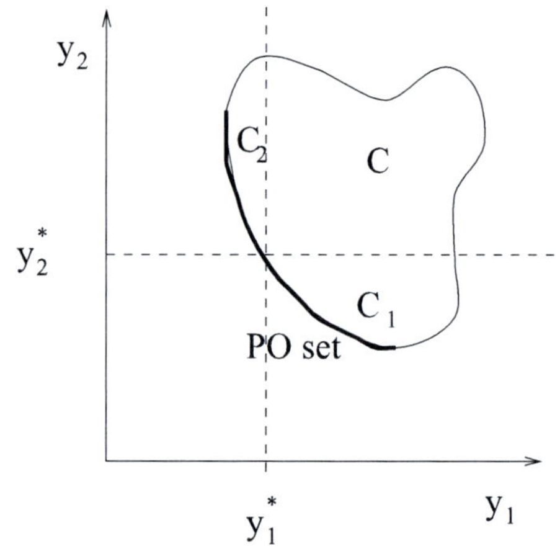

Fig. 1.7 Schmitendorf's theorem: two objective function space f_1, f_2. Adapted from [143]

1.10.4 Ideal and Nadir Design Solutions

The *ideal design variable vector* can be obtained by minimising each of the objective functions f_i individually, subject to the constraints. The ideal design variable vector is represented by a point (called also "utopia point" [152]) in the objective function space (see Fig. 1.7). The ideal design variable vector corresponds generally to a not feasible design solution. It is only a reference design variable vector that constitutes the lower bounds of the Pareto-optimal set for each objective function.

An important reference square matrix is the one whose rows are, respectively, composed by objective function vectors which are obtained by minimising one single objective function at a time. The first row contains the objective function vector in which $f_1(\mathbf{x})$ is minimised. The second row contains the objective function vector in which $f_2(\mathbf{x})$ is minimised and so on for the n_{of}.

$$\begin{pmatrix} f_{1\ \min}(\mathbf{x}_{f_{1\ \min}}) & f_2(\mathbf{x}_{f_{1\ \min}}) & \cdots & f_{n_{of}}(\mathbf{x}_{f_{1\ \min}}) \\ f_1(\mathbf{x}_{f_{2\ \min}}) & f_{2\ \min}(\mathbf{x}_{f_{2\ \min}}) & \cdots & f_{n_{of}}(\mathbf{x}_{f_{2\ \min}}) \\ \vdots & \vdots & \ddots & \vdots \\ f_1(\mathbf{x}_{f_{n_{of}\ \min}}) & f_2(\mathbf{x}_{f_{n_{of}\ mim}}) & \cdots & f_{n_{of}\ \min}(\mathbf{x}_{f_{n_{of}\ min}}) \end{pmatrix} \tag{1.15}$$

Obviously, the ideal objective vector is the main diagonal of the matrix (1.15). The maximum value of each column i of the matrix is a rough estimate of the upper bound of the objective function f_i over the Pareto-optimal set. The upper bounds of the Pareto-optimal set constitute the *nadir design variable vector*.

The row vectors in the matrix (1.15) are Pareto-optimal if they are unique.

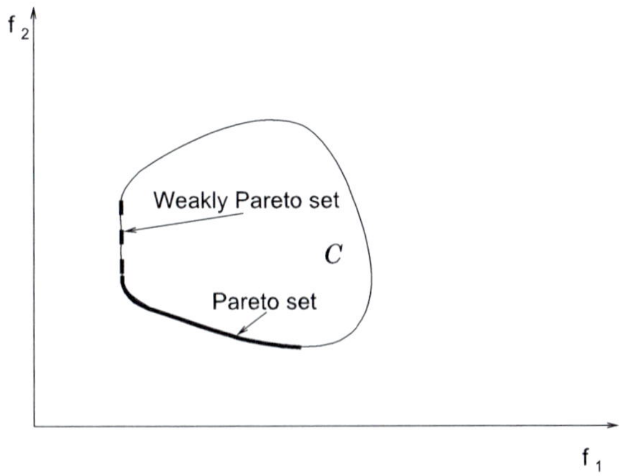

Fig. 1.8 Weak Pareto-optimal set in a two objective function space f_1, f_2. Adapted from [143]

1.10.5 Related Concepts

1.10.5.1 Weak Pareto-optimality

Weak Pareto-optimality: a vector is *weakly Pareto-optimal* if there does not exist any other vector for which all the components are better (Fig. 1.8).

The Pareto-optimal set is a subset of the weak Pareto-optimal set.

1.10.5.2 Proper Pareto-optimality

Let $\mathbf{f}(\mathbf{x}_1)$ and $\mathbf{f}(\mathbf{x}_2)$ be two objective function vectors evaluated, respectively, at two given design variable vectors $\mathbf{x}_1$ and $\mathbf{x}_2$. We denote as *trade-off* the ratio

$$\Lambda_{ij}(\mathbf{x}_1, \mathbf{x}_2) = \frac{f_i(\mathbf{x}_1) - f_i(\mathbf{x}_2)}{f_j(\mathbf{x}_1) - f_j(\mathbf{x}_2)} \tag{1.16}$$

If both $\mathbf{x}_1$ and $\mathbf{x}_2$ belong to the Pareto-optimal set, the *trade-off* is the ratio between the increment of one objective function $\mathbf{f}_i(\mathbf{x})$ and the decrease of other objective functions $\mathbf{f}_j(\mathbf{x})$, $j = 1, \ldots, n_{of} \quad i \neq j$.

Proper Pareto-optimal solutions are defined with reference to trade-off. A Pareto-optimal solution with very high or very low trade-offs is not properly Pareto-optimal.

1.10.6 Basic Problems and Capabilities of Multi-objective Optimisation

Some of the basic problems of MOO (and thus of MOP) are briefly explained below.

- There are objective functions and there are constraints. The difference between them can be fuzzy and some of them will move from objective functions to constraints or vice versa. Some constraints are hard, some not; some will change or disappear, while others may be introduced as the problem knowledge base expands.
- In many cases, the design variable ranges are also fuzzy and flexible and there is a requirement for exploration outside of the default regions. The reason is that the real bounds and limits are not always known from the beginning.
- The solution of the design should contain both optimal solutions and suggestions for extending ranges and/or inclusion/removals of constraints.
- A set of results is required which the engineer can analyse. (That means that the engineer should be able to input those results to some other programs or to consult some database or persons for different aspects of given solutions).
- The designer should not be confused by the number of parameters and by the possibilities optimisation offers, cognitive overload must be avoided.

The problems of conceptual design relate to the fuzzy nature of initial design concepts. Computers should be able to help in exploration of those variants while suggesting some others as well.

An important capability that MOO offers is that of comparing different concept design solutions. Often engineers are in trouble if a well-defined comparison between two systems is requested. Actually, they sometimes fall into the fatal mistake to consider—and compare—one non-optimal system with another which is optimal. The result of the comparison is thus influenced by the non-optimal choice of one system. The conclusion can be favourable to the optimal system, just because the first non-optimal system was considered. MOO allows to take into account *only* optimal systems, thus the comparison is correctly 'restricted' to optimal systems only. This allows to actually choose the best solution after a comprehensive optimal design process.[10]

---

[10]For example, if two concept designs of respectively two cars (one with front wheel drive transmission and the other with rear wheel drive transmission) should be compared, we should take into account *only* the Pareto-optimal design solutions representative of the two different concept designs.

Chapter 2
Introduction to the Optimal Design of Complex Mechanical Systems

The following chapter introduces, with an example, the major topics of the optimal design of complex mechanical systems. An application to structural design is dealt with. The example has been derived from [143] and adapted to this book.

When a designer is able to simulate the physical behaviour of a system by means of a validated mathematical model, the subsequent task is that of defining the system model parameters (also called design variables see Chap. 1) in order to obtain the desired system performances (also called objective functions see Chap. 1).

Often such performances are conflicting, i.e. improving one implies the worsening of another, so a compromise has to be reached.

When the system model parameters are more than four or five and the system's objective functions are more than five or six, both the definition of parameters and the balancing of conflicting performances may become cumbersome or even impossible unless a special approach is adopted.

This chapter will deal with the above-stated problem, i.e. the definition of system model parameters of complex systems, when conflicting performances have to be balanced. The design of complex systems will be accomplished by exploiting Multi-Objective Programming (MOP), an optimisation theory pertaining to Operational Research (OR).

2.1 On the Optimal Design of Complex Systems

In order to introduce to the reader what is meant exactly by optimisation of a complex system, let us resort to an example.

F. M. Ballo et al., *Optimal Lightweight Construction Principles*,
https://doi.org/10.1007/978-3-030-60835-4_2

Let us imagine that a cantilever has to be optimally designed.[1] The example actually deals with a simple system, however, in this apparently simple problem, the main features pertaining to the optimisation of complex systems are present.

The cantilever, shown in Fig. 2.1, has a rectangular cross section and a force acts at the free end. Let us assume that the optimisation problem to be solved is

- *finding the values of the design variables (length b and length h) defining the cantilever cross section*[2]

in order to

- *minimise the cantilever mass*
- *minimise the cantilever deflection at its free end,*

subject to the following conditions

- *the maximum stress at the fixed end must be little than (or equal to) the admissible stress*
- *elastic stability must be guaranteed (i.e. buckling must be avoided).*

A designer should choose the values defining the cross section of the cantilever in order to get it as light and stiff as possible, avoiding both a failure (due to too high stress) and elastic instability.

In mathematical form, the above optimisation problem may be stated as follows.

■ *Given*

l	the cantilever length	[m]
b	the beam cross section width	[m]
h	the beam cross section height	[m]
J	$= \frac{1}{12}bh^3$ the flexural moment of inertia of the section (n–n axis)	[m^4]
F	the force applied at the cantilever free end (see Fig. 2.1)	[N]
σ_E	the material yield stress	[MPa]
η	safety coefficient (≥ 1)	[–]
σ_{adm}	$= \sigma_E/\eta$ the admissible stress at the cantilever fixed end	[MPa]
E	the material modulus of elasticity (Young's modulus)	[MPa]
G	the material modulus of tangential elasticity	[MPa]
ρ	the material density	[kg/m^3]

[1] As an effective *design process* should always produce an optimal solution, *optimisation* and *design* are reputed to be the *same process* by the Authors.

[2] b and h may vary, respectively, within two well-defined ranges.

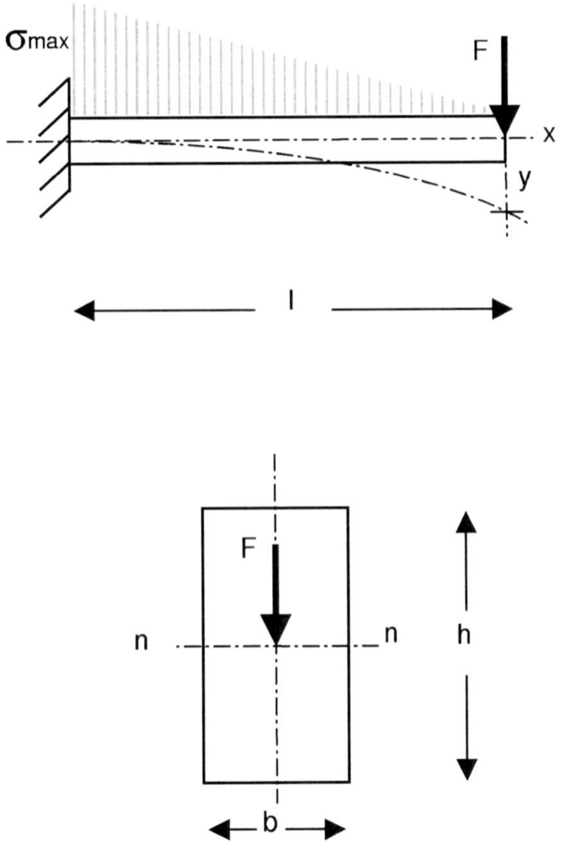

Fig. 2.1 Cantilever whose rectangular cross section is to be defined in order to minimise both the cantilever mass and the cantilever deflection at the free end

- *and defining*

$m = \rho bhl$	the cantilever mass	[kg]
$y = \frac{1}{3}\frac{Fl^3}{EJ} = 4\frac{Fl^3}{Ebh^3}$	the deflection at the free end due to load F	[m]
$\sigma_{max} = 6\frac{Fl}{bh^2}$	the maximum stress located at the top of the cross section at the fixed end of the cantilever	[MPa]
$F_{cr} = \frac{k_1 b^3 h}{l^2}\sqrt{\left(1 - k_2\frac{b}{h}\right) EG}$	the critical load (see [250])	[N]

- *find b and h such that*

$$b_{min} \le b \le b_{max}$$
$$h_{min} \le h \le h_{max}$$

- *and such that*

$$\min \begin{pmatrix} m(b,h) \\ y(b,h) \end{pmatrix} = \begin{pmatrix} \rho bhl \\ 4\frac{Fl^3}{Ebh^3} \end{pmatrix} \tag{2.1}$$

- *subject to*

$$\sigma_{max} = 6\frac{Fl}{bh^2} \le \sigma_{adm} = \sigma_E/\eta \tag{2.2}$$

$$F < F_{cr} = \frac{k_1 b^3 h}{l^2}\sqrt{\left(1 - k_2\frac{b}{h}\right) EG} \tag{2.3}$$

The optimisation problem aims to minimise[3] the *objective functions* $(m(b,h), y(b,h))$ by selecting properly the *design variables*[4] (b,h) and by satisfying the constraints on F_{cr} and σ_{adm}.

In order to solve the problem, one may compute $m(b,h)$ and $y(b,h)$ as function of all possible combinations of b and h.

This is a naive but reasonable approach. In other words, if a couple of values b, h are selected, then, correspondingly, a couple of values $m(b,h)$, $y(b,h)$ can be computed. b and h have to be varied, respectively, within the prescribed ranges $b_{min} \le b \le b_{max}$ and $h_{min} \le h \le h_{max}$. The inequalities (2.2) and (2.3) have to be evaluated for all possible combinations of b and h: if the inequalities are satisfied, the couple of values $m(b,h)$, $y(b,h)$ is kept, otherwise it is discarded.

In Fig. 2.2, the results of such a computation are shown. At one point in the rectangle on the plane b–h corresponds a point in the manifold on the plane y–m. The rectangular manifold on the plane b–h is transformed into the manifold on the plane y–m. In other words, a transformation is established between the domain of design variables and the domain of objective functions.

As all the possible combinations of design variables b, h have been used to generate $m(b,h)$ $y(b,h)$ (having verified that $m(b,h)$ $y(b,h)$ do satisfy the constraints (2.2) and (2.3)), the question is now how to find the values of b, h which minimise concurrently the mass m and the deflection y at the free end of the cantilever.

---

[3]If, for a generic optimisation problem, the objective functions were to be *maximised*, changing their signs would transforms the maximisation problem into a minimisation one, so the presented example dealing with minimisation is quite general.

[4]b and h are precisely *parameters*, with the property that they have to be varied, thus they are *variable parameters* which is a nonsense as parameters do have fixed values. To avoid misunderstanding these variable parameters are named *design variables* to distinguish them from actual parameters having fixed values.

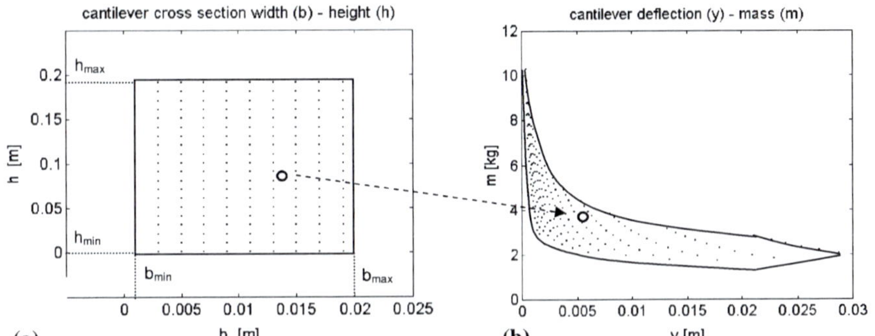

Fig. 2.2 **a** The cantilever cross section width b and height h that are considered for the optimisation. **b** cantilever mass $m(b, h)$ and cantilever deflection at the free end $y(b, h)$. The manifold in b accounts for the constraints on the maximum stress (3.30) and on the critical load (3.31). Data: (symbols referring to (3.30), (3.31) and Fig. 3.5) $b_{min} = 0.001$ m, $b_{max} = 0.020$ m, $h_{min} = 0.001$ m, $h_{max} = 0.200$ m, $\rho = 2700$ kg/m^3 $E = 70000$ MPa, $G = 27000$ MPa, $\sigma_E = 160$ MPa, $\eta = 1$, $F = 1000$ N, $l = 1$ m, $k_1 = 0.669$, $k_2 = 0.63$

The definition of the solution of the addressed optimisation problem is not straightforward and requires a special reasoning.

Let us consider the couple of values $m_A(b_A, h_A)$, $y_A(b_A, h_A)$ referring to point A in Fig. 2.3. The couple of values b_A, h_A defines a cantilever cross section which *should not* be considered by a designer, that is, the cantilever A is a wrong solution for the addressed optimisation problem.

The reason for this is readily explained. Let us consider point $B(m_B(b_B, h_B), y_B(b_B, h_B))$. It is $m_A(b_A, h_A) = m_B(b_B, h_B)$ but $y_B(b_B, h_B) < y_A(b_A, h_A)$: the two cantilevers have the same mass m but the deflection of the cantilever B is smaller than that of the cantilever A. Thus, the cantilever B is better than cantilever A. Similarly, for cantilever C, $m_C(b_C, h_C) < m_A(b_A, h_A)$ and $y_A(b_A, h_A) = y_C(b_C, h_C)$. Also the cantilever C is better than cantilever A because they have the same deflection but the mass of the cantilever C is smaller than that of the cantilever A. Point A is said to be *dominated* by B and C, i.e. B and C *dominate* A. By inspection of Fig. 2.3, A is also dominated by all the design solutions between B and C on the bold line, i.e. the solutions on the bold line between B and C are better at least in one objective function than A.

Given a (wrong) solution A, there exist (right) solutions which are better than A at least in one objective function. All of the solutions corresponding to points in Fig. 2.3 which do not lay on the bold line (defined by the two end points y_{min} and m_{min}) are *wrong solutions* to be discarded by a designer. Conversely, the good or *optimal solutions* are those and only those which are represented by points laying on the said bold line, the set of these solutions is called *Pareto-optimal set*.[5]

[5] Vilfredo Pareto (1848–1923) was an Italian engineer who later became a famous sociologist and economist at the University of Lausanne.

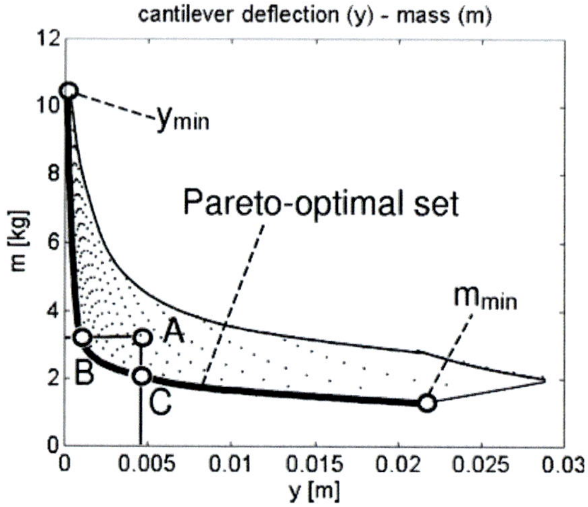

Fig. 2.3 Definition of the Pareto-optimal set in the objective functions domain. Data in Fig. 2.2

The task of the designer is that of choosing a solution from the Pareto-optimal set and only from this set.

It is evident that the designer can choose among an unlimited number of solutions, i.e. the solution is not unique. This is due to the fact that the addressed optimisation problem required to minimise concurrently $m(b, h)$ and $y(b, h)$, i.e. *the minimisation of a vector function* had to be performed.

The minimisation of a vector function is an issue typical in the optimisation of complex systems. Presently, it does not seem to be very well known by designers, even if the theory was developed more than one century ago. On the contrary, the minimisation of a *scalar function*[6] of one or more variables is generally reputed as a relatively simple task.

In order to explain what does it mean 'minimising a vector function', let us start to minimise a scalar function, i.e. let us separately minimise $m(b, h)$ and $y(b, h)$. These two scalar functions are obviously to be minimised by taking into account that constraints (2.2) and (2.3), i.e. a *constrained minimisation* has to be performed.

If only one objective function were considered, e.g. $m(b, h)$, the problem could be formulated mathematically as:

- *Find b and h*
- *such that*

$$b_{min} \leq b \leq b_{max}$$
$$h_{min} \leq h \leq h_{max}$$

- *and such that*

$$\min m(b, h) = \min \rho bhl$$

[6]E.g. minimising only $m(b, h)$ or only $y(b, h)$).

- *subject to*

$$\sigma_{max} = 6\frac{Fl}{bh^2} \leq \sigma_{adm} = \sigma_E/\eta$$

$$F < F_{cr} = \frac{k_1 b^3 h}{l^2}\sqrt{\left(1 - k_2\frac{b}{h}\right) EG}$$

The solution to this optimisation problem in which only one objective function is to be minimised can be denoted by

$$m_{m_{min}}(b_{m_{min}}, h_{m_{min}}),\ y_{m_{min}}(b_{m_{min}}, h_{m_{min}})$$

Similarly, by minimising $y(b, h)$ the solution can be written as

$$m_{y_{min}}(b_{y_{min}}, h_{y_{min}}),\ y_{y_{min}}(b_{y_{min}}, h_{y_{min}})$$

It has to be noticed that if a scalar function has to be minimised, the solution is unique

$$\min m(b, h) \rightarrow m_{m_{min}}(b_{m_{min}}, h_{m_{min}}),\ y_{m_{min}}(b_{m_{min}}, h_{m_{min}}) \rightarrow 1 \text{ solution}$$

$$\min y(b, h) \rightarrow m_{y_{min}}(b_{y_{min}}, h_{y_{min}}),\ y_{y_{min}}(b_{y_{min}}, h_{y_{min}}) \rightarrow 1 \text{ solution}$$

but, if a vector function has to be minimised, the solution is not unique

$$\min \begin{pmatrix} m(b, h) \\ y(b, h) \end{pmatrix} \rightarrow \text{Pareto} - \text{optimal set} \rightarrow \infty^1 \text{optimal solutions}$$

So the concurrent optimisation of two or more objective functions involves finding infinite solutions.

This occurrence is very important and maybe reputed as the key issue of MOP.

The Pareto-optimal set is the set which contains all the infinite solutions coming from the minimisation of a vector function.

The designer has to select one solution from the Pareto-optimal set, and this selection is inherently somewhat subjective. The Pareto-optimal set does contain (by definition) all the best compromise solutions between conflicting objective functions. All of the Pareto-optimal solutions do have the status of the best compromise solutions. The degree or level of the compromise varies in a fuzzy way among solutions. Selecting the desired level of the compromise is the primary or ultimate task of the designer.

The designer has to act as a *decision-maker*, actually, by selecting a solution from the Pareto-optimal set, he decides his desired best compromise between conflicting objective functions.

In our example, the designer has to choose, (from the Pareto-optimal set, and from this set only), the preferred couple of values $m(b, h)$, $y(b, h)$. It is clear that the designer will disregard all of the so-called dominated solutions (the solutions corresponding to points laying inside the manifold in Fig. 2.2b), and will choose one solution from the dominating ones (belonging to the Pareto-set).

In the literature [73, 214], many attempts have been made to help the designer to make a choice. The results are often questionable, mainly because it is assumed that the designer knows perfectly the desired compromise to be reached among conflicting objective functions. This is not always the case, or better, in the design of complex systems, it is *never* the case.

For example, even in our case, it is not easy to answer the questions *how small the deflection should be*, and *how light the cantilever should be*? It is a matter which might involve additional considerations which were not taken into account at the stage of problem formulation.

What actually happens during the optimisation of a complex system is that the designer, before making his choice, acquires an in-depth knowledge of the system performances under investigation. Particularly, he understands which are the limit performances of the system.

For example, in our case, he will realise that an objective function defined by point D in Fig. 2.4 will never be attained, unless the system is changed. Also y_{min} and m_{min} are important boundaries to the performances (i.e. objective functions). These boundaries, depending on the problem under investigation, could be changed by varying the design variable ranges (i.e. the ranges into which the design variables are varied during the optimisation process).

An important reference point is the *utopia or ideal point*. In our example, it is the point whose coordinates are (m_{min}, y_{min}). Obviously, a cantilever having such performances (i.e. objective functions) does not exist, however, if it existed, it would be the best solution. The utopia point can be used to choose Pareto-optimal solutions. Empirical reasoning suggests to choose those Pareto-optimal solutions which are closer to the utopia point.

If the system is complex (tenth of objective functions and tenth of design variables) the knowledge and understanding process (necessary for an effective optimisation) maybe heavy and time consuming. Sometimes the time for optimising a complex system may require as much time as that which was spent to develop the validated mathematical model of the system under investigation. When a designer has at his disposal a validated mathematical model of a system, he might be only halfway (or less!) from the complete (optimal) design of the system.

As the designer has chosen the preferred compromise among conflicting objective functions by selecting a solution from the Pareto-optimal set, then the problem is to define or identify the values of the design variables related to those objective functions. In other words, the designer has to define the design variables that allow the system to perform according to his wishes.

So we have to consider that, having a set of Pareto-optimal solutions, what we really want are the values of design variables corresponding to those Pareto-optimal

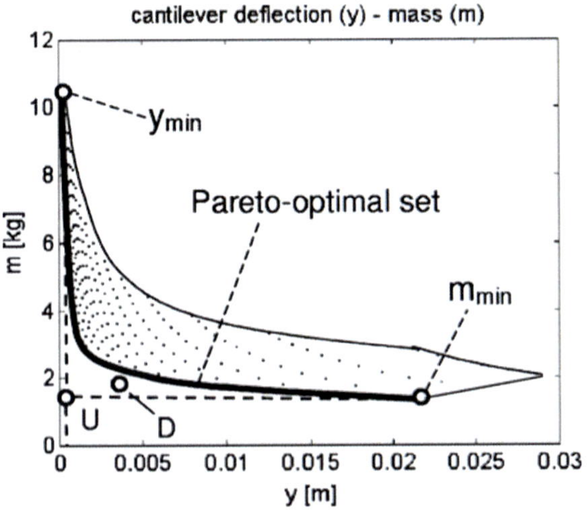

Fig. 2.4 Definition of the utopia or ideal point U (coordinates are m_{min}, y_{min}). The objective functions (m, y) referring to point D cannot be attained by the considered system. Data in Fig. 2.2

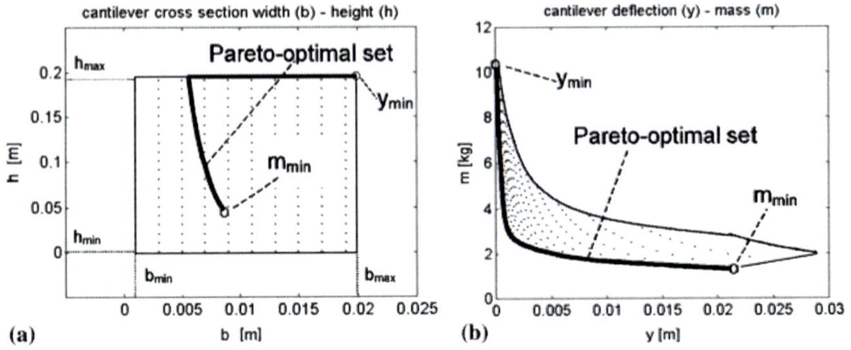

Fig. 2.5 The two Pareto-optimal sets defined, respectively, in the design variable domain (**a**) and in the objective function domain (**b**). Data in Fig. 2.2

solutions. It can be useful to represent the Pareto-optimal solutions both in the objective functions domain or conversely, in the design variables domain (see Fig. 2.5).

With reference to our optimisation problem, in Fig. 2.5, the Pareto-optimal set defined in the design variables domain is shown. As it could be expected, the stiffer cantilever has the highest cross section height h.

There is a direct relationship between the points of the two Pareto-optimal sets represented in Fig. 2.5. It is obvious that the designer will always select the design variables from the Pareto-optimal set represented in Fig. 2.5a.

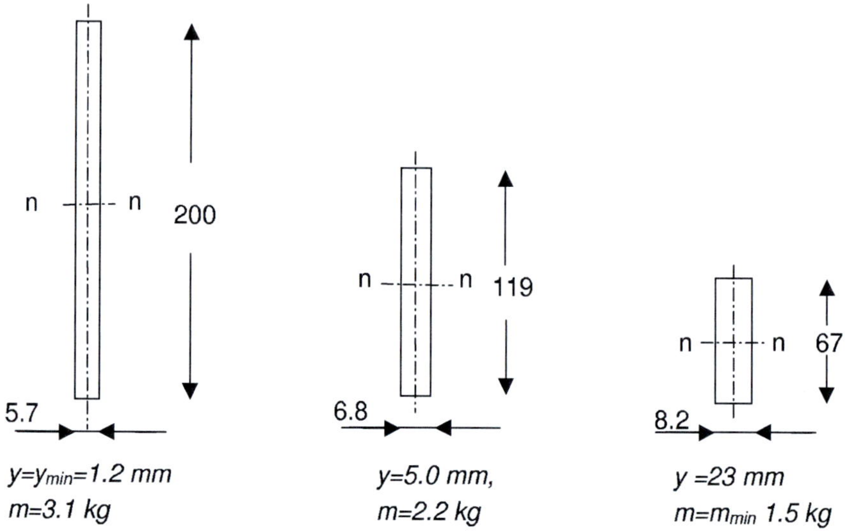

Fig. 2.6 Cross sections of cantilevers belonging to the Pareto-optimal set. Dimensions in [mm]. Left: minimum deflection cantilever (point y_{min} in Fig. 2.5). Centre: a generic Pareto-optimal cantilever. Right: minimum mass cantilever (point m_{min} in Fig. 2.5). Data are shown in Fig. 2.2

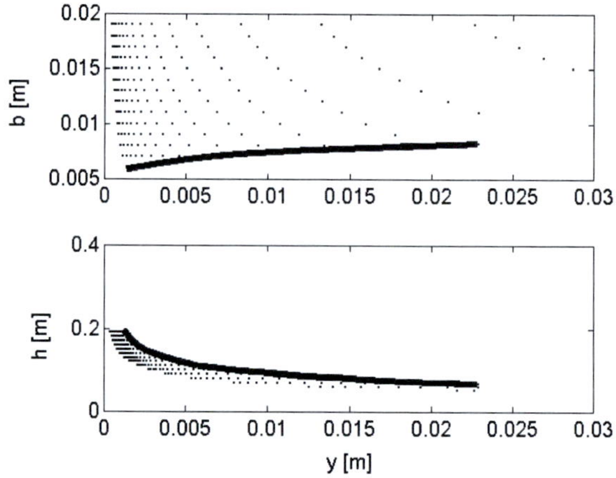

Fig. 2.7 Pareto-optimal cross section width b and Pareto-optimal cross section height h as function of Pareto-optimal deflection y. By these two graphs, the direct relationship between the points of the two Pareto-optimal sets in Fig. 2.5 can be obtained

The shapes of the cross sections of three Pareto-optimal cantilevers are shown in Fig. 2.6. Two of them refer, respectively, to points m_{min}, and y_{min}. The intermediate cantilever in Fig. 2.6 has both the minimum mass m and the minimum deflection y, according to the definition of best compromise addressed before (Figs. 2.5 and 2.7).

2.2 Finding the Pareto-Optimal Sets

From the above considerations, it is clear that a designer should always try to find the Pareto-optimal sets (both in the design variables space and in the objective functions space) to obtain the best solutions. The existence of the Pareto-optimal set is stated by a theorem which guarantees non-empty Pareto-optimal sets under broad conditions,[7] holding for the majority of engineering problems.

In the engineering practice, very rarely problems are solved by resorting to the computation of the Pareto-optimal sets. In fact, the computations to find the Pareto-optimal sets are rather involved and time-consuming.[8] However, in the last years, the ever-increasing power of computers has allowed to attempt the optimal design of complex systems on the basis of Pareto theory.

2.2.1 *Exhaustive Method*

There are many methods to find Pareto-optimal sets [143]. The one, called exhaustive method, which has been used in the presented example referring to the cantilever, is the simplest, but heavier due to the huge amount of computations required. To explain why the exhaustive method is unmanageable, let us resort to an example.

Let us imagine that a complex system is defined by 10 design variables n_{dv} and that 30 objective functions n_{of} have to be taken into account. Each design variable may assume $n_v = 10$ different values within its definition range. Let us assume that, given a combination of design variables, the time t_s for performing a simulation to obtain the value of one single objective function is 1 s. By the exhaustive method, the number of all possible combinations of design variables values is

$$n_{cdv} = n_v^{n_{dv}}$$

and the total time t_t for computing all of the objective functions as function of all possible combinations of design variables is

$$t_t = t_s n_{of} n_v^{n_{dv}} \tag{2.4}$$

substituting the numerical values

$$t_t = t_s n_{of} n_v^{n_{dv}} = 1 \cdot 30 \cdot 10^{10}\,\text{s} = 3 \cdot 10^{11}\,\text{s} = 9645\,\text{years} \tag{2.5}$$

---

[7]The assumptions are that the design variables space is closed and the objective functions are continuous functions of the design variables [152].

[8]This seems the main reason why, presently, very few designers are instructed to apply the optimisation theory based on Pareto-optimality.

the total time t_t for computing all of the objective functions is clearly a time too long to solve an engineering problem (10000 years seems to be the age of *homo sapiens sapiens*). Additionally if, for some practical reason, the number of design variables should be increased by two (from 10 to 12), the total time requested for simulations would be 100 times longer

$$t_t = t_s n_{of} n_v^{n_{dv}} = 1 \cdot 30 \cdot 10^{10+2}\,\text{s} = 100 \cdot 9645\,\text{years} = 964500\,\text{years}$$

These figures are startling and state evidently that an exhaustive construction[9] of the Pareto-optimal set is, in general, not possible, unless the system under consideration is very simple.

The critical factors in Eq. (2.4) are the number of design variables n_{dv}, the number of values the design variables may assume within their respective definition ranges n_v, and the simulation time for computing one single objective function t_s. Often, the number of objective functions n_{of} is not critical.

2.2.2 *Uniformly Distributed Sequences and Random Search*

There have been conceived many methods to reduce the above addressed total number of simulations $n_v{}^{n_{dv}}$, in fact, it is not absolutely necessary to explore all of the possible design configurations to construct the Pareto-optimal sets [143].

Sometimes it is sufficient to reduce (to a considerable extent) the number of simulations to estimate properly the Pareto-optimal set both in the space of objective functions and in the space of design variables.

For example, in Fig. 2.8a, a regular grid representing an ordered combination of design variables values is compared with two different combinations (centre and right) which are still ordered (in the sense that points are not randomly distributed) but the number of points is greatly reduced with respect to the previous combination. Points related to the combination in Fig. 2.8 (centre) are somehow 'equally' distributed along vertical planes. Every square of the three vertical planes contains one point. Points along these planes are somehow "well distributed". For each vertical plane, one out of the three coordinates of a point is fixed. So—in order to reconstruct the relationships between the design variables and the objective functions—the informative contribution given by the design variable which is kept fixed might be nearly the same for all points laying on the same plane. This causes a computational inefficiency as many simulations will be performed at a given (fixed) value of a design variable.

The distribution in Fig. 2.8c overcomes this problem, the information on the relationships between the objective functions and the design variables is complete and

[9]When all of the simulations have been made on the basis of all possible combinations of design variables, the construction of all Pareto-optimal sets is performed by selecting the so-called dominating solutions from the dominated ones [143].

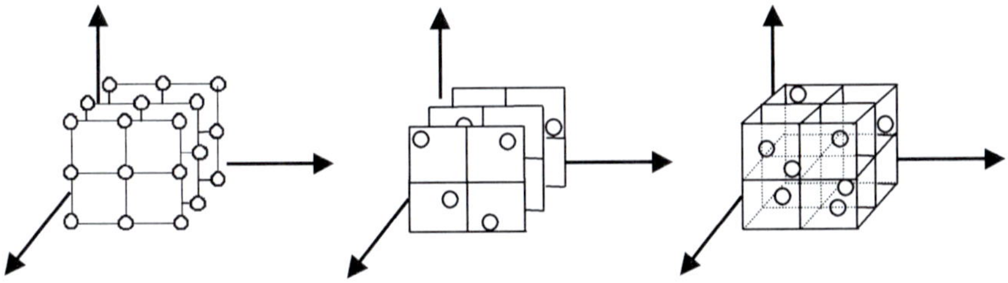

Fig. 2.8 Three different combinations of design variables values. Number of design variables: $n_{dv} = 3$. **a**: Regular grid, number of values the design variables may assume within their respective definition ranges: $n_v = 3$, total number of design variable combinations: 27. **b**: Low discrepancy grid, total number of design variable combinations: 12. **c**: Low discrepancy grid, total number of design variable combinations: 8

less redundant. From a mathematical point of view, the above explanation is very rough; anyway, it gives some hint on how and why simulations can be reduced to estimate the Pareto-optimal sets.

Orthogonal Arrays and *Low Discrepancy Sequences* [143] are particularly suited to be used to reduce as much as possible the number of combinations of design variables used as input data for simulations. These sequences are called *Uniformly Distributed Sequences* and are said to have *low discrepancy* as the points are uniformly distributed in the space. In multi-dimensional spaces (>3 up to 30 or more dimensions), the low discrepancy placement of points is a peculiar mathematical matter which requires a special theoretical background.

2.2.3 Genetic Algorithms

Another well-known method to optimise complex systems is based on Genetic Algorithms [143]. This approach mimics what has happened during the evolution of living creatures. Living creatures evolve by adapting as much as possible to the environment. It is assumed that

- individuals have a chance to reproduce themselves according to their fitness (reproduction step)
- an individual cross its genes with the ones of its partner to generate a new individual (cross-over step)
- some mutation of the genes always acts during the previous cross-over process (mutation step)

The generated individuals start a new three-step generation process (reproduction, cross-over, mutation). A number of generations are necessary to select fit individuals belonging to Pareto-optimal sets.

A design project evolves by adapting as much as possible both to the designer aims and to the design constraints.

An individual corresponds to a design solution, the genes correspond to a design variable combination which identifies a single design solution. A direct mathematical relationship is established between genes and design variables: the design variables values are expressed in binary form and these binary strings correspond to the genes. Each design variable can be converted into its binary equivalent, and thereby mapped into a fixed length of zeros and ones.

An individual/design solution is selected for cross-over according to its fitness. The fitness is related to the values of the objective functions. If the fitness is high, the individual/design solution will have more chances to reproduce, i.e. crossing its genes with those of a partner. The cross-over process just mix the binary strings pertaining to genes/design variables of two parents to generate one new individual. The genes/design variables of the new individual/design solution may slightly vary due to mutation.

Genetic algorithms do have many variants which are often developed to solve particular optimisation problems.

Genetic algorithms are very useful, especially when the design variables values take discrete values. Unfortunately, they are not very simple to be used (with respect to other methods such as Global Approximation, see next subsection), especially when dealing with the optimisation of (very) complex systems. In fact, they often require the exact definition of some algorithm parameters that influence the efficiency of the search.

2.2.4 *Comparison of Broadly Applicable Methods to Solve Optimisation Problems*

In Table 2.1, three general and broadly applicable methods to solve optimisation problems are presented together with their optimisation properties and performances. All of the three methods have been introduced in the preceding subsections. They allow a vectorial formulation, i.e. they permit the optimisation of all of the objective functions concurrently (direct derivation of Pareto-optimal sets).

Computational efficiency refers to how fast an optimisation can be. For simple problems (number of design variables less than 5 or 6), all the three presented methods work well. Complex problems are handled well by genetic algorithms, on the contrary, the exhaustive method is absolutely unsuited for this kind of problems.

The accuracy in the definition of the Pareto-sets is the best for the exhaustive method in which design variables may vary in non-continuous ranges (discrete values). All of the three methods in Table 2.1 do have the following important properties

- they can deal with design variables which vary within non-continuous ranges (discrete values)
- objective functions can be non-continuous functions of the design variables.

Table 2.1 General and broadly applicable methods for solving optimisation problems. Other methods very efficient for specific cases are reported in Table 2.3 (– – very bad; – bad; + good; very good + +)

	Computational efficiency (time)		Accuracy		Discrete design variables values allowed	Objective functions need to be continuous functions
	Simple problem	Complex problem				
Exhaustive	–	– –	++ *(discrete design variables)*	– *(continuous design variables values)*	Yes	No
Uniformly distributed sequences	+	–	–		Yes	No
Genetic algorithms	+	+	+		Yes	No

2.2.5 Global Approximation—Artificial Intelligence—Machine Learning

The MOO of very complex systems may require still a prohibitive simulation effort, even reducing the number of combinations of design variables by means of the mentioned techniques.

To optimise the performances of complex systems, the relationship between design variables and objective functions can be *approximated* by means of a purely mathematical model. The parameters of the purely mathematical model (which can be either an artificial neural network or a piecewise quadratic function, or others) are defined on the basis of a limited number of simulations performed by means of the originally validated mathematical model of the system under consideration [143].

In other words, there are two models, the first which is the original validated mathematical model that the designer uses to simulate the actual physical behaviour of the system to be optimised, and the second mathematical model which is a purely mathematical model, able to approximate the outputs of the first model.

Obviously, it is supposed, as it always happens in actual applications, that the first model requires time-consuming simulations and the second model provides quick simulation outputs. Typically the simulation time of the second model is a small fraction (1/10–1/10000) of the simulation time requested by the first model.

The accuracy in the approximation of the outputs depends on the approximation model employed. In Table 2.2, known methods for Global approximation are presented. Linear and quadratic interpolation are well suited to approximate locally,

Table 2.2 Approximation methods for solving optimisation problems. (− bad; + good; very good + +)

	Approximation domain	Evaluation accuracy		Tuning effort
		Simple model	Complex model	
Linear interpolation	Local	+	−	++
Quadratic interpolation	Local	+	−	++
Radial basis function neural networks	Local/global	++	+	+
Multi-layer perceptron neural networks	Global	++	++	−
Statistical approximation	Global	++	+	−

i.e. in the neighbourhood of a single point, the objective functions. They are suited for very simple optimisation problems, and the tuning (i.e. the process for defining the parameters of the pure mathematical model) is relatively easy. The evaluation accuracy, i.e. the ability to reproduce the original values of objective functions given the values of design variables, is not very high. Response surface methodologies use linear and quadratic approximation models.

Radial basis function neural networks seem to be accurate, easy to be tuned with appropriate algorithms and efficient in the evaluation of objective functions. Multi-layer perceptron neural networks perform better than previous radial basis function neural networks but require more effort in tuning. These approximation methods are particularly suited for complex design optimisation problems (Fig. 2.9).

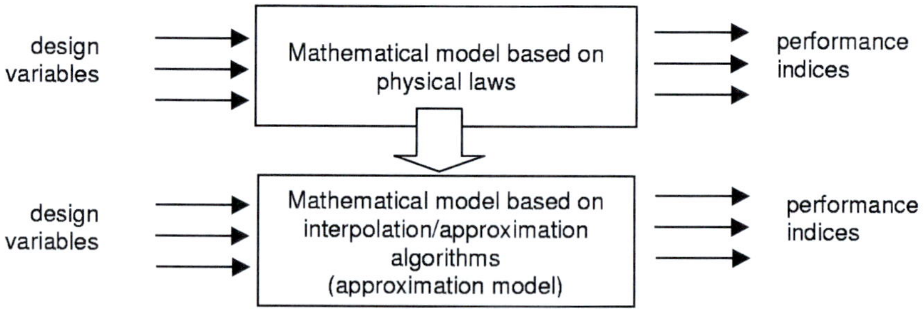

Fig. 2.9 Global Approximation approach to solve optimisation problems. The original model based on physical laws is substituted by another purely mathematical model based on interpolation/approximation algorithms

2.2.6 *Multi-objective Programming via Non-linear Programming*

Historically, MOO problems were numerically solved resorting to non-linear programming (NLP), i.e. by transforming the original vector problem into a scalar one [143]. Obviously, this transformation is still effective but it is not recommended any longer for solving general complex optimisation problems, i.e. finding Pareto-optimal sets.

A well-known method for transforming a vector optimisation problem into a scalar one is the *Constraints Method*. Given the MOO problem as a vector function to be minimised

$$\min \mathbf{f}(\mathbf{x}) = \min \begin{Bmatrix} f_1(\mathbf{x}) \\ f_2(\mathbf{x}) \\ \dots \\ f_{n_{of}}(\mathbf{x}) \end{Bmatrix} \qquad \mathbf{x} = \begin{Bmatrix} x_1 \\ x_2 \\ \dots \\ x_{n_{dv}} \end{Bmatrix}$$

subject to n_c constraints

$$\mathbf{g}(\mathbf{x}) = \begin{Bmatrix} g_1(\mathbf{x}) \\ g_2(\mathbf{x}) \\ \dots \\ g_{n_c}(\mathbf{x}) \end{Bmatrix} \leq 0$$

the original vector problem is transformed into the new scalar one

$$\min f_1(\mathbf{x}) \qquad \mathbf{x} = \begin{Bmatrix} x_1 \\ x_2 \\ \dots \\ x_{n_{dv}} \end{Bmatrix}$$

subject to the $n_c + (n_{of} - 1)$ constraints

$$\mathbf{g}(\mathbf{x}) = \begin{Bmatrix} f_2(\mathbf{x}) - \varepsilon_2 \\ \dots \\ f_{n_{of}}(\mathbf{x}) - \varepsilon_{n_{of}} \\ g_1(\mathbf{x}) \\ g_2(\mathbf{x}) \\ \dots \\ g_{n_c}(\mathbf{x}) \end{Bmatrix} \leq 0$$

By varying the values of the elements of the vector

$$\varepsilon = \begin{Bmatrix} \varepsilon_2 \\ \dots \\ \varepsilon_{n_{of}} \end{Bmatrix} \tag{2.6}$$

the Pareto-optimal set can be found.

This scalarisation can be applied to any kind of optimisation problem. It is not straightforward how to vary ε, so this method is only used to refine a Pareto-optimal solution. In other words, if a Pareto-optimal solution is known approximately, by applying the constraints method in the neighbourhood of the approximately known solution, the approximation can be significantly improved to the desired extent.

Another widespread method for scalarisation is the weight method. Given the MOO problem as a vector function to be minimised

$$\min \mathbf{f}(\mathbf{x}) = \min \begin{Bmatrix} f_1(\mathbf{x}) \\ f_2(\mathbf{x}) \\ \dots \\ f_{n_{of}}(\mathbf{x}) \end{Bmatrix} \qquad \mathbf{x} = \begin{Bmatrix} x_1 \\ x_2 \\ \dots \\ x_{n_{dv}} \end{Bmatrix}$$

subject to n_c constraints

$$\mathbf{g}(\mathbf{x}) = \begin{Bmatrix} g_1(\mathbf{x}) \\ g_2(\mathbf{x}) \\ \dots \\ g_{n_c}(\mathbf{x}) \end{Bmatrix} \leq 0$$

the original vector problem is transformed into the new scalar one

$$\min(w_1 f_1(\mathbf{x}) + w_2 f_2(\mathbf{x}) + \dots + w_{n_{of}} f_{n_{of}}(\mathbf{x})) \qquad \mathbf{x} = \begin{Bmatrix} x_1 \\ x_2 \\ \dots \\ x_{n_{dv}} \end{Bmatrix}$$

subject to the n_c constraints

$$\mathbf{g}(\mathbf{x}) = \begin{Bmatrix} g_1(\mathbf{x}) \\ g_2(\mathbf{x}) \\ \dots \\ g_{n_c}(\mathbf{x}) \end{Bmatrix} \leq 0$$

by varying the elements of the vector

$$\mathbf{w} = \begin{Bmatrix} w_1 \\ w_2 \\ \dots \\ w_{n_{of}} \end{Bmatrix}$$

the Pareto-optimal set can be found both in the space of objective functions and in the space of design variables, provided that the Hessian of $w_1 f_1(\mathbf{x}) + w_2 f_2(\mathbf{x}) + \cdots + w_{n_{of}} f_{n_{of}}(\mathbf{x})$ is a semidefinite positive matrix in the design variables space (a sufficient condition [152] for this is that $w_1 f_1(\mathbf{x}) + w_2 f_2(\mathbf{x}) + \cdots + w_{n_{of}} f_{n_{of}}(\mathbf{x})$ is globally convex in the design variables space).

This scalarisation can be applied to any kind of convex optimisation problem, for other problems it might fail in finding all Pareto-optimal solutions.

2.2.7 *Algorithms to Solve Optimisation Problems in Scalar Form*

In Table 2.3, the most used algorithms for solving MOO problems via scalar formulation are presented and compared. Solving the optimisation problem in scalar form deals with the minimisation of one function of many design variables subject to constraints (on design variables).

Obviously, the algorithms that could solve the optimisation problem in vector form are still available for solving the scalar problem.

The *Simplex Algorithm* performs well and does not require the computation of the derivatives of the function.

Sequential Unconstrained Minimisation Technique (SUMT) is a kind of gradient method and requires the function to be continuous together with its first derivative. Discrete values are not permitted because the search is based on gradient evaluation.

Sequential Quadratic Programming (SQP) performs better than SUMT, and requires the same conditions on the functions to be minimised.

The main disadvantage of this method is that they are all local methods. So the finding of global minima are all influenced by the starting point that must be close to the global solution.

2.3 Understanding Pareto-Optimal Solutions

After Pareto-optimal sets have been computed, the designer can make a choice and select from these sets the preferred solution featuring the desired compromise among objective functions.

This process can be preceded and even accelerated by special analyses allowing the designer to have an insight into the physical phenomena he/she is trying to analyse. In other words, there are some analyses by which the designer may understand the reason why a Pareto-optimal solution requires such a design variable combination. In particular, these analyses may show which is

- the relationship between two Pareto-optimal objective functions,
- the relationship between two Pareto-optimal design variables,

Table 2.3 Methods for solving MOO problems via Non-Linear Programming (scalar formulation). (− − very bad; − bad; + good; very good + +)

	Computational efficiency		Accuracy		Discrete design variables values allowed	Objective functions continuous functions
	Simple problem	Complex problem				
Exhaustive	−	− −	++ (*discrete design variables values*)	− (*continuous design variables values*)	Yes	No
Uniformly distributed sequences	+	−	−		Yes	No
Genetic algorithms	+	+	+		Yes	No
Simplex	+	−	+		Yes	No
Sequential unconstrained minimisation technique	+	−	+		No	Yes
Sequential quadratic programming	+	+	+		No	Yes

- which is the relationship between a Pareto-optimal design variable and a Pareto-optimal objective function

In Fig. 2.10, these different relationships are shown. The Pareto-optimal set is defined in a n_{of} dimensional domain, so the Pareto-optimal values projected onto a bi-dimensional domain (f_i, f_j) appear in a picture like the one in Fig. 2.10. Thus, the addressed analyses can be performed on the basis of a statistical approach possibly combined with the plotting of graphs like those represented in Fig. 2.10. The *Spearman rank-order correlation coefficient* can be used to assess the above-introduced relationships and thus is particularly suited to discover significant relationships. This coefficient is able to identify monotonic non-linear relationships and thus is particularly suited to discover significant relationships that may suggest important engineering solutions. If the relationships are not monotonic a partition of the domain could be performed as shown in Fig. 2.11.

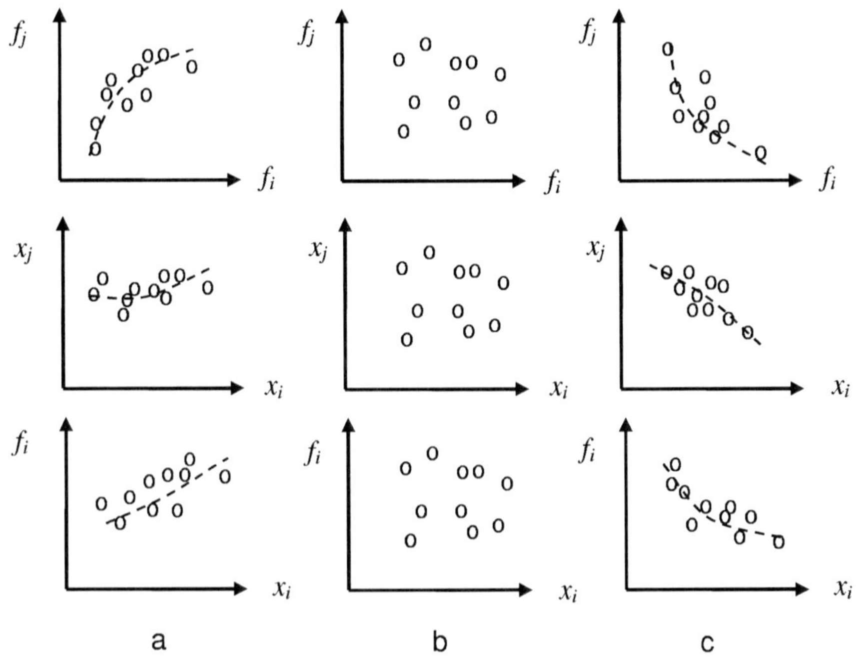

Fig. 2.10 Relationships between Pareto-optimal objective functions (f_i–f_j), between Pareto-optimal design variables (x_i–x_j), and between Pareto-optimal objective functions and Pareto-optimal design variable (f_i–x_i). The values can be either directly correlated (column (**a**)), uncorrelated (column (**b**)), or indirectly correlated (column (**c**))

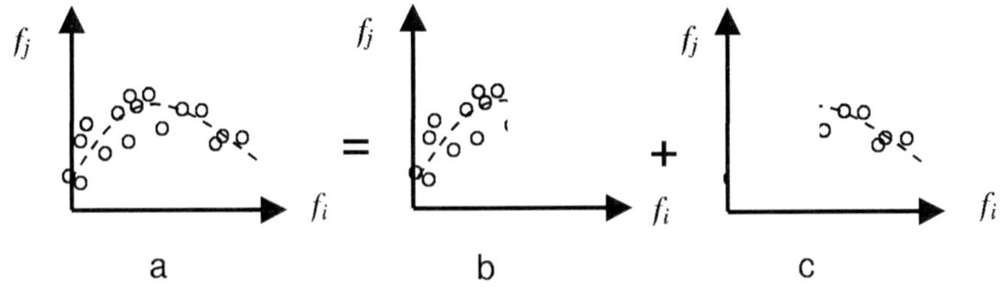

Fig. 2.11 Partition of the f_i domain to obtain two monotonic interpolation functions f_i. The Spearman rank-order correlation coefficient can be computed separately for cases **b** and **c**

By computing the Spearman rank-order correlation coefficient, very useful information on the system under consideration can be gained. In fact, a kind of sensitivity analysis focused on the Pareto-optimal solutions can be performed. In conventional sensitivity analysis, the relationships between objective functions and design variables are computed in the neighbourhood of one simple reference design solution. The analysis does not distinguish dominated from non-dominated solutions, i.e. the analysis is *not* restricted to Pareto-optimal solutions. So very poor information can be gained with conventional parameter sensitivity analysis. An analysis restricted to the very special Pareto-optimal values only may give significant hints to the designer. Actually, the relationships between Pareto-optimal objective functions and Pareto-optimal design variables can be highlighted.

Chapter 3
Analytical Derivation of the Pareto-Optimal Set with Application to Structural Design

Often at the earliest stage of an engineering project, a preliminary optimisation is useful, in order to allow the designer to ascertain the envisaged performance of the system under development.

Providing an efficient (i.e. analytical) tool to quickly define the Pareto-optimal set could be an extremely valuable chance to make the right design decision at the right time.

The procedure proposed here to obtain the Pareto-optimal set in analytical form refers mostly to design problems described by a limited number of design variables and a limited number of objective functions and constraints.

In the first part of this chapter, the analytical derivation of the expression of the Pareto-optimal set for MOO problems is dealt with.

According to the knowledge of the authors, in the literature, very few papers exist on this topic and related issues. A survey of current continuous MOO concepts and methods is presented in [141]. Some relevant contributions are given in [109, 142] in which some new formulations of the Fritz John first-order conditions are proposed and analysed. In [248] first and second-order conditions are proposed for a convex multi-objective problem via scalarisation and in [2] some second-order conditions are analysed in detail. In [2, 109, 142, 186, 248, 256] necessary and/or sufficient conditions are discussed but no procedures are introduced nor mentioned to allow the analytical derivation of the Pareto-optimal set. A slightly different formulation of the Fritz John conditions is used in [12], where a symbolic algorithm for finding the Pareto front is discussed.

The procedure proposed in this book is based on the reformulation of the Fritz John conditions for Pareto-optimality (first-order conditions). Necessary conditions (a relaxed form of the Fritz John ones) are introduced and used to define the procedure to find analytically the Pareto-optimal set [127].

In the second part of the chapter, basic engineering examples are used to show the effectiveness of the proposed procedure.

F. M. Ballo et al., *Optimal Lightweight Construction Principles*,
https://doi.org/10.1007/978-3-030-60835-4_3

First, the Fonseca and Fleming problem with two design variables and two objective functions has been addressed.

Second, the diameter of two spheres pressed one against the other has been designed to obtain minimum mass, minimum compliance with the constraint of structural integrity.

Third, the ideal cantilever discussed in Chap. 2 has been designed in order to minimise both mass and deflection in presence of maximum stress and buckling design constraints. In this chapter, a full insight into the analytical expressions of the Pareto-optimal solution of the cantilever is provided. Analytical solutions for optimal beam design are described in [168] for a single-objective problem (min compliance) where the design variable is the area of the beam. In [28, 70, 129, 143], the multi-objective problem referring to the derivation of an optimal cantilever beam has been introduced and solved by applying numerical optimisation methods.

Fourth, the Fonseca and Fleming problem with three design variables, two objective functions and two design constraints has been addressed.

3.1 The Fritz John Necessary Conditions

Let us consider the following MOO problem [152]

$$\begin{array}{ll} minimise & \mathbf{f}(\mathbf{x}) = \mathbf{f}(f_1(\mathbf{x}), f_2(\mathbf{x}, \ldots, f_{n_{of}}(\mathbf{x})) \\ subject\ to & \mathbf{x} \in \mathcal{X} = \{\mathbf{x} \in \mathbf{R}^{n_{dv}} \mid \mathbf{g}(\mathbf{x}) = \big(g_1(\mathbf{x}), g_2(\mathbf{x}), \ldots, g_{n_c}(\mathbf{x})\big) \leq \mathbf{0}\} \end{array} \tag{3.1}$$

The problem (3.1) has n_{dv} design variables $(x_1, \ldots, x_{n_{dv}})$, n_{of} objective functions $(f_1, \ldots, f_{n_{of}})$, and n_c constraint functions $(g_1, \ldots, g_{n_c})$.

The Fritz John necessary conditions for Pareto-optimality are reported below (see [152])

Fritz John conditions. Let the objective and constraint functions of problem (3.1) be continuously differentiable at a decision vector $\mathbf{x}^* \in \mathcal{X}$. A necessary condition for $\mathbf{x}^*$ to be Pareto-optimal is that vectors must exists reading $\mathbf{0} \leq \boldsymbol{\lambda} \in \mathbf{R}^{n_{of}}$ and $\mathbf{0} \leq \boldsymbol{\mu} \in \mathbf{R}^{n_c}$. If $(\boldsymbol{\lambda}, \boldsymbol{\mu}) \neq (\mathbf{0},\mathbf{0})$

$$\begin{cases} \sum_{i=1}^{n_{of}} \lambda_i \nabla f_i(\mathbf{x}^*) + \sum_{j=1}^{n_c} \mu_j \nabla g_j(\mathbf{x}^*) = \mathbf{0} \\ \\ \mu_j g_j(\mathbf{x}^*) = 0 \quad \forall j = 1, \ldots, n_c. \end{cases} \tag{3.2}$$

The Fritz John conditions (3.2) become the well-known Karush–Kuhn–Tucker (KKT) (1.7), sufficient conditions for Pareto-optimality if $\boldsymbol{\lambda} \neq \mathbf{0}$ and if objective and constraint functions are convex. Dealing with convex MOO problems, every local Pareto-optimal solution is also global Pareto-optimal, see [152].

Convexity can be easily checked by considering the Hessian H of a function $f(\mathbf{x})$. If the Hessian is positive semi-definite $(H(f(\mathbf{x})) \geq 0)$ then $f(\mathbf{x})$ is convex [9].

Actually, the convexity assumption can be relaxed and the KKT sufficient conditions are also valid if the objective functions are pseudo-convex and the constraint functions quasi-convex, see [109, 142].

A function $f(\mathbf{x})$ is quasi-convex if $f(\beta \mathbf{x}^I + (1-\beta)\mathbf{x}^{II}) \leq \max[f(\mathbf{x}^I),\ f(\mathbf{x}^{II})]$ for all $0 \leq \beta \leq 1$ and for all $\mathbf{x}^I,\ \mathbf{x}^{II}$.

A differentiable function $f(\mathbf{x})$ is pseudo-convex if for all $\mathbf{x}^I,\ \mathbf{x}^{II}$ such that $\nabla f(\mathbf{x}^I)^T\ (\mathbf{x}^{II} - \mathbf{x}^I) \geq 0$, we have $f(\mathbf{x}^{II}) \geq f(\mathbf{x}^I)$. Every pseudo-convex function is also quasi-convex [152].

Pseudo-convexity can be checked for a twice differentiable function $f(\mathbf{x})$ by means of the first and second partial derivatives arranged into the bordered determinant

$$|\mathbf{B}| = \begin{vmatrix} 0 & f_1 & f_2 & \dots & f_n \\ f_1 & f_{11} & f_{12} & \dots & f_{1n} \\ f_2 & f_{21} & f_{22} & \dots & f_{2n} \\ \dots & \dots & \dots & \dots & \dots \\ fn & f_{n1} & f_{n2} & \dots & f_{nn} \end{vmatrix} \tag{3.3}$$

along with its leading principal minors

$$|\mathbf{B}_1| = \begin{vmatrix} 0 & f_1 \\ f_1 & f_{11} \end{vmatrix} \quad |\mathbf{B}_2| = \begin{vmatrix} 0 & f_1 & f_2 \\ f_1 & f_{11} & f_{12} \\ f_2 & f_{21} & f_{22} \end{vmatrix} \quad \dots \quad |\mathbf{B}_n| = |\mathbf{B}| \tag{3.4}$$

where $f_i \equiv \partial f / \partial x_i$ and $f_{ij} \equiv \partial^2 f / \partial x_i \partial x_j$.

If $|\mathbf{B}_1| < 0$, $|\mathbf{B}_2| < 0$, ..., $|\mathbf{B}_n| < 0$ for all $\mathbf{x}$ then $f(\mathbf{x})$ is pseudo-convex [211, 222] (sufficient condition).

Just to give a hint, let us consider two functions $f_1(x) = x + x^3$ and $f_2(x) = x^3$. Both the two functions are continuous, differentiable and monotonically increasing; none of them is convex. In fact, it can be easily verified that neither $\dfrac{\partial^2 f_1}{\partial x^2}$ nor $\dfrac{\partial^2 f_2}{\partial x^2}$ are positive $\forall x \in R$.

Actually, function f_1 is pseudo-convex, since $\nabla f_1 = 1 + 3x^2 > 0\ \forall x \in R$ and the product $\nabla f_1 \left(x^I\right) \cdot \left(x^{II} - x^I\right) \geq 0\, \forall\, x^I, x^{II} \in R\ s.t.\ x^{II} \geq x^I$. As the function is strictly monotonically increasing, it is clear that $f_1(x^{II}) \geq f_1(x^I)\ \forall\ x^I, x^{II} \in R\ s.t.\ x^{II} \geq x^I$.

On the other hand, the condition of pseudo-convexity is not verified for function f_2. In fact, if the two values $x^I = 0$ and $x^{II} = -1$ are selected, we can easily verify that $\nabla f_2\,(0) \cdot (-1-0) = 0$, but $f_2\,(-1) < f_2\,(0)$.

Actually, it can be demonstrated that function f_2 is quasi-convex. In fact, for the same values $x^I = 0$ and $x^{II} = -1$, the condition $f(\beta \cdot 0 + (1-\beta) \cdot (-1)) \leq \max[0,\ -1]$ is verified $\forall\ 0 \leq \beta \leq 1$, being the function monotonically increasing. Additionally, as f_2 is monotonically increasing $\forall\, x\ \in\ R$, the validity of the condition can be extended to any pair of values of x^I and $x^{II} \in\ R$.

3.2 The L Matrix

Let $\nabla\mathbf{f}$ and $\nabla\mathbf{g}$ be the matrices of the gradients (Jacobian matrices) of objective and constraint functions, respectively

$$\nabla\mathbf{f} = \left[\nabla f_1 \ \nabla f_2 \ldots \nabla f_{n_{of}}\right] = \begin{bmatrix} \frac{\partial f_1}{\partial x_1} & \cdots & \frac{\partial f_{n_{of}}}{\partial x_1} \\ \vdots & \ddots & \vdots \\ \frac{\partial f_1}{\partial x_{n_{dv}}} & \cdots & \frac{\partial f_{n_{of}}}{\partial x_{n_{dv}}} \end{bmatrix} \quad [n_{dv} \text{ x } n_{of}]$$
$$\nabla\mathbf{g} = \left[\nabla g_1 \ \nabla g_2 \ldots \nabla g_{n_c}\right] = \begin{bmatrix} \frac{\partial g_1}{\partial x_1} & \cdots & \frac{\partial g_{n_c}}{\partial x_1} \\ \vdots & \ddots & \vdots \\ \frac{\partial g_1}{\partial x_{n_{dv}}} & \cdots & \frac{\partial g_{n_c}}{\partial x_{n_{dv}}} \end{bmatrix} \quad [n_{dv} \text{ x } n_c] \tag{3.5}$$

Let $\mathbf{O}$ be the null matrix $[n_c \text{ x } n_{of}]$ and

$$\mathbf{G} = diag(g_1, g_2, \ldots, g_{n_c}) \qquad [n_c \text{ x } n_c] \tag{3.6}$$

a diagonal matrix which has on the main diagonal the constraint functions.

Let $\mathbf{L}$ be the matrix

$$\mathbf{L} = \begin{bmatrix} \nabla\mathbf{f} & \nabla\mathbf{f} \\ \mathbf{O} & \mathbf{G} \end{bmatrix} \qquad [(n_{dv} + n_c) \text{ x } (n_{of} + n_c)] \tag{3.7}$$

The Fritz John conditions can be simply written as follows.

Fritz John conditions (matrix form). Let the objective and constraint functions of problem (3.1) be continuously differentiable at a decision vector $\mathbf{x}^* \in \mathcal{X}$. A necessary condition for $\mathbf{x}^*$ to be Pareto-optimal is the existence of a vector $\boldsymbol{\delta} \in \mathbf{R}^{n_{of}+n_c}$ such that

$$\mathbf{L} \cdot \boldsymbol{\delta} = \mathbf{0}$$
$$\text{with} \quad \mathbf{L} = \mathbf{L}(\mathbf{x}^*) \quad \text{and} \quad \boldsymbol{\delta} \geq \mathbf{0} \quad and \quad \boldsymbol{\delta} \neq \mathbf{0}. \tag{3.8}$$

This definition is completely equivalent to definition (3.2) since $\boldsymbol{\delta}$ is the vector

$$\boldsymbol{\delta} = \begin{bmatrix} \boldsymbol{\lambda} \\ \boldsymbol{\mu} \end{bmatrix} \qquad [(n_{of} + n_c) \text{ x } 1] \tag{3.9}$$

Let us notice that no constraints on the respective values of n_{dv}, n_{of}, n_c have been set, so $\mathbf{L}$ is in general rectangular.

3.3 Analytical Derivation of the Pareto-Optimal Set

The matrix form of the Fritz John conditions (3.8) can be employed to derive the analytical expression of the Pareto-optimal set.

All of the following considerations refer to problems in which $n_{dv} \geq n_{of}$, that is the number of design variables is greater or equal than the number of objective functions.

The Fritz John conditions (see Eq. 3.8) can be relaxed by removing the condition $\boldsymbol{\delta} \geq \mathbf{0}$.

The relaxed Fritz John conditions read

$$\mathbf{L} \cdot \boldsymbol{\delta} = \mathbf{0}$$

$$\text{with} \qquad \mathbf{L} = \mathbf{L}(\mathbf{x}^*) \qquad \text{and} \qquad \boldsymbol{\delta} \neq \mathbf{0}. \tag{3.10}$$

This relaxation implies that we are dealing with necessary conditions also in presence of convex objective functions and constraints. So, the analytical expression derived on the basis of the relaxed form contains the actual Pareto-optimal set but also non-Pareto-optimal solutions.

The non-Pareto-optimal solutions have to be eliminated by computing the minimum of each objective function. Such minima obviously define the boundaries of the Pareto-optimal set.

The relaxed Fritz John conditions (Eq. 3.10) are a homogeneous system of linear equations [7]. The trivial solution $\boldsymbol{\delta} = \mathbf{0}$ is obviously of no interest.

If we consider the homogeneous overdetermined system of $(n_{dv} + n_c)$ linear equations in $(n_{of} + n_c)$ variables, $\mathbf{L} \cdot \boldsymbol{\delta} = \mathbf{0}$, there will be non-trivial solutions only when the system has enough linearly dependent equations so that the number of independent equations is at most $(n_{of} + n_c)$. But being $(n_{dv} + n_c) \geq (n_{of} + n_c)$, the number of independent equations (i.e. $rank(\mathbf{L})$) could be as high as $(n_{of} + n_c)$, in which case the trivial solution $\boldsymbol{\delta} = \mathbf{0}$ is the only one. The matrix $\mathbf{L}^\mathbf{T}\mathbf{L}$ is positive definite if and only if all the columns of $\mathbf{L}$ are linearly independent [33], i.e. $rank(\mathbf{L}) = n_{of} + n_c$ and a positive definite matrix is always nonsingular. If $\mathbf{L}^\mathbf{T}\mathbf{L}$ is nonsingular, there is the unique trivial solution $\boldsymbol{\delta} = \mathbf{0}$, but if $\mathbf{L}^\mathbf{T}\mathbf{L}$ is singular there are infinitely many solutions. This approach gives the exact solution when one does exist.

So, Eq. 3.10 admits non-trivial solution if

$$\det(\mathbf{L}^\mathbf{T}\mathbf{L}) = 0. \tag{3.11}$$

For a square matrix **L**, this condition is simply obtained by setting to zero its determinant.

If $n_{dv} < n_{of}$ (the number of design variables is smaller than the number of objective functions), $\det(\mathbf{L}^T\mathbf{L})$ is always zero and the problem is no longer a minimisation problem. The analytical expression of the Pareto-optimal set can be simply derived by directly applying the substitution method [143, 148].

So, we can state the following necessary conditions for Pareto-optimality (relaxed Fritz John conditions)

L-matrix necessary conditions for Pareto-optimality. Let the objective and constraint functions of problem (3.1) be continuously differentiable at a decision vector $\mathbf{x}^* \in \mathcal{X}$ and let be $n_{dv} \geq n_{of}$. A necessary condition for $\mathbf{x}^*$ to be Pareto-optimal is that $\det(\mathbf{L}^T\mathbf{L}) = 0$ with $\mathbf{L} = \mathbf{L}(\mathbf{x}^*)$.

By computing the determinant of the **L** matrix, the analytical relationship between the design variables that represents the Pareto-optimal set in the design variables domain can be obtained.

The flow chart of the proposed procedure to find the Pareto-optimal set in the design variables domain is shown in Fig. 3.1.

To get the analytical expression of the Pareto-optimal set in the objective functions domain, the relation obtained from the solution of Eq. 3.11 has to be substituted into the expressions of the objective functions. Referring to the flow diagram of Fig. 3.2, after this substitution (block (6)), the n_{of} objective functions are related to the remaining $n_{dv} - 1$ independent design variables.

Once at this point, a set of $n_{of} - 1$ auxiliary functions t_l $\left(with\ l = 1, \ldots, n_{of} - 1\right)$ have to be defined (block (7) in Fig. 3.2). The auxiliary functions $t_l\left(x_1, \ldots x_j, \mid j \neq i\right)$ depend on the $n_{dv} - 1$ independent design variables, but are defined in such a way to parametrise the objective functions on t_l only. Obviously, this operation, may be extremely cumbersome and even impossible if the analytical expressions are too complex.

Now that all the objective functions are expressed in terms of the auxiliary functions t_l only, the inverse functions have to be computed on a set of $n_{of} - 1$ objective functions (block (8) in Fig. 3.2).

Finally, by substituting the computed $n_{of} - 1$ inverse functions in the last expression of the objective function, the analytical formula of the Pareto-optimal set in the objective functions domain is obtained.

For a number of special problems, the L-matrix necessary conditions for Pareto-optimality can be further simplified as described in the following subsections.

3.3.1 Unconstrained Problem

Clearly, if the problem (3.1) is unconstrained, that is there are no constraint functions, the L matrix is simply $\mathbf{L} = \nabla\mathbf{f}$. Hence, the L-matrix condition is

$$\det(\mathbf{L}^T\mathbf{L}) = 0 \quad \Rightarrow \quad \det(\nabla\mathbf{f}^T\nabla\mathbf{f}) = 0 \tag{3.12}$$

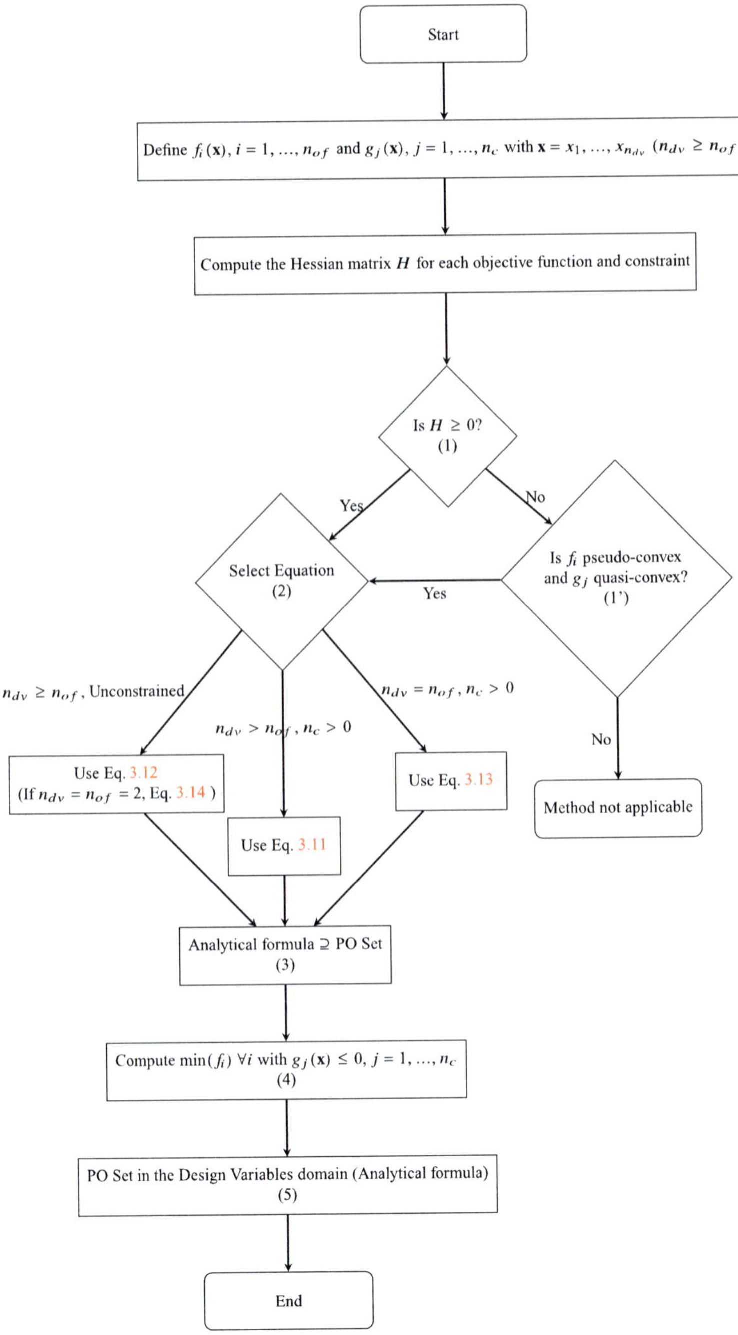

Fig. 3.1 Flow chart of the proposed procedure to find analytically the Pareto-optimal set in the design variables domain

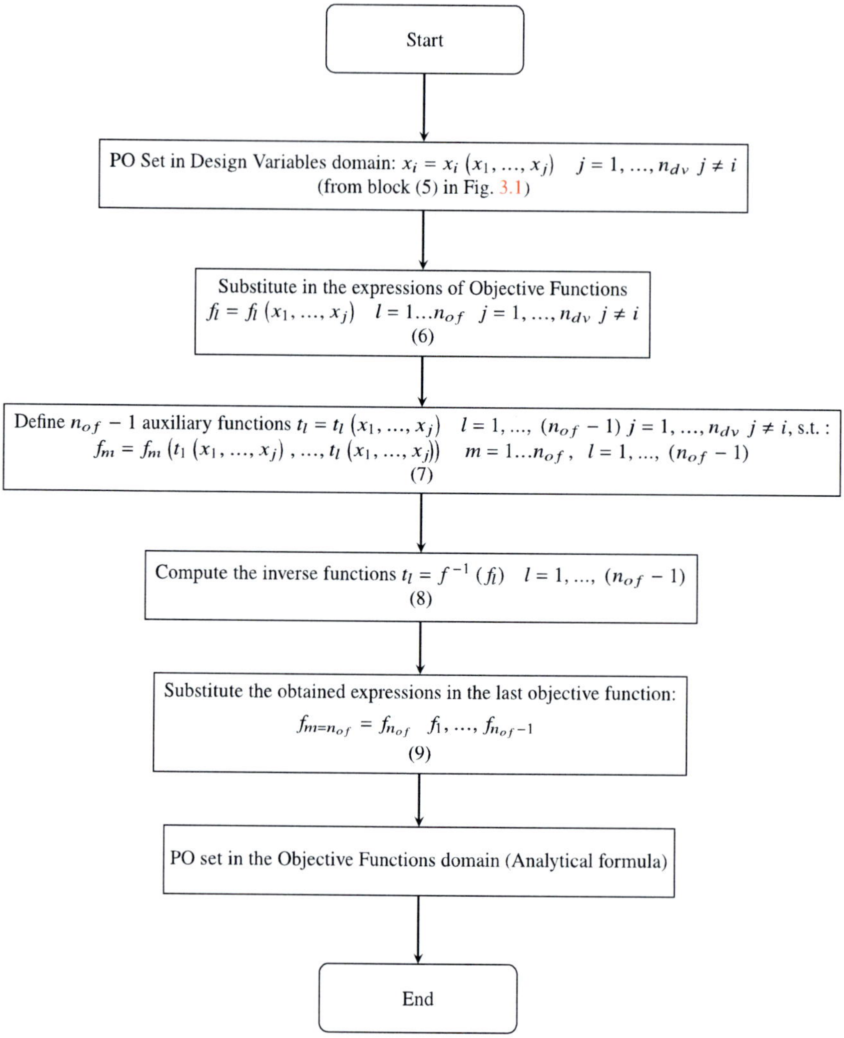

Fig. 3.2 Flow chart of the proposed procedure to find analytically the Pareto-optimal set in the objective functions domain

3.3.2 *Even Number of Design Variables and Objective Functions*

If $n_{dv} = n_{of}$, that is the number of design variables is equal to the number of objective functions, the L matrix is square. So, it is not necessary to multiply it by its transposed matrix and the L-matrix condition is simply, see Eqs. 3.6, 3.7

$$\det(\mathbf{L}^T\mathbf{L}) = 0 \quad \Rightarrow \quad \det(\mathbf{L}) = 0 \quad \Rightarrow \quad \left(\prod_{i=1}^{n_c} g_i\right) \cdot \det(\nabla\mathbf{f}) = 0 \tag{3.13}$$

Hence, the Pareto-optimal set for the constrained problem can be an active constraint and/or the Pareto-optimal set for the unconstrained problem, see Eq. 3.12.

3.3.2.1 Unconstrained Problem with $n_{dv} = n_{of} = 2$

For the simplest multi-objective unconstrained problem with two objective functions and two design variables, the L-matrix condition is

$$\det(\nabla\mathbf{f}^T\nabla\mathbf{f}) = 0 \quad \Rightarrow \quad \det(\nabla\mathbf{f}) = 0 \quad \Rightarrow \quad \frac{\partial f_1}{\partial x_1} \cdot \frac{\partial f_2}{\partial x_2} = \frac{\partial f_1}{\partial x_2} \cdot \frac{\partial f_2}{\partial x_1} \tag{3.14}$$

According to the knowledge of the authors, this formula is original. The formula has been successfully applied by the authors in [89] to solve the actual and very important engineering problem referring to the trade-off between road-holding and comfort of ground vehicles.

3.4 Case Studies

A number of case studies are presented which address the derivation of the Pareto-optimal set in the design variables domain. The Pareto-optimal set in the objective functions domain is derived, when possible, by substituting the obtained expressions of the Pareto-optimal set in the design variables domain into the expressions of the objective functions as described by the flow chart of Fig. 3.2.

3.4.1 Case #1. Two Design Variables, Two Objective Functions, No Constraints

The problem proposed by Fonseca and Fleming in [80] has been selected from the ones presented in the literature [208], and used to find analytically the Pareto-optimal set (see Fig. 3.1).

The problem has two design variables and two objective functions and reads

$$\textit{minimise} \quad \begin{pmatrix} f_1(x_1, x_2) = 1 - e^{-[(x_1-1/\sqrt{2})^2+(x_2-1/\sqrt{2})^2]} \\ \\ f_2(x_1, x_2) = 1 - e^{-[(x_1+1/\sqrt{2})^2+(x_2+1/\sqrt{2})^2]} \end{pmatrix} \tag{3.15}$$

Convexity can be easily verified for the problem given by Eq. 3.15 by computing the Hessian matrix for the two objective functions and checking that $H(f_1(x_1, x_2)) \geq 0$ and $H(f_2(x_1, x_2)) \geq 0$, see Step 1 in Fig. 3.1.

Equation 3.14 (Step 2 in Fig. 3.1) can be directly applied to obtain the analytical expression of the Pareto front, see Step 3 in Fig. 3.1.

$$(x_1 - 1/\sqrt{2})(x_2 + 1/\sqrt{2}) = (x_1 + 1/\sqrt{2})(x_2 - 1/\sqrt{2}) \quad \Rightarrow \quad x_1 = x_2 \tag{3.16}$$

which has to be limited by the minima of the two objective functions considered separately.

The two minima are computed by directly setting to zero the gradients, being both the Hessian matrices for f_1 and f_2 positive semi-definite, see Step 4 in Fig. 3.1.

$$\begin{aligned} \nabla f_1 = 0 \quad (H(f_1) \geq 0) \quad &\Rightarrow \quad x_1 = 1/\sqrt{2}, x_2 = 1/\sqrt{2} \\ \nabla f_2 = 0 \quad (H(f_2) \geq 0) \quad &\Rightarrow \quad x_1 = -1/\sqrt{2}, x_2 = -1/\sqrt{2} \end{aligned} \tag{3.17}$$

So the analytical expression of the Pareto-optimal front in the design variables domain (see Step 5 in Fig. 3.1) reads

$$x_1 = x_2 \quad with \quad -1/\sqrt{2} \leq x_1 \leq 1/\sqrt{2} \quad and \quad -1/\sqrt{2} \leq x_2 \leq 1/\sqrt{2}. \tag{3.18}$$

The Pareto-optimal set in the objective functions domain can be derived analytically by simple substitutions. By following the flowchart of Fig. 3.2, the substitution of Eq. 3.16 in the objective functions expressions gives

$$f_1 = 1 - e^{-2(x_2 - 1/\sqrt{2})^2} \tag{3.19}$$

$$f_2 = 1 - e^{-2(x_2 + 1/\sqrt{2})^2} \tag{3.20}$$

computing the inverse of Eq. 3.20 and removing the non-Pareto solution, we obtain

$$x_2 = \frac{1}{\sqrt{2}} \left(\sqrt{-\log(1 - f_2)} - 1 \right). \tag{3.21}$$

It has to be noticed that in this case, being the functions relatively simple, there was no need to define any auxiliary function (or alternatively the auxiliary function is $t = x_2$).

By substituting Eq. 3.21 into Eq. 3.19, the analytical expression of the Pareto-optimal set in the objective functions domain can be computed as

$$f_1 = 1 + (f_2 - 1)\, e^{-4+4\sqrt{-\log(1-f_2)}} \quad 0 \leq f_2 \leq 0.982. \tag{3.22}$$

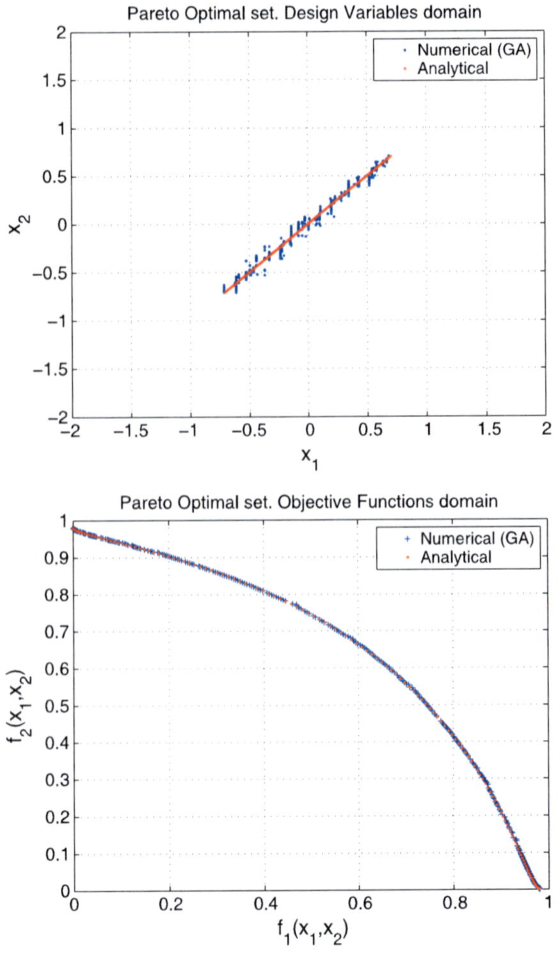

Fig. 3.3 Fonseca and Fleming problem. Analytical versus numerical solution. Pareto-optimal set into the design variables domain (top) and into the objective functions domain (bottom) See Sect. 3.4.1

Even if the two objective functions are convex, the Pareto front is non-convex, so the Pareto-optimal set cannot be computed by applying a simple weighted sum approach [143]. A state of the art multi-objective genetic algorithm (MOGA) [84] has been used to validate the analytical results. The results are shown in Fig. 3.3. The MOGA requires about ten thousand objective functions evaluations to obtain a reasonable level of accuracy.

3.4.2 Case #2. Two Design Variables, Two Objective Functions, One Constraint

Let us consider the case of two spheres pressed one against the other. The problem has two design variables, i.e. the diameters of the two spheres D_1 and D_2. Two objective functions are to be minimised, namely, the total mass M and the relative displacement of the two spheres along the axis of loading y. The design constraint is the maximum stress in one (or both) of the spheres σ_{max}. The problem could refer to the basic design of sintered materials [45]. The elastic stress and deformation produced by the pressure between two spheres have been modelled on the basis of Hertz's theory [250].

The optimisation problem reads

$$minimise \quad \begin{pmatrix} M(D_1, D_2) = \frac{1}{6}\pi\,\rho\left({D_1}^3 + {D_2}^3\right) \\ \\ y(D_1, D_2) = k_y \sqrt[3]{\frac{P^2\,(D_1+D_2)}{D_1\,D_2\,E^2}} \end{pmatrix}$$

$$\sigma_{max} = k_\sigma \sqrt[3]{\frac{P\,E^2\,(D_1 + D_2)^2}{D_1^2\,D_2^2}} \leq \sigma_{adm} \tag{3.23}$$

given ρ the material density, E the material modulus of elasticity, σ_{adm} the material admissible stress, $k_\sigma = 0.616$, $k_y = 1.55$ [250] and P the external force.

Pseudo-convexity can be easily verified for the problem in Eq. 3.23 by applying Eq. 3.4. $M(D_1, D_2)$ and $y(D_1, D_2)$ are pseudo-convex, being $|\mathbf{B}_1| < 0$, $|\mathbf{B}_2| < 0$ for all D_1, D_2, see Step 1' in Fig. 3.1.

Equation 3.13 (Step 2 in Fig. 3.1) can be directly applied to obtain the analytical expression of the Pareto front, see Step 3 in Fig. 3.1.

$$P^2\,k_y\,\pi\,\rho\,\left({D_1}^4 - {D_2}^4\right)\left(\sigma_{adm} - k_\sigma \sqrt[3]{\frac{E^2\,P\,(D_1 + D_2)^2}{{D_1}^2\,{D_2}^2}}\right) = 0 \tag{3.24}$$

The Pareto-optimal front for the constrained problem is coincident with the Pareto-optimal set for the unconstrained problem, see Eq. 3.12, up to the activation of the maximum stress constraint $\sigma_{max} = \sigma_{adm}$ and it reads

$$D_1 = D_2 \quad for \quad D_2 \geq 2\,E\,\sqrt{\frac{P\,k_\sigma^3}{\sigma_{adm}^3}} \tag{3.25}$$

which has to be limited by the minima of the two objective functions considered separately.

The two minima read, see Step 4 in Fig. 3.1.

$$\begin{aligned} &min \quad M(D_1, D_2) \quad D_1 = D_2 \rightarrow 0 \\ &min \quad y(D_1, D_2) \quad D_1 = D_2 \rightarrow \infty \end{aligned} \tag{3.26}$$

Equations 3.26 don't affect the results given in Eq. 3.25, so the analytical expression of the Pareto-optimal set in the design variables domain (see Step 5 in Fig. 3.1) reads

$$D_1 = D_2 \quad for \quad D_2 \geq 2\,E \sqrt{\frac{P\,k_\sigma^3}{\sigma_{adm}^3}} \tag{3.27}$$

In the objective functions domain, the analytical expression of the Pareto-optimal set (obtained by simple substitution) reads

$$y = k_y \sqrt[3]{\frac{2\,P^2}{E^2\,(\frac{3\,M}{\pi\,\rho})^{\frac{1}{3}}}} \tag{3.28}$$

A numerical procedure has been used to validate the analytical results. The results are shown in Fig. 3.4.

3.4.3 Case #3. Two Design Variables, Two Objective Functions, Two (four) Constraints

Let us consider the design of a cantilever beam already introduced in Chap. 2. The cantilever, shown in Fig. 3.5, has a rectangular cross section and a force acts at the free end. Let us assume that the optimisation problem to be solved is to find the values of the design variables (length b and length h) defining the cantilever cross section[1] in order to minimise both the cantilever mass and the cantilever deflection at its free end. The two constraints refer, respectively, to the maximum stress at the fixed end (the stress must be less than (or equal to) the admissible stress) and to elastic stability (buckling to be avoided).

Again, as shown in Chap. 2, we choose the values defining the cross section of the cantilever in order to get it as light and stiff as possible, avoiding both a failure (due to too high stress) and elastic instability.

In mathematical form, the above optimisation problem may be stated as follows:

[1] b and h may vary, respectively, within two well-defined ranges.

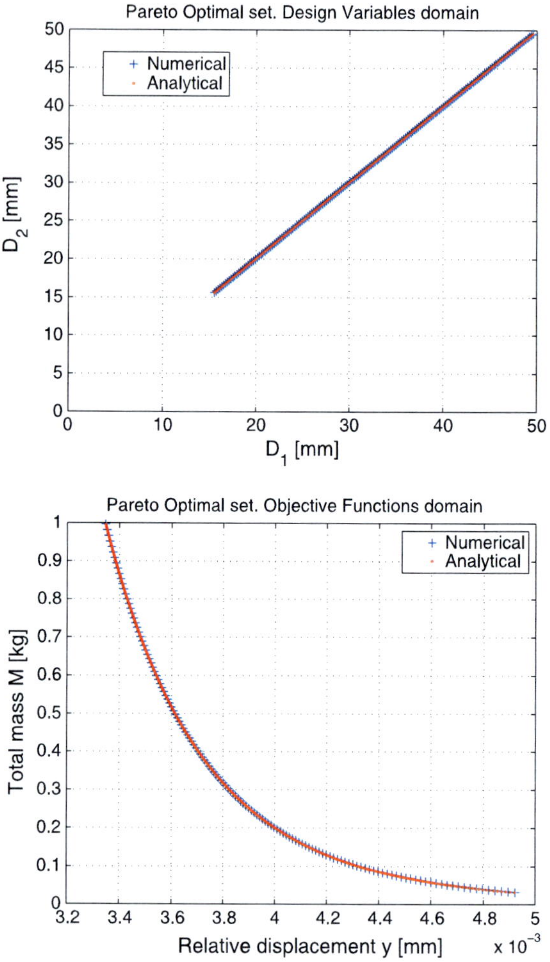

Fig. 3.4 Analytical versus numerical solution. Pareto-optimal set into the design variables (D_1, D_2) domain (top) and into the objective functions (y, M) domain (bottom) for the problem 3.23. Material 100Cr6 Steel, $\rho = 7800\,\text{kg/m}^3$, $E = 210000\,\text{MPa}$, $\sigma_{adm} = 2500\,\text{MPa}$, $F = 100\,\text{N}$, $k_\sigma = 0.616$, $k_y = 1.55$

Given

l	the cantilever length	[m]
b	the beam cross section width	[m]
h	the beam cross section height	[m]
J	$= \frac{1}{12}bh^3$ the flexural moment of inertia of the section (n–n axis)	[m^4]
F	the force applied at the cantilever free end (see Fig. 3.5)	[N]
σ_E	the material yield stress	[MPa]
η	safety coefficient ($\geq$1)	[–]
σ_{adm}	$= \sigma_E / h$ the admissible stress at the cantilever fixed end	[MPa]
E	the material modulus of elasticity (Young's modulus)	[MPa]
G	the material modulus of tangential elasticity	[MPa]
ρ	the material density	[kg/m^3]

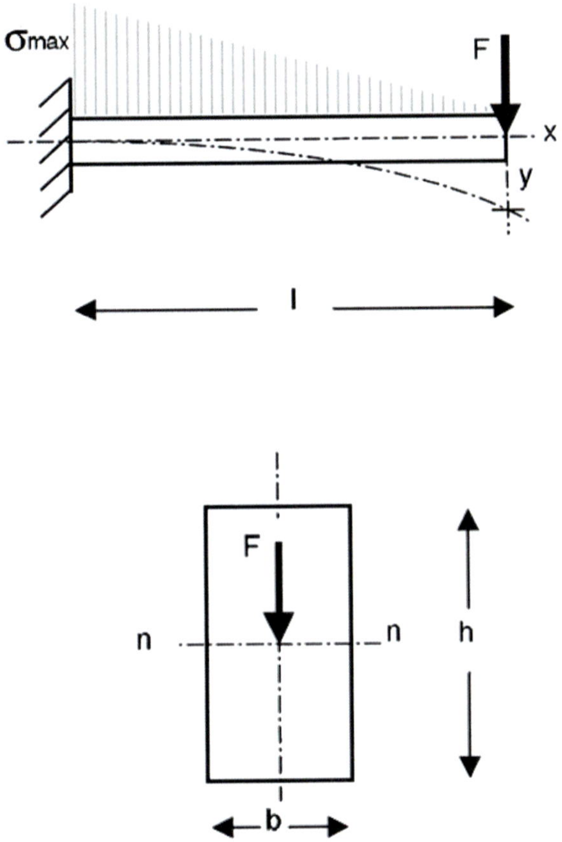

Fig. 3.5 Cantilever whose rectangular cross section has to be defined in order to minimise both the cantilever mass and the cantilever deflection at the free end

and defining

$$m = \rho bhl$$

the cantilever mass [kg]

$$y = \frac{1}{3}\frac{Fl^3}{EJ} = 4\frac{Fl^3}{Ebh^3}$$

the deflection at the free end due to load F [m]

$$\sigma_{max} = 6\frac{Fl}{bh^2}$$

the maximum stress located at the top of the cross section at the fixed end of the cantilever [MPa]

$$F_{cr} = \frac{k_1 b^3 h}{l^2}\sqrt{\left(1 - k_2\frac{b}{h}\right) EG}$$

the critical load (see [86, 250]) [N]

find b and h such that

$$b_{min} \leq b \leq b_{max}$$
$$h_{min} \leq h \leq h_{max}$$

and such that

$$\min \begin{pmatrix} m(b,h) \\ y(b,h) \end{pmatrix} = \min \begin{pmatrix} \rho bhl \\ 4\frac{Fl^3}{Ebh^3} \end{pmatrix} \tag{3.29}$$

subject to

$$\sigma_{max} = 6\frac{Fl}{bh^2} \leq \sigma_{adm} = \sigma_E/\eta \tag{3.30}$$

$$F < F_{cr} = \frac{k_1 b^3 h}{l^2}\sqrt{\left(1 - k_2\frac{b}{h}\right) EG} \tag{3.31}$$

The analytical solution of the stated optimisation problem can be found by first considering the problem as unconstrained, i.e. admissible maximum stress and elastic stability constraints are removed from the problem formulation. The objective functions are continuous and differentiable. The beam deflection y is convex [9]

$$\mathbf{H} = 4Fl^3 \begin{bmatrix} \frac{2}{b^3 h^3} & \frac{3}{b^2 h^4} \\ \frac{3}{b^2 h^4} & \frac{12}{bh^5} \end{bmatrix} \tag{3.32}$$

being the eigenvalues of matrix $\mathbf{H}$ both real and positive.

The beam mass m is not convex, but only pseudo-convex, being m strictly monotonically increasing for any feasible value of b and h ($b > 0,\ h > 0$).

The L-matrix defined by Eq. 3.7 with $\nabla \mathbf{G} = \mathbf{0}$ (no constraints present), reads

$$\mathbf{L} = \begin{bmatrix} h\,l\,\rho & -\frac{4\,F\,l^3}{E\,b^2\,h^3} \\ b\,l\,\rho & -\frac{12\,F\,l^3}{E\,b\,h^4} \end{bmatrix} \tag{3.33}$$

By applying Eq. 3.14, we have

$$-\frac{8\,F\,l^4\,\rho}{E\,b\,h^3} = 0 \tag{3.34}$$

which has solution only if $bh^3 \to \infty$. Such solution not belonging to the set of the positive and finite numbers has no physical meaning. The nonexistence of the Pareto-optimal set can be proved by considering the ϵ-constraint method [100] and by applying the monotonicity principles [164]. The first monotonicity principle is violated for this optimisation problem. In fact, if we apply the ϵ-constraint method and we consider the beam mass (m) as objective function and the beam deflection ($y(b,h)$) as (active) design constraint, the objective function ($m(b,h)$) is mono-

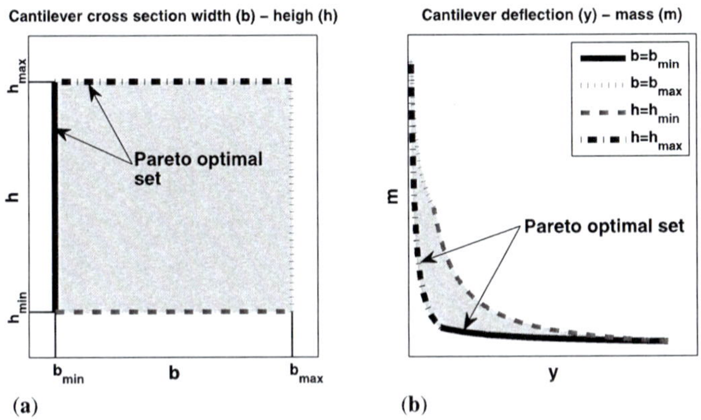

Fig. 3.6 Optimal design of a cantilever beam. The two Pareto-optimal sets defined, respectively, in the design variables domain (**a**) and in the objective functions domain (**b**). Unconstrained problem with $b_{min} \leq b \leq b_{max}$ and $h_{min} \leq h \leq h_{max}$. Data in Table 3.1

tonically increasing by considering both the design variables (b, h), but the design variables are not bounded below by the active constraint on beam deflection $(y(b, h))$.

By including upper and lower bounds for b and h, the Pareto-optimal set for the unconstrained problem is given by the two boundaries on the design variables domain ($b = b_{min}$ and $h = h_{max}$), as shown in Fig. 3.6. The Pareto-optimal set is limited between $\min m = m(b_{min}, h_{min})$ and $\min y = y(b_{max}, h_{max})$.

If a sufficiently large design space is considered, the two design constraints read

$$\sigma = \frac{6\,F\,l}{b\,h^2} \leq \sigma_{max} \tag{3.35}$$

$$F_{cr} = \frac{k_1 b^3 h}{l^2}\sqrt{\left(1 - k_2\frac{b}{h}\right) EG} \geq F \tag{3.36}$$

that for any given b can be rewritten as

$$h \geq \sqrt{\frac{6\,F\,l}{b\,\sigma_{max}}} \tag{3.37}$$

$$h \geq \frac{b\,\mathrm{k}_2}{2} + \frac{\sqrt{E^2\,G^2\,b^8\,\mathrm{k}_1{}^2\,\mathrm{k}_2{}^2 + 4\,E\,F^2\,G\,l^4}}{2\,E\,G\,b^3\,\mathrm{k}_1} \tag{3.38}$$

Equation 3.37 represents the limit on the admissible maximum stress, it is a monotonic decreasing function proportional to $1/\sqrt{b}$ and it has limits $h \to \infty$ as $b \to 0$ and $h \to 0$ as $b \to \infty$. Equation 3.38 represents the limit on buckling. It has limits

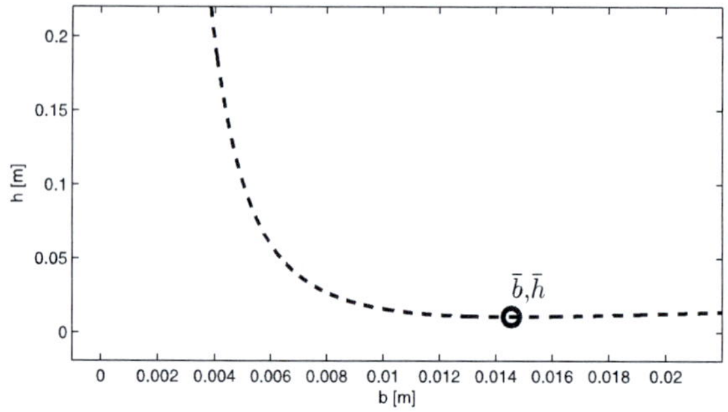

Fig. 3.7 Optimal design of a cantilever beam. Equation (3.38) represents the buckling limit. It has a minimum at $\bar{b}, \bar{h}$. For $b > \bar{b}$ the buckling constraint Eq. (3.38) has no physical meaning

$h \to \infty$ as $b \to 0$ and $h \to \infty$ as $b \to \infty$. Function (3.38), plotted in Fig. 3.7, is non-monotonic and it has a minimum at

$$\bar{b} = \sqrt[8]{\frac{36\, F^2\, l^4}{7\, E\, G\, k_1^2\, k_2^2}}, \qquad \bar{h} = \sqrt[8]{\frac{7^7\, F^2\, l^4\, k_2^6}{6^6\, E\, G\, k_1^2}} \tag{3.39}$$

It must be noticed that for $b > \bar{b}$ the buckling constraint Eq. (3.38) has no physical meaning, i.e. the buckling constraint has to be considered only for $b \leq \bar{b}$. For $b \to 0$, the equation of the buckling constraint is proportional to $1/b^3$ and, thus, is more binding than the constraint on the maximum stress. For $b = \bar{b}$ and $h = \bar{h}$ the maximum stress on the cantilever beam can be computed by replacing Eq. 3.39 into the expression of the maximum stress (3.35). By considering reasonable values for the system parameters, the computed stress level for $b = \bar{b}$ and $h = \bar{h}$ is unacceptable for any engineering material. Therefore at $b = \bar{b}$ (and above) the constraint on the maximum stress is active.

Since for $b \to 0$ the buckling constraint is active and for $b = \bar{b}$ the constraint on the maximum stress is active and the two constraints functions are continuous and monotonic in the range $0 < b < \bar{b}$, there must be a design solution given by equation

$$\frac{b\, \mathrm{k}_2}{2} + \frac{\sqrt{E^2\, G^2\, b^8\, \mathrm{k}_1{}^2\, \mathrm{k}_2{}^2 + 4\, E\, F^2\, G\, l^4}}{2\, E\, G\, b^3\, \mathrm{k}_1} - \sqrt{\frac{6\, F\, l^2}{b\, \sigma_{max}}} = 0 \tag{3.40}$$

in which the active constraint switches from buckling to maximum stress.

By substituting Eq. 3.37 into the expressions of the mass and of the deflection of the cantilever beam, the following equation representing the maximum stress constraint in the objective functions domain can be obtained:

$$m = \frac{9\, \rho\, F\, E}{\sigma_{\max}{}^2}\, y \tag{3.41}$$

Equation 3.41 is a straight line with positive angular coefficient, so the buckling constraint is the active one up to the maximum acceptable stress level, and the L-matrix of Eq. 3.7 reads

$$\mathbf{L} = \begin{bmatrix} h\,l\,\rho - \frac{4\,F\,l^3}{E\,b^2\,h^3} & \frac{E\,G\,b^2\,\mathrm{k}_1\,(6\,h-7\,b\,\mathrm{k}_2)}{2\,l^2\,\sqrt{\frac{E\,G\,(h-b\,\mathrm{k}_2)}{h}}} \\ b\,l\,\rho - \frac{12\,F\,l^3}{E\,b\,h^4} & \frac{E\,G\,b^3\,\mathrm{k}_1\,(2\,h-b\,\mathrm{k}_2)}{2\,h\,l^2\,\sqrt{\frac{E\,G\,(h-b\,\mathrm{k}_2)}{h}}} \\ 0 \quad 0 & -\frac{F\,l^2-b^3\,h\,\mathrm{k}_1\,\sqrt{\frac{E\,G\,(h-b\,\mathrm{k}_2)}{h}}}{l^2} \end{bmatrix} \tag{3.42}$$

Being $n = k$, we apply Eq. 3.13, so we have

$$h = \frac{b\,\mathrm{k}_2}{2} + \frac{\sqrt{E^2\,G^2\,b^8\,{\mathrm{k}_1}^2\,{\mathrm{k}_2}^2 + 4\,E\,F^2\,G\,l^4}}{2\,E\,G\,b^3\,\mathrm{k}_1} \tag{3.43}$$

Equation 3.43 is the analytical expression of the Fritz John necessary condition when the buckling constraint is active and it is coincident with the expression of the buckling boundary condition itself (see Eq. 3.38).

The equation of this set in the space of the objective functions can be expressed as

$$y = y\,(m) = y\,(m\,(b,\,h)) \tag{3.44}$$

In this case, the derivation of $y = y(m)$ could not be performed as the analytical derivation is cumbersome.

The Pareto-optimal set is the portion of Eq. 3.43 limited by Eq. 3.40.

The first and second derivatives of the function can be found by applying the chain rule

$$\frac{dm}{dy} = \frac{\partial m}{\partial b}\frac{\partial b}{\partial y} + \frac{\partial m}{\partial h}\frac{\partial h}{\partial y} = -\frac{E\,b^2\,h^4\,\rho}{3\,F\,l^2} \tag{3.45}$$

$$\frac{d^2y}{dm^2} = \frac{\partial\left(\frac{dy}{dm}\right)}{\partial b}\frac{\partial b}{\partial m} + \frac{\partial\left(\frac{dy}{dm}\right)}{\partial h}\frac{\partial h}{\partial m} = \frac{5\,E\,b^3\,h^7\,\rho}{18\,F^2\,l^5} \tag{3.46}$$

Equations 3.45 and 3.46 show that this function is decreasing and convex for any value of b and h. Therefore, the points belonging to the buckling boundary condition belong to the Pareto-optimal set [31, 66]. Figure 3.8 shows the Pareto-optimal set. In the design variables domain, the Pareto-optimal set is composed by the curve representing the buckling boundary condition from the point defined by Eq. 3.40 to the point where $h = h_{max}$ and then on the boundary $h = h_{max}$ from this point to $b = b_{max}$.

A numerical validation is reported in Fig. 3.9.

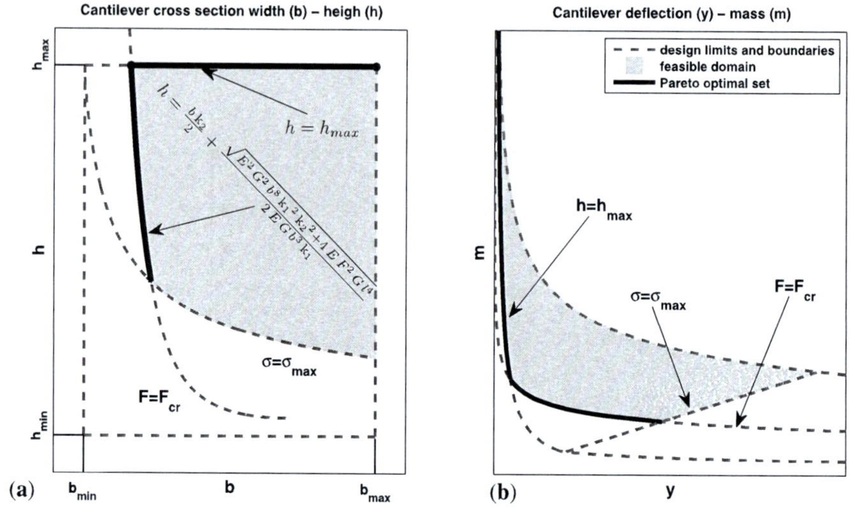

Fig. 3.8 Optimal design of a cantilever beam with a rectangular cross section. The two Pareto-optimal sets defined, respectively, in the design variables domain (**a**) and in the objective functions domain (**b**). The Pareto-optimal set is given by either the buckling condition or the maximum allowed height of the cross section. Data in Table 3.1

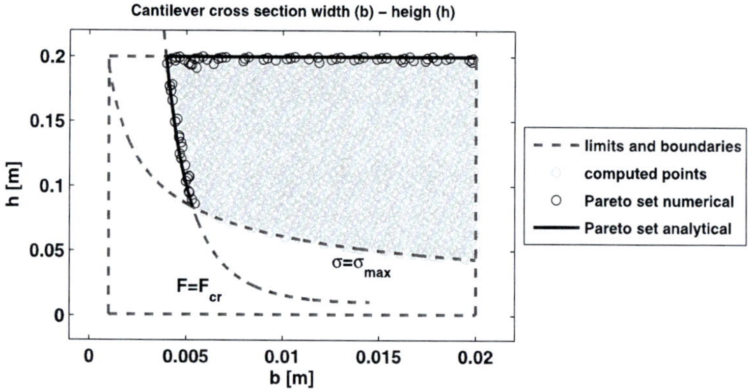

Fig. 3.9 Numerical validation of the Pareto-optimal solutions in the design variables domain for the cantilever beam with rectangular section of Fig. 3.5. Numerical validation. Data in Table 3.1

3.4.4 *Case #4. Three Design Variables, Two Objective Functions, Two Constraints*

The problem proposed by Fonseca and Fleming in [80] with three design variables has been considered.

The problem has three design variables, two objective functions and two constraints

Table 3.1 Structural parameters of the rectangular cantilever beam of Fig. 3.5

Notation	Description	Value	Unit
ρ	Material density	2700	$\frac{\text{kg}}{\text{m}^3}$
L	Beam length	1	m
E	Material modulus of elasticity	$70 \cdot 10^9$	Pa
G	Material modulus of tangential elasticity	$27 \cdot 10^9$	Pa
σ_E	Material yield stress	$160 \cdot 10^6$	Pa
η	Safety factor	1	–
b_{min}	Lower bound for b	0.001	m
b_{max}	Upper bound for b	0.020	m
h_{min}	Lower bound for h	0.001	m
h_{max}	Upper bound for h	0.05	m
F	Load at free end	1000	N

$$
\begin{aligned}
& \textit{minimise} \begin{pmatrix} f_1(x_1, x_2, x_3) = 1 - e^{-[(x_1-1/\sqrt{3})^2+(x_2-1/\sqrt{3})^2+(x_3-1/\sqrt{3})^2]} \\ \\ f_2(x_1, x_2, x_3) = 1 - e^{-[(x_1+1/\sqrt{3})^2+(x_2+1/\sqrt{3})^2+(x_3+1/\sqrt{3})^2]} \end{pmatrix} \\
& \frac{1}{2} \leq x_1 \leq \frac{1}{2}
\end{aligned} \tag{3.47}
$$

Convexity can be easily verified for the problem given by Eq. 3.47 by computing the Hessian matrix for the two objective functions and checking that $H(f_1(x_1, x_2, x_3)) \geq 0$ and $H(f_2(x_1, x_2, x_3)) \geq 0$, see Step 1 in Fig. 3.1.

Equation 3.11 (Step 2 in Fig. 3.1) can be applied to obtain the analytical expression of the Pareto front, see Step 3 in Fig. 3.1.

$$
\begin{aligned}
&(x_1 - x_2)(x_2 - x_3)\left(16x_1^4 + 8x_1^2 + 5\right) - (x_1 - x_3)(x_2 - x_3)\left(16x_1^4 + 8x_1^2 + 5\right) \\
&- (x_1 - x_2)(x_1 - x_3)(2x_1 - 1)^2(2x_1 + 1)^2 = 0
\end{aligned} \tag{3.48}
$$

which gives

$$
\begin{aligned}
x_1 = x_2 = x_3 \quad & if\ \frac{1}{2} \leq (x_1 = x_2 = x_3) \leq \frac{1}{2} \\
x_1 = \frac{1}{2},\ x_2 = x_3 \quad & if\ (x_2 = x_3) > \frac{1}{2} \\
x_1 = \frac{1}{2},\ x_2 = x_3 \quad & if\ (x_2 = x_3) < -\frac{1}{2}
\end{aligned} \tag{3.49}
$$

which has to be limited by the minima of the two objective functions taking into account the design constraints.

The two minima have been computed, see Step 4 in Fig. 3.1.

$$
\begin{aligned}
\min f_1 \quad &\Rightarrow \quad x_1 = \frac{1}{2},\ x_2 = 1/\sqrt{3},\ x_3 = 1/\sqrt{3} \\
\min f_2 \quad &\Rightarrow \quad x_1 = -\frac{1}{2},\ x_2 = -1/\sqrt{3},\ x_3 = -1/\sqrt{3}
\end{aligned}
\tag{3.50}
$$

So the analytical expression of the Pareto-optimal front in the design variables domain (see Step 5 in Fig. 3.1) reads

$$
\begin{aligned}
x_1 = x_2 = x_3 \quad & if\ -\frac{1}{2} \leq (x_1 = x_2 = x_3) \leq \frac{1}{2} \\
x_1 = \frac{1}{2},\ x_2 = x_3 \quad & if\ \frac{1}{2} < (x_2 = x_3) \leq 1/\sqrt{3} \\
x_1 = -\frac{1}{2},\ x_2 = x_3 \quad & if\ -1/\sqrt{3} \leq (x_2 = x_3) < -\frac{1}{2}
\end{aligned}
\tag{3.51}
$$

A numerical procedure has been used to validate the analytical results. The results are shown in Fig. 3.10.

3.5 Summary

A procedure has been proposed to find—when possible—the Pareto-optimal set in the analytical form in the design variables domain. Both the objective and constraint functions are assumed to be twice differentiable and convex or pseudo-convex. The Fritz John necessary condition for the Pareto-optimality has been re-formulated in matrix form. This formulation has been employed to derive a new necessary condition (the L-matrix necessary condition) that is a relaxed form of the Fritz John one.

If the number of design variables is greater than (or equal to) the number of objective functions, the L-matrix condition can be applied for the analytical derivation of the Pareto-optimal set.

In particular, when two design variables and two objective functions define an optimisation problem, the Pareto-optimal set can be computed quite easily by applying the simple formula 3.14 which seems original and is based on simple partial derivatives.

If the number of design variables equals the number of objective functions, the Pareto-optimal set in the design variables domain can be found after the product of the constraint functions times the determinant of the Jacobian of the objective functions.

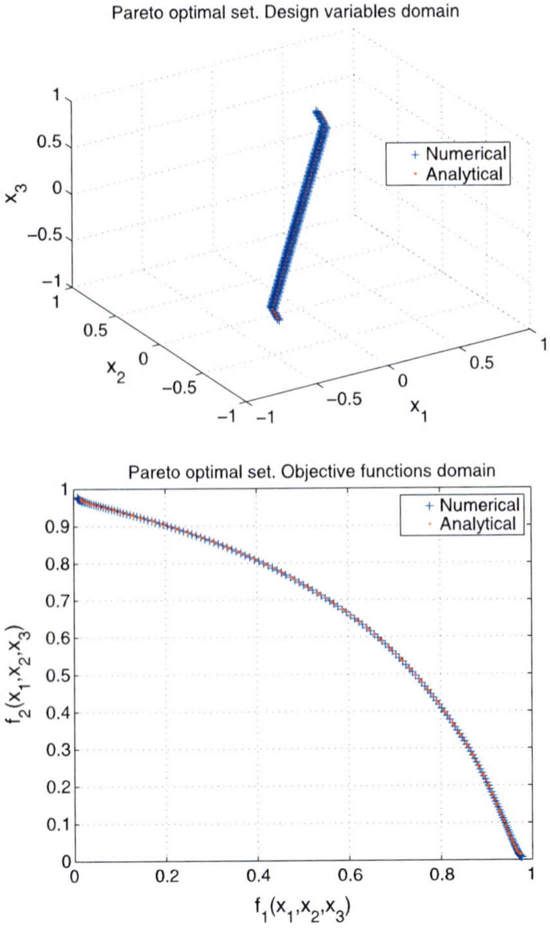

Fig. 3.10 Fonseca and Fleming problem. Analytical versus numerical solution. Pareto-optimal set into the design variables (x_1, x_2, x_3) domain (top) and into the objective functions (f_1, f_2) domain (bottom). See Sect. 3.4.4

After the Pareto-optimal set has been found in the design variables domain, the Pareto-optimal set in the objective functions domain can be derived, provided that a compact analytical form exists.

The proposed procedure for the analytical derivation of Pareto-optimal sets appears to be quite general and can be easily applied to problems with low dimensionality. Obviously, a numerical check of the derived analytical Pareto-optimal set is quite useful. Nonetheless, the analytical formulation of the Pareto-optimal set provides a strong reference for designers.

An attempt to prove the effectiveness of the proposed procedure to find analytically the Pareto-optimal set has been performed. A number of basic engineering problems have been addressed. First, the test problem proposed by Fonseca and Fleming with two design variables has been solved analytically. Second, the respective radii of two spheres pressed one against the other have been defined by minimising both the total mass and the total deflection, with the constraint to preserve their structural integrity. The result is that the diameter of the two spheres must be the same to obtain Pareto-optimal solutions. Third, the (constrained) dimensions of the rectangular cross section of a cantilever beam subject to bending have been defined by minimising both the mass and the deflection, with the constraint to preserve the structural integrity (referring to both maximum stress and buckling). The result is that the Pareto-optimal set in the design variables domain is defined by either the buckling or the maximum height of the rectangular cross section. Sections approaching the buckling constraint, i.e. sections with high height/width ratio, represent the best compromise. Fourth, the test problem proposed by Fonseca and Fleming with three design variables and two inequality constraints has been solved analytically.

Chapter 4
Bending of Beams of Arbitrary Cross Sections—Optimal Design by Analytical Formulae

At the very early stage of the design of a structure, the designer has to make a number of choices that will affect the whole project and that could be extremely costly and time-consuming to be modified in a later design stage. Many possible design solutions are initially available. A preliminary optimisation can be useful to the designer to get insight into the problem, to estimate the attainable performances.

The conceptual model of the structure, at this initial stage, is generally very simple and a few design variables have to be considered. In Chap. 3, a simple and efficient tool is presented to derive the Pareto-optimal set for design problems described by a limited number of design variables and objective functions. The availability of analytical expressions for the Pareto-optimal set allows the designer to quickly select many possible solutions and to choose the most promising system configurations for the subsequent refined design.

Referring to the simple, but very common, case of the design of a beam under bending, the first decision that has to be made is the material selection. The material selection is not trivial as it involves many conflicting requirements. The beam should be able to sustain the load without failing (structural safety) or becoming unstable (elastic stability). Additionally, mass and compliance should be minimised to obtain a light and stiff structure (lightweight design) [9, 86].

The comparison of different materials cannot be separated from the analysis of the cross sections that the beam can assume [10]. In [10], the effects of the cross-sectional shape on the material selection for beams and axles are analysed. Procedures are presented in the book [10] for the correct choice of material and shape. The Pareto-optimality theory is applied to balance cost and performance.

The analytical solution for a single objective optimisation (minimum compliance) of a beam under different and separated loads is derived in [168]. The considered design variable is the area of the beam. In [21] the shape optimisation of beams under bending or torsion is performed analytically for the different cases of stress, stiffness or stability-driven design. The optimisation of cylindrical bar cross sections with regular polygonal contours under stiffness and strength constraints is discussed in [22]. The problem of finding the minimum area of a thin-walled closed symmetrical cross

F. M. Ballo et al., *Optimal Lightweight Construction Principles*,
https://doi.org/10.1007/978-3-030-60835-4_4

section subject to bending with prescribed constant thickness and bending stiffness is solved in [175]. In [28, 70, 129, 143], the multi-objective problem referring to the derivation of an optimal beam has been introduced and solved by applying numerical optimisation methods. In [90], the multi-objective optimisation of a beam of fixed cross section (rectangular) is solved analytically.

Despite the many contributions mentioned above, in the known literature a comprehensive analytical method to solve the basic problem of designing a beam subjected to bending was never proposed. The chapter of this book aims at covering this gap.

4.1 Bending of a Beam

In this section, the problem of the design of a beam subjected to bending is stated.

Figure 4.1 shows a beam subjected to bending. The geometry of the cross section is characterised by the three quantities A, I and y_G representing, respectively, the area, the moment of inertia and the maximum distance from the neutral axis of a point of the cross section border, respectively. The cross section is constant along the length of the beam. L is an arbitrary length and, for sake of generality, in the following will be considered unitary. By this assumption, any computed value will represent the corresponding quantity per unit length. The beam material is described by its density ρ and its elastic modulus E. According to this notation, with $L = 1$, mass m and compliance c of the beam can be expressed as

$$m = \rho A \quad [\text{kg/m}] \tag{4.1}$$

$$c = \frac{\theta}{L} = \frac{M}{EI} \quad [\text{rad/m}] \tag{4.2}$$

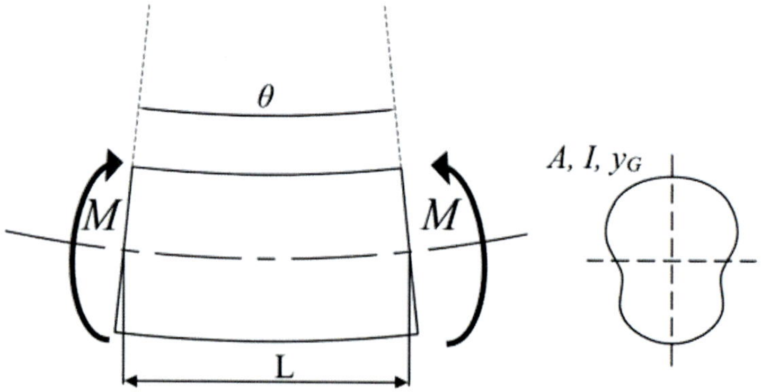

Fig. 4.1 Cantilever beam subjected to bending moment

Mass and compliance have to be minimised.

The structural safety of the design is represented by the maximum admissible stress acting on the structure. The constraint can be written as

$$\sigma_{max} \leq \sigma_{adm} \tag{4.3}$$

where σ_{adm} is the admissible stress and σ_{max} is the maximum stress in the structure. It can be computed as [250]

$$\sigma_{max} = \frac{M y_G}{I} = \frac{M}{Z} \tag{4.4}$$

where Z is the section modulus.

Referring to buckling, a general formula taken from [10] will be considered.

4.1.1 Shape Factor for the Elastic Bending

Equation 4.2 shows that the compliance of a beam due to bending is defined by the elastic modulus E of the material and the moment of inertia of the section I. Considering a square section of edge length b_0 and area $A = b_0^2$, the moment of inertia I_0 of the section reads

$$I_0 = \frac{b_0^4}{12} = \frac{A^2}{12} \tag{4.5}$$

Let us consider the ratio between the compliance c_0 of the square section and the compliance c of an arbitrary cross section of the moment of inertia I [10]

$$\phi^e = \frac{c_0}{c} = \frac{M}{EI_0} \cdot \frac{EI}{M} = \frac{I}{I_0} = \frac{12I}{A^2} \tag{4.6}$$

ϕ^e is the *shape factor for elastic bending* (called shape factor in the following) of the cross section and represents its efficiency, in terms of compliance, with respect to the square section with the same area A (and therefore with the same mass per unit of length). Elongated shapes in the direction orthogonal to the neutral axis have $\phi^e > 1$.

4.1.2 Stress Factor for Elastic Bending

Be Z_0 the section modulus of a square section

$$Z_0 = \frac{b_0^3}{6} = \frac{A^{3/2}}{6} \tag{4.7}$$

the *stress factor for elastic bending* (called stress factor in the following) ϕ^f can be defined as the ratio between the section modulus of a arbitrary section Z and the section modulus of the square section Z_0

$$\phi^f = \frac{Z}{Z_0} = \frac{6Z}{A^{3/2}} \tag{4.8}$$

ϕ^f represents the efficiency of a cross section, in terms of maximum stress, with respect to the square section.

By combining Eqs. 4.8, 4.6 and 4.4, ϕ^f can be related to ϕ^e as

$$\phi^f = \frac{\phi^e}{2y_G}\sqrt{A} \tag{4.9}$$

where y_G is the maximum distance from the neutral axis of a point laying on the border of the beam cross section.

4.1.3 Buckling

Several papers can be found providing accurate buckling formulae for different beams [181, 182, 250]. Buckling is often considered to be an elastic phenomenon, in many practical applications buckling can be caused by local defects [10]. Standards [69] and manuals [10] suggest simplified formulae that consider, besides the elastic properties of the materials, also the limit stress for the computation of the instability limit. In general, higher resistance materials are more prone to local effects, thus the instability limit decreases as the admissible stress increases. A formulation suitable for an early stage design is given in [10] and reads

$$\phi^e \leq \phi^e_{cr} \simeq 2.3\sqrt{\frac{E}{\sigma_{adm}}} \tag{4.10}$$

Equation 4.10 is well in accordance with standards such as Eurocode [69]. Moreover, in [10] it is shown that often commercially available semi-finished rods comply with the limit of Eq. 4.10. In Table 4.1, upper bounds for shape and stress efficiency factors for some materials taken from [10] and based on the analysis of actually available cross sections are reported.

Table 4.1 Upper limits for the shape factor (data from [10])

Material	ϕ^e_{cr}
Structural steel	65
6061 aluminium alloy	44
Glass fibre reinforced polymer (GCFRP)	39
Carbon fibre reinforced polymer (CFRP)	39
Nylon	12
Hard wood	5
Elastomers	<6

4.1.4 Optimisation Problem Formulation

Introducing the shape factor ϕ^e into Eq. 4.2 and the stress efficiency factor ϕ^f into Eq. 4.3, the problem of the mass and compliance optimisation of a beam of arbitrary shape subjected to bending can be stated in mathematical form as follows:

Given

M	the applied moment (see Fig. 4.1)	[Nm]
σ_{adm}	the admissible stress of the material	[MPa]
E	the material modulus of elasticity (Young's modulus)	[MPa]
ρ	the material density	[kg/m^3]

and defining

A	the beam cross section area	[m^2]
ϕ^e	the beam cross-sectional shape factor	[–]
ϕ^f	the beam cross-sectional stress factor	[–]
m	the beam mass (per unit of length)	[kg/m]
c	the compliance of the cantilever (rotation per unit of length) due to the moment M	[rad/m]
σ_{max}	the maximum stress	[MPa]
ϕ^e_{cr}	the maximum shape factor for stability	[–]

find A and ϕ^e such that

$$\min \begin{pmatrix} m(A, \phi^e) \\ c(A, \phi^e) \end{pmatrix} = \min \begin{pmatrix} \rho A \\ \frac{12M}{E\phi^e A^2} \end{pmatrix} \tag{4.11}$$

where

$$\begin{aligned} m &= \rho A \\ c &= \frac{12M}{E\phi^e A^2} \end{aligned} \tag{4.12}$$

subject to

$$A_{min} \leq A \leq A_{max} \tag{4.13}$$

$$\phi^e_{min} \leq \phi^e \leq \phi^e_{max} \tag{4.14}$$

$$\sigma_{max} = \frac{6M}{\phi^f A^{3/2}} \leq \sigma_{adm} \tag{4.15}$$

$$\phi^e \leq \phi^e_{cr} = 2.3\sqrt{\frac{E}{\sigma_{adm}}} \tag{4.16}$$

By inspecting Eqs. 4.12, 4.15 and 4.16, it can be observed that the objective functions actually depend on the two design variables A and ϕ^e, but one of the constraint (Eq. 4.15) depends also on the additional variable ϕ^f (and thus on y_G). Table 4.2 reports the expressions of A, ϕ^e and ϕ^f for some commonly used beam cross sections.

Referring to the different cross sections of Table 4.2, three main groups can be identified.

Group (1) Only one parameter is required to define the cross section of the beam. In this case, the number of design variables is less than the number of objective functions. If A is the free parameter, ϕ^e and ϕ^f can be computed as functions of A (i.e. $\phi^e = \phi^e(A)$ and $\phi^f = \phi^f(A)$). In this case, the solution can be found by simply substituting the constraints expressions into the objective functions (see [143, 148]). Design optimisation is not required. This case will not be further discussed in this chapter.

Group (2) Two parameters are required to define the cross section of the beam. In this case, the number of design variables is equal to the number of objective functions. A and ϕ^e are the free parameters, ϕ^f can be computed as functions of A and ϕ^e (i.e. $\phi^f = \phi^f(A, \phi^e)$).

Group (3) Three (or more) parameters are required to define the cross section of the beam. In this case, the number of design variables is still equal to the number of the objective functions since only A and ϕ^e appear in the expressions of the objective functions m and c in Eq. 4.12. However, for this group of cross sections, the maximum distance from the neutral axis y_G is an additional parameter that changes the structural safety constraint of the problem (Eqs. 4.4 and 4.15) and $\phi^f = \phi^f(A, \phi^e, y_G)$ (Eq. 4.9). In other words, from the analytical point of view, the number of parameters that define the beam cross sections of Group 3 is higher than the number of design variables (A and ϕ^e) used in the problem of Eq. 4.12. In this case, each Pareto-optimal solution can be therefore obtained by more than one combination of the cross-sectional parameters. In Sect. 4.3, this scenario is analysed with reference to a I-shaped cross section beam.

Table 4.2 Formulas for some commonly used beam cross sections. As explained in Sect. 4.1.1, the expressions of ϕ^e are obtained with reference to a square cross section with same area A of the section shape under consideration

	#	Section shape	A	I	y_G	ϕ^e	ϕ^f
Group 1	1	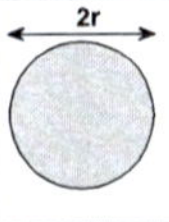	πr^2	$\frac{\pi r^4}{4}$	r	$\frac{3}{\pi}$	$\frac{3}{2\sqrt{\pi}}$
Group 2	2	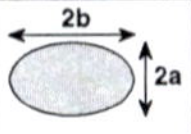	πab	$\frac{\pi a^3 b}{4}$	a	$\frac{3a}{\pi b}$	$\frac{3\sqrt{a}}{2\sqrt{\pi b}}$
	3	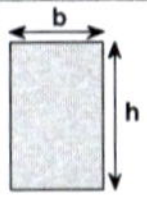	bh	$\frac{bh^3}{12}$	$\frac{h}{2}$	$\frac{h}{b}$	$\sqrt{\left(\frac{h}{b}\right)}$
	4	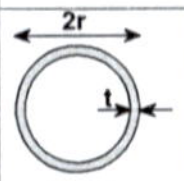 r >> t	$2\pi rt$	$\pi r^3 t$	r	$\frac{3r}{\pi t}$	$\frac{3\sqrt{r}}{\sqrt{2\pi t}}$
Group 3	5	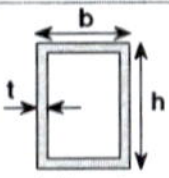 b,h >> t	$2t\,(h+b)$	$\frac{h^2 t(h+3b)}{6}$	$\frac{h}{2}$	$\frac{h^2(h+3b)}{2t(h+b)^2}$	$\frac{h(h+3b)}{\sqrt{2t(h+b)^3}}$
	6	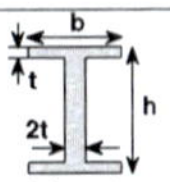 b,h >> t	$2t\,(h+b)$	$\frac{h^2 t(h+3b)}{6}$	$\frac{h}{2}$	$\frac{h^2(h+3b)}{2t(h+b)^2}$	$\frac{h(h+3b)}{\sqrt{2t(h+b)^3}}$
	7	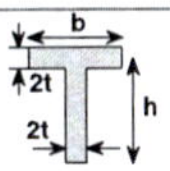 b,h >> t	$2t\,(h+b)$	$\frac{h^3 t(h+4b)}{6(h+b)}$	$\frac{h(h+2b)}{2(h+b)}$	$\frac{h^3(h+4b)}{2t(h+b)^3}$	$\frac{\sqrt{2}h^2(h+4b)}{2\sqrt{t(h+b)^3(h+2b)}}$
	8	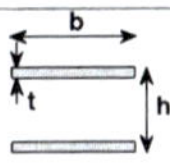 b,h >> t	$2bt$	$\frac{bth^2}{2}$	$\frac{h}{2}$	$\frac{3h^2}{2bt}$	$\frac{3h}{\sqrt{2bt}}$

4.2 Pareto-Optimal Set for the Beam of Arbitrary Cross-Sectional Shape Subjected to Bending

For the analytical solution of the stated optimisation problem, let us start by considering the unconstrained problem, i.e. structural and stability constraints (Eqs. 4.15 and 4.16) are removed from the problem formulation.

The two objective functions are continuous and differentiable. The beam mass m, being function only of A and monotonically increasing, is pseudo-convex [90]. The

Hessian matrix of the beam compliance c reads

$$\mathbf{H} = \frac{24M}{A^2\phi^e}\begin{bmatrix} \frac{3}{A^2} & \frac{1}{A\phi^e} \\ \frac{1}{A\phi^e} & \frac{1}{\phi^{e2}} \end{bmatrix} \tag{4.17}$$

and has both eigenvalues real and positive. Thus, the beam compliance c is convex [9]. For an unconstrained problem with the same number of design variables and objective functions, Eq. 3.14 can be applied and reads

$$det\,(\nabla \mathbf{f}) = det\left(\begin{bmatrix} \rho & -\frac{24M}{EA^3\phi^e} \\ 0 & -\frac{12M}{EA^2\phi^{e^2}} \end{bmatrix}\right) = -\frac{12M\rho}{EA^2\phi^{e^2}} = 0 \tag{4.18}$$

which has solution only if $A^2\phi^{e^2} \rightarrow \infty$. Such solution not belonging to the set of the positive and finite numbers has no physical meaning. The nonexistence of the Pareto-optimal set can be proved by considering the ϵ-constraint method [152] and by applying the monotonicity principles [164]. In fact, if we apply the ϵ-constraint method and we consider the beam mass (m) as objective function and the beam compliance (c) as (active) design constraint, the unconstrained multi-objective minimisation problem of Eq. 4.11 is transformed into

$$\begin{aligned} &min\ (\rho A) \quad subject\ to: \\ &\frac{12M}{E\phi^e A^2} \leq \epsilon \end{aligned} \tag{4.19}$$

with ϵ the constraint bound. The objective function in the first line of Eq. 4.19 is monotonically increasing with A. The constraint of the second line of Eq. 4.19 states that $A \geq \sqrt{\frac{12M}{E\phi^e\epsilon}}$. It is immediate to understand that, to minimise $A, \phi^e \rightarrow \infty$, which violates the second monotonicity principle (ϕ^e is not properly bounded) [164]. So, there must exist at least one active constraint [164] and the Pareto-optimal set is defined by the active constraints.

This fact is conceptually very important. In the early design of a beam subjected to bending, constraints play a crucial role.

In other words, the optimal design of a beam subjected to bending is always a constrained optimal solution.

When the buckling constraint is active, the shape factor ϕ^e reads (see Eq. 4.16)

$$\phi^e = 2.3\sqrt{\frac{E}{\sigma_{adm}}} \tag{4.20}$$

By replacing Eq. 4.20 into Eq. 4.12 and eliminating A, the expression of the beam mass m as function of the beam compliance c can be derived as

Table 4.3 Computed values of k and r_G for the cross sections of Table 4.2. Section numbers refer to Table 4.2

Section #	$k = \frac{\phi^f}{\sqrt{\phi^e}}$	$r_G = \sqrt{\frac{I}{A}}$
1	$\frac{\sqrt{3}}{2}$	$\frac{r}{2}$
2	$\frac{\sqrt{3}}{2}$	$\frac{a}{2}$
3	1	$\frac{h}{2\sqrt{3}}$
4	$\sqrt{\frac{3}{2}}$	$\frac{r}{\sqrt{2}}$
5	$\sqrt{\frac{h+3b}{h+b}}$	$\frac{h}{2\sqrt{3}}\sqrt{\frac{(h+3b)}{(h+b)}}$
6	$\sqrt{\frac{h+3b}{h+b}}$	$\frac{h}{2\sqrt{3}}\sqrt{\frac{(h+3b)}{(h+b)}}$
7	$\frac{\sqrt{h(h+4b)}}{h+2b}$	$\frac{h}{2\sqrt{3}}\sqrt{\frac{h(h+4b)}{(h+b)^2}}$
8	$\sqrt{3}$	$\frac{h}{2}$

$$m = 2\rho\sqrt[4]{\frac{9M^2\sigma_{adm}}{2.3^2E^3}}\frac{1}{\sqrt{c}} \tag{4.21}$$

Equation 4.21 is always decreasing and convex, thus, when active, the buckling constraint is part of the Pareto-optimal set. On the buckling constraint, $m \to \infty$ if $c \to 0$ and $m \to 0$ if $c \to \infty$.

Referring to the stress constraint, the problem of how to relate ϕ^f to the design variables has to be addressed. For a generic cross section, A, ϕ^e and y_G are independent. By considering A, ϕ^e and y_G as independent, a functional relationship between A, ϕ^e, y_G and ϕ^f can be assumed.

The considered relationship between ϕ_f and A, ϕ_e and y_G reads

$$\phi^f = k\sqrt{\phi^e} \tag{4.22}$$

where k is a positive expression and depends on the cross-sectional shape (see Table 4.3). For simple cross-sectional shapes, k is a constant. For more complex cross-sectional shapes, k is function of the cross-sectional parameters.

By replacing Eq. 4.22 in Eq. 4.15, for any A, ϕ^e is limited by the stress constraint

$$\phi^e \geq \frac{36M^2}{k^2\sigma_{adm}^2A^3} \tag{4.23}$$

By substituting Eq. 4.23 in Eq. 4.12 and eliminating out A, the expression of the stress constraint, i.e. beam mass m as function of the beam compliance c, can be obtained

$$m \geq \frac{3\rho ME}{k^2\sigma_{adm}^2}c \tag{4.24}$$

The constraint boundary of Eq. 4.24 can be obtained by replacing the "≥" with "="

$$m = \frac{3\rho M E}{k^2 \sigma_{adm}^2} c = B \frac{c}{k^2} \tag{4.25}$$

where $B = \dfrac{3\rho M E}{\sigma_{adm}^2}$ is a positive constant. In case of a constant value of k (e.g. for cross section numbers 1, 2, 3, 4 and 8 in Table 4.3), Eq. 4.25 is a straight line with positive slope. The boundary condition in Eq. 4.25 intersects the buckling constraint.

The intersection between buckling and stress constraints represents the extremum of the Pareto-optimal set for the minimum value of mass.

If k is function of the cross-sectional parameters, the slope of Eq. 4.25 has to be studied.

By considering Eq. 4.9 and the definition of ϕ^e (Eq. 4.6), ϕ^f can be rewritten as

$$\phi^f = \sqrt{\phi^e} \frac{\sqrt{\phi^e}}{2 y_G} \sqrt{A} = \sqrt{\phi^e} \cdot \sqrt{3} \frac{r_G}{y_G} \tag{4.26}$$

where $r_G = \sqrt{\frac{I}{A}}$ is the radius of gyration of the cross section and its expressions for the considered cross sections are reported in Table 4.3.

By replacing Eq. 4.26 into Eq. 4.22, k can be rewritten as

$$k = \frac{\phi^f}{\sqrt{\phi^e}} = \sqrt{3} \frac{r_G}{y_G} \tag{4.27}$$

In general, the following approximations of y_G and r_G can be considered (see Tables 4.2 and 4.3)

$$\begin{aligned} y_G &= \tfrac{h}{2} + \xi \mathrel{\underset{\sim}{\propto}} h \\ r_G &\mathrel{\underset{\sim}{\propto}} h \end{aligned} \tag{4.28}$$

where ξ is the distance between the neutral axis and the middle of the height of the cross section.

From Eqs. 4.27 and 4.28, it follows that k is almost constant and the ratio $\frac{c}{k^2}$ is increasing in the plane $c - m$. This boundary condition intersects the buckling constraint. For the purpose of the preliminary design, it is reasonable to assume that the intersection between buckling and stress constraints represents the extremum of the Pareto-optimal set for the minimum value of mass. Given the sign of the inequality in Eq. 4.15, only the portion of the buckling constraint above the stress constraint gives feasible solutions.

The intersection between the buckling and structural safety constraints can be easily computed by simply using Eqs. 4.21 and 4.24. The coordinates of the intersection in the objective function domain read

$$c = \frac{1}{E}\sqrt[3]{\frac{4k^4\sigma_{adm}^4}{3M\phi_{cr}^e}} \tag{4.29}$$

$$m = \rho\sqrt[3]{\frac{36M^2}{k^2\phi_{cr}^e\sigma_{adm}^2}} \tag{4.30}$$

and in the design variables domain read

$$A = \sqrt[3]{\frac{36M^2}{k^2\phi_{cr}^e\sigma_{adm}^2}} \tag{4.31}$$

$$\phi^e = \phi_{cr}^e \tag{4.32}$$

From the designer perspective, not all materials can be employed for all the possible shapes. For each material, the minimum attainable mass can be estimated by Eq. 4.30. As shown in Table 4.3, for Group 3 cross-sectional shapes, k depends on the cross-sectional parameters. For this group of cross sections, the maximum attainable value of k has to be considered to obtain the solution with the minimum attainable mass.

The last boundary constraint is related to the maximum available room for the cross section. To discuss this constraint, the limits A_{min} and A_{max} of Eq. 4.13 and of ϕ_{min}^e and ϕ_{max}^e of Eq. 4.14 have to be analysed.

For each cross section, A_{max} represents the maximum area of the cross section that can be fitted in the allowable room. In fact, from an engineering point of view, the maximum allowable room is not described by the area, but, in general, by the maximum height and the maximum width that the section can assume. Depending on the shape of the section, a maximum value of area can be derived. The minimum value of A (A_{min}) is given by the minimum dimensions that can be realised (depending for instance on material and technological constraints). Each of these two limits is related to a value of ϕ^e given by the corresponding values of the section parameters.

In a similar way, the values of ϕ_{min}^e and ϕ_{max}^e are given by the possible combinations of the parameters of each cross section. A corresponding value of A can be computed for these two values. Figure 4.2 depicts this situation for a I-shaped cross section.

If complying with the other constraints, the two solutions P_1 (minimum area) and P_2 (maximum area) of Fig. 4.2 are the points at minimum mass and minimum compliance, respectively, and are the extrema of the Pareto-optimal set for the considered shape.

According to Eq. 4.18, the Pareto-optimal set can be only on the border of the design variables space. The Pareto-optimal set is connected [31, 155]. Therefore, to connect P_1 to P_2, only two options are available. The Pareto-optimal set is one of the two border lines connecting P_1 and P_2. From Eq. 4.12, it can immediately be seen that the upper border line dominates the bottom border line of Fig. 4.2 and therefore represents the Pareto-optimal set.

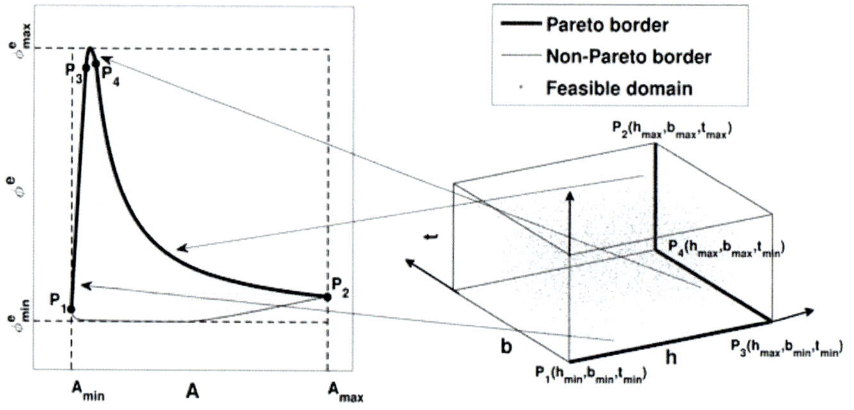

Fig. 4.2 Pareto-optimal set in the design variables domain for a I-shaped cross section when only the available room boundary constraint is considered. Left: $A - \phi^e$ plane. Right: cross-sectional parameters plane. Parameters definition in Table 4.2

Another consequence of Eq. 4.18 is that in the n-dimensional space of the cross-sectional parameters (i.e. the actual design parameters of any cross section), the Pareto-optimal set can be only on the edges of the domain. This can be easily proved by fixing all the parameters but one. To obtain an unbounded value of the expression $A^2\phi^{e^2}$, the remaining parameter has to become infinity or zero. But, being each parameter bounded, it follows that the considered parameter must assume either its minimum or maximum value. In other words, on the Pareto-optimal set cross-sectional parameters vary one at a time from their minimum to their maximum value (or vice versa).

If the problem has solutions for a given cross section, point P_2 must be reachable for that cross section. At P_2, the area of the cross section is at its maximum. It is of clear engineering interest the knowledge of the only cross section parameter to be varied starting from P_2 to reduce the area of the cross section (on the Pareto-optimal set, the cross-sectional parameters vary one at a time).

In order to remain on one edge of the domain and on the upper bound in the plane $A - \phi^e$, the parameter to change is the one that minimises the derivative of ϕ^e with respect to A at P_2. Calling $\mathbf{p}$ the vector of the parameters of the cross section, the minimum of the derivative can be computed as

$$\min\left(\left.\frac{d\phi^e\left(A\left(\mathbf{p}\right)\right)}{dA\left(\mathbf{p}\right)}\right|_{@P_2}\right) = \min\left(\left[\left.\frac{\frac{\partial \phi^e(\mathbf{p})}{\partial p_i}}{\frac{\partial A(\mathbf{p})}{\partial p_i}}\right|_{@P_2}\right]_{i=1\ldots n} \cdot \mathbf{v}\right) \tag{4.33}$$

where p_i is one of the n parameters of the cross section and $\mathbf{v}$ is a unit direction vector in the space of the cross-sectional parameters. Since from Eq. 4.18 the Pareto-optimal set can be only on the edges of the domain, the minimum of the derivative can be found only when $\mathbf{v}$ coincides with one of such directions. Therefore, $\mathbf{v}$ can assume only combinations where a single component is one and all the other components are zeros. By replacing this condition in Eq. 4.33, the minimum of the derivative of ϕ^e with respect to A at P_2 can be computed as

$$\min\left(\left.\frac{d\phi^e\left(A\left(\mathbf{p}\right)\right)}{dA\left(\mathbf{p}\right)}\right|_{@P_2}\right) = \min_{i=1\ldots n}\left(\left.\frac{\frac{\partial\phi^e(\mathbf{p})}{\partial p_i}}{\frac{\partial A(\mathbf{p})}{\partial p_i}}\right|_{@P_2}\right) \tag{4.34}$$

Equation 4.34 states the condition to find the unique cross-sectional parameter that has to be changed to move from the point at minimum compliance along with the Pareto-optimal set.

It can be observed that if we call h the parameter describing the height of the section, both $\frac{d\phi^e}{dh}$ and $\frac{dA}{dh}$ are positive for any combination of the other parameters. Therefore, $\frac{d\phi^e}{dh}/\frac{dA}{dh}$ is positive and the height of the cross section is never the first parameter to be changed. Actually, being this ratio always positive, starting from P_2, the height of the cross section can be reduced only after the maximum value of ϕ^e has been reached.

In practical problems, a different constraint (elastic stability or structural safety) is usually reached before ϕ^e gets to its maximum values. The condition $h = h_{max}$ (being h_{max} the maximum value of h) is always part of the Pareto-optimal set. In most cases, the switch between the available room constraint and another constraint happens with $h = h_{max}$ still active. This conclusion is general, being always possible to parametrise a section in order to have its height as a parameter.

In Table 4.4 the derivatives of ϕ^e with respect to A at point P_2 of maximum mass and minimum compliance for the cross sections of Table 4.2 are reported.

Referring to different cross-sectional shapes, the absolute minimum value of compliance, corresponding to the maximum possible value of mass, is given by a rectangular section with height coincident with the maximum crosssection height and width coincident with the maximum width. The rigorous way to compute the Pareto-optimal set at low values of compliance is to order all the possible cross sections for decreasing levels of efficiency up to the rectangular cross section. A reasonable way of approximating this curve, is to linearly connect the point at minimum compliance of the most efficient cross section to the absolute minimum of the compliance (rectangular cross section).

In Fig. 4.3, the Pareto-optimal set in the objective functions domain is shown for a steel beam (data in Table 4.5) considering shapes 3, 6 and 8 of Table 4.2. The applied moment is 1000 Nm and the allowable space is a square with side length 0.15 m. A maximum section thickness of 0.01 m is considered.

The Pareto-optimal set of Fig. 4.3 can be divided into three regions:

(1) Buckling constraint: from point **A** to point **B**. In this region, the buckling constraint is active. The curve is the same for all shapes, just the limits vary (points $\mathbf{A_R}$ for the rectangular section, $\mathbf{A_I}$ for the I section and $\mathbf{A_S}$ for the sandwich-like section, see Eq. 4.30 and Table 4.3). The minimum absolute mass (point **A**) is given by the shape with the highest value of the constant k in Table 4.3 (in this

Table 4.4 Computed values of the derivatives of ϕ^e with respect to A for the solution with maximum mass and minimum compliance for the cross sections reported in Table 4.2. Section numbers refer to Table 4.2

Section #	Parameters	Derivatives
1	r	$\left.\frac{d\phi^e}{dr} / \frac{dA}{dr}\right\rvert_{@P_2} = 0$
2	a, b	$\left.\frac{d\phi^e}{da} / \frac{dA}{da}\right\rvert_{@P_2} = \frac{3}{\pi^2 b_{max}^2}$ $\left.\frac{d\phi^e}{db} / \frac{dA}{db}\right\rvert_{@P_2} = -\frac{3}{\pi^2 b^2}$
3	h, b	$\left.\frac{d\phi^e}{dh} / \frac{dA}{dh}\right\rvert_{@P_2} = \frac{1}{b_{max}^2}$ $\left.\frac{d\phi^e}{db} / \frac{dA}{db}\right\rvert_{@P_2} = -\frac{1}{b^2}$
4	r, t	$\left.\frac{d\phi^e}{dr} / \frac{dA}{dr}\right\rvert_{@P_2} = \frac{3}{2\pi^2 t_{max}^2}$ $\left.\frac{d\phi^e}{dt} / \frac{dA}{dt}\right\rvert_{@P_2} = -\frac{3}{2\pi^2 t^2}$
5 & 6	h, b, t	$\left.\frac{d\phi^e}{dh} / \frac{dA}{dh}\right\rvert_{@P_2} = \frac{h\left(6b_{max}^2+3b_{max}h+h^2\right)}{4t_{max}^2 (h+b_{max})^3}$ $\left.\frac{d\phi^e}{db} / \frac{dA}{db}\right\rvert_{@P_2} = -\frac{h_{max}^2 (3b-h_{max})}{4t_{max}^2 (h_{max}+b)^3}$ $\left.\frac{d\phi^e}{dt} / \frac{dA}{dt}\right\rvert_{@P_2} = -\frac{h_{max}^2 (3b_{max}+h_{max})}{4t^2 (h_{max}+b_{max})^3}$
7	h, b, t	$\left.\frac{d\phi^e}{dh} / \frac{dA}{dh}\right\rvert_{@P_2} = \frac{h^2\left(12b_{max}^2+4b_{max}h+h^2\right)}{4t_{max}^2 (h+b_{max})^3}$ $\left.\frac{d\phi^e}{db} / \frac{dA}{db}\right\rvert_{@P_2} = -\frac{h_{max}^3 (8b-h_{max})}{4t_{max}^2 (h_{max}+b)^4}$ $\left.\frac{d\phi^e}{dt} / \frac{dA}{dt}\right\rvert_{@P_2} = -\frac{h_{max}^3 (4b_{max}+h_{max})}{4t^2 (h_{max}+b_{max})^4}$
8	b, t	$\left.\frac{d\phi^e}{db} / \frac{dA}{db}\right\rvert_{@P_2} = -\frac{3h_{max}^2}{4b^2 t_{max}^2}$ $\left.\frac{d\phi^e}{dt} / \frac{dA}{dt}\right\rvert_{@P_2} = -\frac{3h_{max}^2}{4b_{max}^2 t^2}$

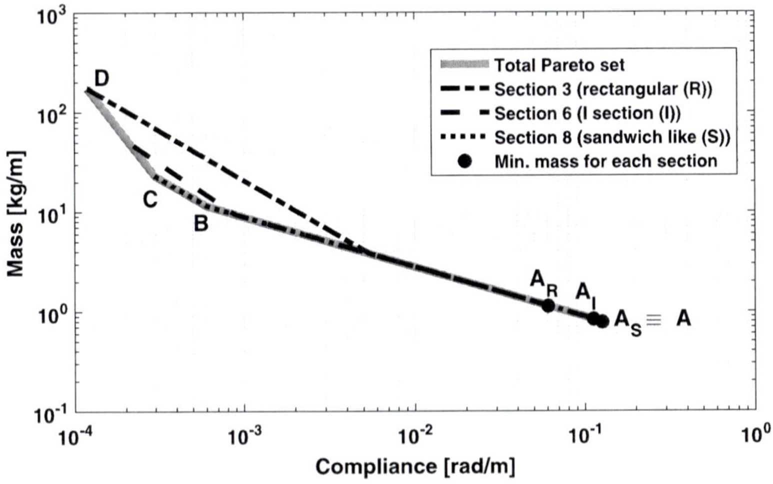

Fig. 4.3 Pareto-optimal set in the objective functions domain for a mild steel beam subjected to bending moment. A-B: constraint due to buckling. B-C: constraint due to the available room for the most efficient cross-sectional shape. C-D: linear connection with the cross-sectional shape with the lowest attainable compliance. Data in Table 4.5. Section numbers refer to Table 4.2. Applied moment 1000 Nm, $h_{max} = b_{max} = 0.15\,\text{m}$, $t_{max} = 0.01\,\text{m}$

Table 4.5 Material data

Material	Limit stress (MPa)	Elastic modulus (MPa)	Density (Mg/m^3)	ϕ_{cr}^{e} *
Structural steel	500	200000	7.8	44
6061 aluminium alloy	200	70000	2.7	43
Hard wood	60	13500	0.9	5

(*) minimum between the value computed by Eq. 4.16 and the practical limits of Table 4.1

case shape 8, see Table 4.2, i.e. $\mathbf{A} \equiv \mathbf{A_S}$). Solution **B** refers to the most efficient shape (in this case shape 8, see Table 4.2) when the maximum section height is reached.

(2) Available room constraint on the most efficient cross section: from point **B** to point **C**. The curve is given by the most efficient shape along the maximum height curve up to the buckling constraint (in this case shape 8, see Table 4.2). At point B, the height of the cross section has reached its maximum value. In moving from point B to point C, the other parameters of the cross section are progressively increased one at a time (we have demonstrated that, when the volume constraint is active, the Pareto-set is on the border of the design variables space). Point **C** represents the solution with minimum compliance, and maximum mass, for the most efficient shape, i.e. shape 8 of Table 4.2 when all the cross-sectional parameters reach their maximum value.

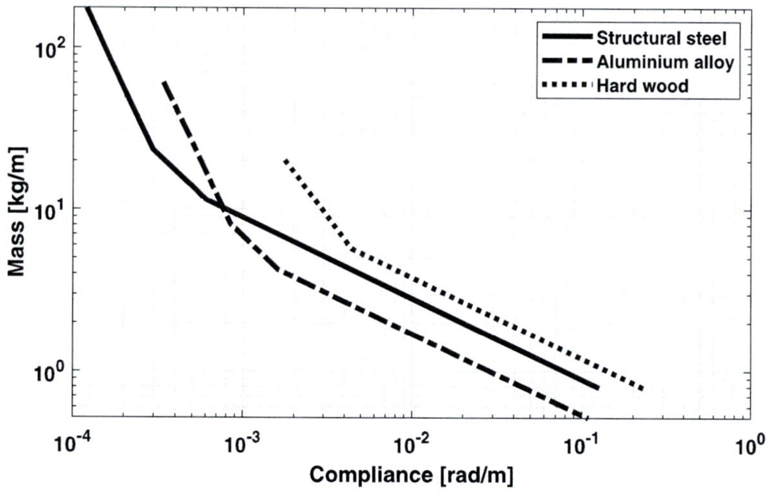

Fig. 4.4 Pareto-optimal set in the objective functions domain for mild steel, aluminium alloy and wood beams subjected to bending moment. Data in Table 4.5. Applied moment 1000 Nm, $h_{max} = b_{max} = 0.15\,\text{m}$, $t_{max} = 0.01\,\text{m}$

(3) Approximation of the available room constraint on other possible cross sections: from point **C** to point **D**. The curve linearly connects the point at minimum compliance on the most efficient cross section to the absolute minimum of compliance (rectangular section, shape 3 in Table 4.2). This curve roughly interpolates the points at minimum compliance of the other cross sections. Not all points of this curve can be actually obtained since only a discrete number of cross section shapes are available. However, this part of the Pareto-optimal set is of small interest due to the limited practical use, being this design region very close to the maximum mass. In this region, a small decrement of the compliance costs a large increase in mass.

In Fig. 4.4, the Pareto-optimal sets referring to three different materials (structural steel, aluminium alloy and hardwood, data in Table 4.5) for the same load and geometrical limits of the previous example are reported. For the wood, only rectangular and I-shaped cross sections have been considered (the beam profiles that are commonly available on the market).

As expected, aluminium shows better performances than the other considered materials for low values of mass, while steel is the best choice for low levels of compliance. Hardwood, although has the highest ratio E/ρ, shows the worst performance among the considered materials. Topologically different cross-sectional configurations are allowed by using metals.

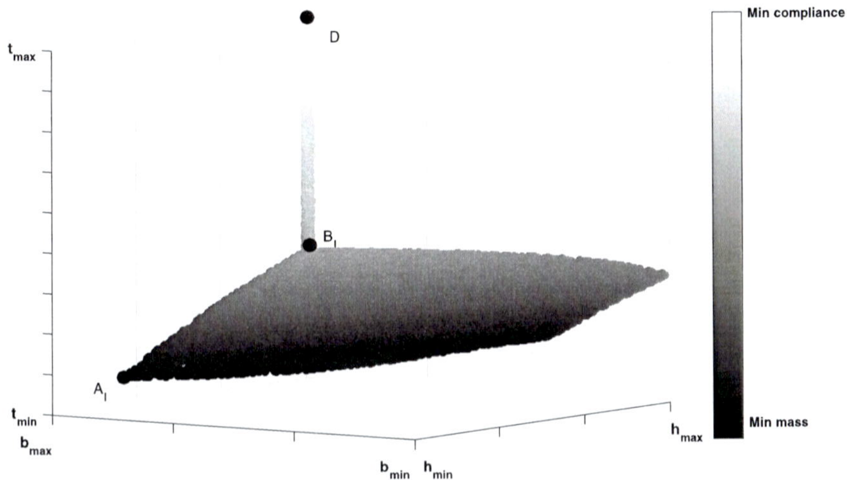

Fig. 4.5 Pareto-optimal set for a steel I beam in the cross-sectional parameters space. Parameters definition in Table 4.2. Material data in Table 4.5. Applied moment 1000 Nm, $h_{max} = b_{max} = 0.15$ m, $t_{max} = 0.01$ m

4.3 An Application to the Design of an I-Shaped Cross Section

An I-shaped cross-sectional (Group 3 in Table 4.2) steel beam is considered in the following.

The Pareto-optimal solutions in the objective functions domain are reported in Fig. 4.7. Each optimal solution in the objective functions domain corresponds biunivocally to a single solution A, ϕ^e in the design variables domain, as shown in Fig. 4.6. Being the I-shaped cross section defined by three parameters, namely, h, b, t (see Table 4.2), each Pareto-optimal solution in the design variables domain can be obtained by more than one combination of the cross-sectional parameters. This is clearly shown in Fig. 4.5. The optimal design solutions with the same level of grey are defined by different values of h, b, t, but they have exactly the same performance in terms of mass m and compliance c (and the same value of A and ϕ^e).

Point A_I, in Figs. 4.5, 4.6 and 4.7 represents the solution with minimum mass; point B_I, the switching point between the buckling constraint and the available room constraint; point D, the solution with minimum compliance.

4.4 Summary

In this chapter, Pareto-optimality theory has been applied to the study of a beam subjected to bending.

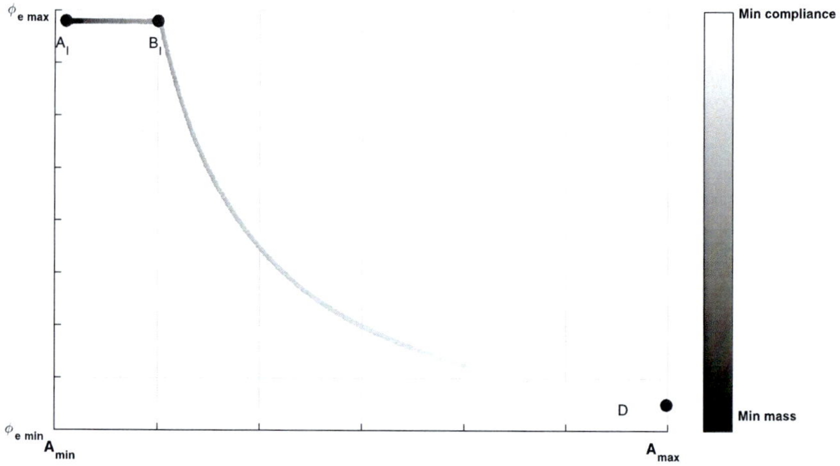

Fig. 4.6 Pareto-optimal set in the design variables domain for an I shaped steel cross section. Material data in Table 4.5. Applied moment 1000 Nm, $h_{max} = b_{max} = 0.15\,\text{m}$, $t_{max} = 0.01\,\text{m}$

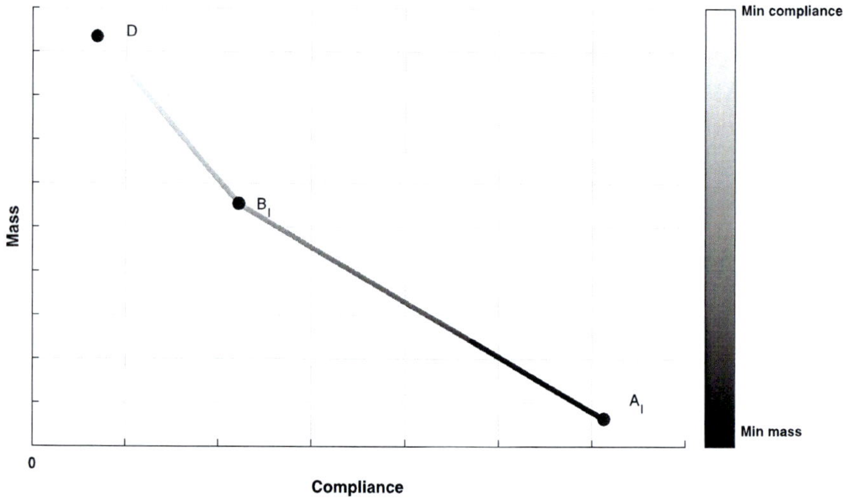

Fig. 4.7 Pareto-optimal set in the objective functions domain for an I shaped steel cross section. Material data in Table 4.5. Applied moment 1000 Nm, $h_{max} = b_{max} = 0.15\,\text{m}$, $t_{max} = 0.01\,\text{m}$

The Pareto-optimal set for a beam of arbitrary cross-sectional shape and material has been derived analytically under the constraints of allowable room, structural safety (maximum stress) and elastic stability (buckling). The analytical derivation has proved that the Pareto-optimal set is composed of two connected regions. The first region is given by the elastic stability constraint, while the second region is given by the available room constraint. In particular, the limit on the maximum height of the beam section is the leading parameter when the room constraint is active.

The structural safety constraint is not part of the Pareto-optimal set. Structural safety limits the minimum mass of the beam. In fact, the point at minimum mass is given by the intersection between the elastic stability and the structural safety constraints.

For elastic stability, a simplified buckling formulation that includes also manufacturing considerations has been chosen. This formulation is quite conservative. In case it is required to exploit the limit performances of a material, the approach used in this paper can still be applied by changing the buckling formulation.

The proposed approach can be effectively used by the designer at a very early stage of the project when the material and the shape of the structure has still to be defined.

Chapter 5
Bending of Lightweight Circular Tubes—Optimal Design

The aim of this chapter is that of providing a sound theoretical guideline for a rigorous, and maybe ultimate, concept design of thin-walled circular tubes subject to bending. Such tubes can be of any dimension and made from any material. The topic is quite classical (even apparently elementary) but the relevant literature [10, 28, 113, 244] does not address the lightweight sizing of thin-walled circular tubes subject to bending by exploiting MOO theory. In [10], Ashby proposes an excellent method to design beams subject to bending of any cross-sectional shape. In Chap. 4, Ashby's formulae were exploited to derive analytical Pareto-optimal solutions for general beams under bending. However, in this chapter, a more focused design method is proposed, which highlights how the paradigms of Pareto optimality (i.e. MOO) may be used in the addressed case. The design of both big structures such as windmill columns and nano-structures such as nano-tubes may take advantage of a basic theory as the one proposed in this chapter.

General-purpose mechanical designers, during the concept phase, are often engaged in a trial-and-error activity based on the usage of elementary formulae [250] based on elastic theory. By changing parameters (namely geometrical dimensions and material properties), a specific solution is found which, more or less, fulfils the technical specifications on mass and compliance. No information is generally given on why such a solution has been chosen instead of another, and whether or not the preferred solution is the optimal one [103].

A paradigm change of the design process for circular tubes subject to bending is proposed in the chapter. Provided that the designer knows in advance the mass or the compliance to be attained, the optimal geometrical parameters of the thin-walled tube (together with the material properties) are promptly suggested by the analytical formulae proposed in the chapter. In other words, the chapter is aimed to give designers a broad overview of the lightweight design problem of thin-walled tubes subject to bending, in order to allow a conscious choice of an optimal solution.

The ever-existing problem of lightweight design is dealt with here by making use of MOO. MOO seems the proper theoretical tool for dealing with the sizing of the cross section of a structural member [21, 99, 129, 143, 152, 164, 185]. The reader

F. M. Ballo et al., *Optimal Lightweight Construction Principles*,
https://doi.org/10.1007/978-3-030-60835-4_5

who is not familiar with MOO can refer to [143] and to Chaps. 2–3 of this book for a tutorial on such a topic.

MOO allows, mainly for sizing problems, to consider the minimisation of both mass and compliance (i.e. maximisation of the stiffness), provided that a number of constraints are satisfied, namely, safety (admissible stress), stability (buckling), available room (maximum radius of the tube), thickness of the tube (arising from technological issues). Mass and compliance are objective functions, i.e. performance indices, function of the design variables (see Sect. 1.5). The design variables (Sect. 1.6) are the geometrical dimensions of the tube (and—if needed—Young's modulus). The constraints (Sect. 1.7) are given as functions of the design variables.

The solution of (simple) MOO problems can be derived analytically (see e.g. Chap. 3) for a number of engineering problems. Analytical formulae, whenever available, can be quite useful for providing paradigms for concept design [143, 152, 164]. The derivation of analytical expression(s) for the addressed optimal design of thin-walled tubes, can be obtained by employing the method described in Chap. 3. The method refers to a matrix formulation (the so-called **L** matrix) of a relaxed form of the Fritz John necessary condition.

According to authors' knowledge, few papers on the concept design of beams subject to bending can be found in the literature. [168] addressed the minimum compliance problem applied to Euler–Bernoulli and Timoshenko beams. The cross-sectional area was chosen as a design variable, several constraint conditions, load distributions and cross-sectional shapes have been investigated. In this chapter we will concentrate on one single cross section (thin-walled tube) and we will consider not only compliance but also limit stress and buckling.

In [10], a set of design formulae for beams subjected to bending are given to quickly compare different materials. Stiffness or strength-oriented designs are considered separately and a different set of formulae is derived for each of the two situations. In this chapter, both strength and stiffness are considered in a unique formulation.

Specific topics are dealt with in the following papers. In [76], thin-walled tubes filled with a special foam are considered, Pareto-optimal design is used to design efficient structures, we focus instead on a simpler and propedeutical concept design analysis of thin-walled tubes. In [98] CFRP hollow section beams are studied, the problem is too complex to be addressed by the method we presented in this book, however the concepts presented in this chapter can be used for a preliminary analysis. In [138], advanced topologies of hollow beam structures are addressed, the paper could be considered a subsequent step with respect to our contribution. In [163], FE analysis is applied for the study of thin-walled carbon nano-tubes which inspired us to think of our method as a very preliminary step towards the design of nano-tubes.

The chapter addresses at first the lightweight design problem, then the optimal solutions are found both in the design variable domain and in the objective function domain. Based on the targets referring to mass and compliance, the proper material to be used is discussed.

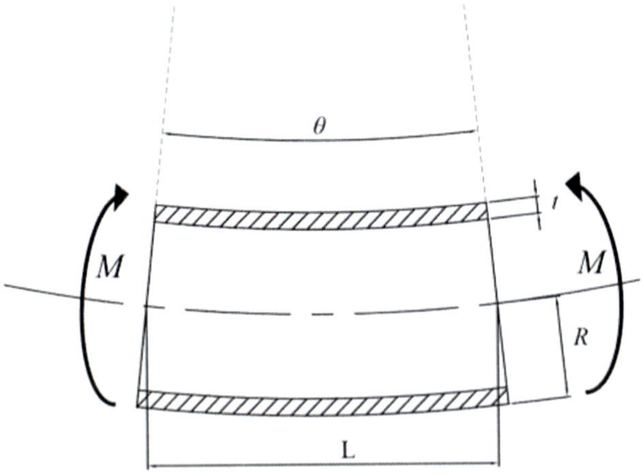

Fig. 5.1 Thin-walled tube subjected to bending moment. $t \leq R/10$

5.1 Equations Describing the Bending of a Thin-Walled Tube

In this section, the lightweight design problem of a thin-walled tube subject to bending is defined. This elementary problem has had, since the solution was discovered in the nineteenth century, relevant engineering applications (see, for instance, [120]) and is a classical problem any mechanical or civil designer faces during his or her career.

In Fig. 5.1, the bending of a thin-walled tube is depicted. The geometric dimensions of the tube are the external radius R, the thickness t and the length L. M is the applied bending moment. The optimal design aims to minimise the mass and maximise the stiffness (i.e. minimise the compliance) of the tube in Fig. 5.1. Mass and compliance are the *objective functions* of the problem [143, 164].

For sake of a general formulation of the problem, both the mass of the tube m_t and the relative rotation of the two end cross sections of the tube θ will be divided by the axial length of the tube L. We will briefly refer to mass m and compliance c as

$$m = m_t/L = 2\pi\rho t R \quad [kg/m] \tag{5.1}$$

$$c = \theta/L = \frac{M}{\pi E t R^3} \quad [rad/m] \tag{5.2}$$

where ρ is the density of the material (which is considered omogeneous and isotropic), E is the elastic (bulk) modulus of the material. Equations 5.1 and 5.2 are, respectively, functions of the geometrical dimensions of the tube (R and t). R and t are the *design variables* of the problems. The designer can achieve different performance of the structure (mass, stiffness) by changing the value of these two geometric dimensions. R and t are physical quantities and are limited by attainable minimum or maximum values, respectively [164]

$$R_{min} \leq R \leq R_{max} \tag{5.3}$$

$$t_{min} \leq t \leq t_{max} \tag{5.4}$$

The maximum value of the radius R_{max} can be seen as a constraint on the available room, while the minimum thickness t_{min} as a technological constraint.

The safety of the structure is related to a constraint on the maximum admissible stress σ_{adm}. The constraint expression reads

$$\sigma_{max} \leq \sigma_{adm} \tag{5.5}$$

where σ_{max} is the maximum stress in the structure and can be computed as [250]

$$\sigma_{max} = \frac{M}{\pi t R^2} \tag{5.6}$$

When thin-walled cross sections are employed, failures due to local buckling can occur. Local buckling phenomena depend on many parameters (cross-sectional geometry, material...) but mainly on the wall thickness (see [181, 182] for updated buckling studies that highlight the accuracy limit of our approach). As the thickness of the cross section is decreased, the structure gets more exposed to local buckling phenomena. Calling M_{cr} the critical value of bending moment, the following constraint has to be considered

$$M \leq M_{cr} \tag{5.7}$$

where M_{cr} has expression [23, 250]

$$M_{cr} = k_{cr} E t^2 R \tag{5.8}$$

where $k_{cr} = k/a$ where k $= 0.99$ is a constant (theoretical value) and $a = 1 - \nu^2$ with ν the Poisson coefficient of the material.

5.2 Optimal Design of Thin-Walled Tubes Subject to Bending

In this section, the lightweight problem is formulated mathematically according to the theory on MOO ([143, 152, 164]) described in Sect. 3.1.

The MOO problem reads
Find

$$
\begin{aligned}
&min \begin{bmatrix} c(t,R) = \frac{M}{\pi E t R^3} \\ m(t,R) = 2\pi \rho t R \end{bmatrix} \quad so\ that \\
&\sigma_{max} \leq \sigma_{adm} = \frac{M}{\pi t R^2} \\
&M \leq M_{cr} = k_{cr} E t^2 R \\
&t_{min} \leq t \leq t_{max} \\
&R_{min} \leq R \leq R_{max}
\end{aligned}
\tag{5.9}
$$

Let us notice that we require the minimum of both *objective functions* m and c. The solution of such a problem (in the Pareto-optimal sense, see [143]) is a set that can be described by one or more (piecewise) analytical functions.

By applying the theory reported in Chap. 3, *the necessary conditions* for sizing a thin-walled tube subject to bending can be derived. Given the objective functions and the constraints listed in Eq. 5.9, the $\mathbf{L}$ matrix (see Sect. 3.1) is constructed and reported in Eq. 5.10.

$$
\mathbf{L} = \begin{bmatrix}
2R\pi\rho & -\frac{M}{E R^3 \pi t^2} & -2ERk_{cr}t & -\frac{M}{R^2\pi t^2} & 0 & 0 & 1 & -1 \\
2\pi\rho t & -\frac{3M}{E R^4 \pi t} & -Ek_{cr}t^2 & -\frac{2M}{R^3\pi t} & 1 & -1 & 0 & 0 \\
0 & 0 & M - ERk_{cr}t^2 & 0 & 0 & 0 & 0 & 0 \\
0 & 0 & 0 & \frac{M}{R^2\pi t} - \sigma_{adm} & 0 & 0 & 0 & 0 \\
0 & 0 & 0 & 0 & R - R_{max} & 0 & 0 & 0 \\
0 & 0 & 0 & 0 & 0 & R_{min} - R & 0 & 0 \\
0 & 0 & 0 & 0 & 0 & 0 & t - t_{max} & 0 \\
0 & 0 & 0 & 0 & 0 & 0 & 0 & t_{min} - t
\end{bmatrix}
\tag{5.10}
$$

By applying the necessary optimality condition on $\mathbf{L}$ which reads

$$
\det(\mathbf{L}) = 0 \tag{5.11}
$$

after symbolic manipulation and removing the constraints which are not active (see Appendix 5.5 for the analytical proofs), the result reads

$$
\begin{aligned}
&(R - R_{max})^2 \cdot (t - t_{min})^2 \cdot (M + \\
&\quad - ERk_{cr}t^2)^2 \cdot (M - R^2 \pi \sigma_{adm} t)^2 = 0
\end{aligned}
\tag{5.12}
$$

This analytical formula is a necessary condition for the Pareto-optimal set in the design variable domain. In other words, Eq. 5.12 contains all of the Pareto-optimal solutions of the problem addressed in Eq. 5.9.

By the inspection of Eq. 5.12, we see that, as expected from MOO theory, solutions have to satisfy the constraint conditions. Actually, the equation we found is the product of the respective constraint conditions on available room, thickness of the tube, buckling and admissible stress (safety).

In Eq. 5.12, we see that exclusively (i.e. only) constraint conditions appear. No other conditions appear, in other words, no other analytical combination of design variables appear. This can be explained as follows.

Let us consider the simplified problem in which the constraints in Eq. 5.9 are removed.

$$min \begin{bmatrix} c(t,R) = \frac{M}{\pi E t R^3} \\ m(t,R) = 2\pi \rho t R \end{bmatrix} \tag{5.13}$$

Since $n = k = 2$, Eq. 3.14 can be employed for solving the MOO problem. According to Eq. 3.14 the gradient of the objective functions reads

$$\nabla \mathbf{F} = \begin{bmatrix} \frac{\partial c}{\partial R} & \frac{\partial m}{\partial R} \\ \frac{\partial c}{\partial t} & \frac{\partial m}{\partial t} \end{bmatrix} = \begin{bmatrix} -\frac{3M}{\pi E R^4 t} & 2\pi \rho t \\ -\frac{-M}{\pi E R^3 t^2} & 2\pi \rho R \end{bmatrix} \tag{5.14}$$

By imposing $det\,(\nabla \mathbf{F}) = 0$, we obtain

$$-\frac{4M\rho}{E R^3 t} = 0 \tag{5.15}$$

which has solution for $R^3 t \to \infty$. Such solution—not belonging to the set of the positive and finite numbers—has no physical meaning and must be discarded [90, 164].

The Pareto-optimal set is therefore given for *active constraints only*, as shown by Eq. 5.12. We will discuss in the subsequent section how constraints can interact to define the Pareto-optimal set.

5.2.1 Pareto-Optimal Set in the Design Variable Domain and in the Objective Function Domain (Necessary Conditions)

We have produced in Eq. 5.12 the necessary condition for solutions to be Pareto-optimal in the design variable domain. The corresponding Pareto-optimal set in the objective function domain is

$$\left(m - \frac{2M\rho}{E R_{max}^2 c}\right) \cdot \left(m - 2\rho \sqrt[3]{\frac{M\pi^2 t_{min}^2}{Ec}}\right) \cdot \left(m + \right. \\ \left. - 2\rho \sqrt[5]{\frac{\pi^4 M^3}{E^3 k_{cr}^2 c}}\right) \cdot \left(m - \frac{2\rho E M c}{\sigma_{adm}^2}\right) = 0 \tag{5.16}$$

Such an equation has been derived by inverting the objective functions $m = m(t, R)$, $c = c(t, R)$ to obtain $t = t(c, m)$, $R = R(c, m)$, i.e.

$$t = \frac{1}{2\pi\rho}\sqrt{\frac{Em^3c}{2\rho M}} \qquad R = \sqrt{\frac{2\rho M}{Emc}} \tag{5.17}$$

Actually, Eq. 5.16 is—as the dual Eq. 5.12—the product of the constraint conditions on available room, thickness of the tube, buckling and admissible stress (safety), respectively.

In principle, the designer can take the Pareto-optimal solution from either Eq. 5.12 or Eq. 5.16. But, since such equations contain solutions that may be not Pareto-optimal, the designer must check whether or not the preferred solution does satisfy the constraints.

5.3 Sizing of Thin-Walled Tubes with Constraints on Available Room, on Minimum Thickness, on Buckling, on Admissible Stress

In Fig. 5.2, a number of possible cases of Pareto-optimal sets in the design variable domain are shown. Bold lines highlight the Pareto-optimal sets (the reason for this will be explained in the subsequent subsection). Given the lower and upper bounds on t and R, respectively, the feasible solutions have been computed numerically and highlighted in grey. The constraints on buckling and stress drop some solutions. Equation 5.12 holds for *any* of the Pareto-optimal sets that are shown.

In the next subsections, the Pareto-optimal sets depicted in Fig. 5.2 are discussed in detail.

5.3.1 *Case* ①

Let us consider case ① of Fig. 5.2. The Pareto-optimal set in the design variable domain is given by the two expressions $t = t_{min}$ and $R = R_{max}$, as shown in Fig. 5.3. The full analytical expressions of the Pareto-optimal sets, both in the design variable domain and in the objective function domain, are reported in Table 5.1.

Three candidate optimal solutions for concept design are the ones highlighted with rhomb, square and circle in panel ①. Actually, they could be a good initial guess for a design iteration.

General hints for the designer are to occupy as much room as possible and keep the thickness as small as possible. This provides low compliance c with the corresponding minimum mass m.

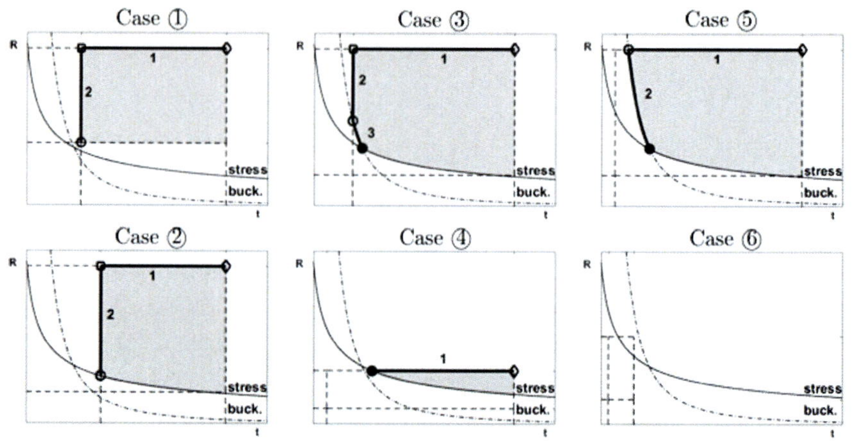

Fig. 5.2 Sizing of thin-walled tubes subject to bending. Pareto-optimal sets in the design variable domain are depicted in bold lines. Influence of the respective upper and lower bounds on t and R are shown. Feasible solutions are highlighted in grey colour Three candidate optimal solutions for concept design are the ones highlighted with rhomb, square and circle in panel ①

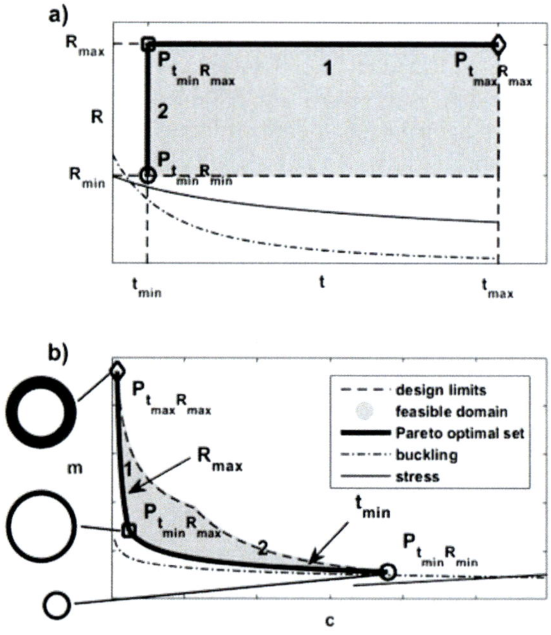

Fig. 5.3 Case 1 of Fig. 5.2, Pareto-optimal sets defined, respectively, in the design variable domain (**a**) and in the objective function domain (**b**). The only active constraints are $t = t_{min}$ and $R = R_{max}$. Analytical expressions of Pareto-optimal sets in Table 5.1

Table 5.1 Case ① of Fig. 5.2. Analytical expressions of the Pareto-optimal sets in both the design variables domain and objective functions domain

Analytical expressions

Pareto optimal set in the design variable domain	Pareto optimal design in the objective functions domain	Boundaries (Fig. 5.3)	
		Pareto optimal subset	Points
$R = R_{max}$	$m = \frac{2\rho M}{E R_{max}^2 c}$	1	$\left[P_{t_{max} R_{max}}, P_{t_{min} R_{max}}\right]$
$t = t_{min}$	$m = 2\rho \sqrt[3]{\frac{\pi^2 M t_{min}^2}{Ec}}$	2	$\left[P_{t_{min} R_{max}}, P_{t_{min} R_{min}}\right]$

Coordinates of points

Point	t	R	c	m
$P_{t_{max} R_{max}}$	t_{max}	R_{max}	$\frac{M}{\pi E t_{max} R_{max}^3}$	$2\pi\rho t_{max} R_{max}$
$P_{t_{min} R_{max}}$	t_{min}	R_{max}	$\frac{M}{\pi E t_{min} R_{max}^3}$	$2\pi\rho t_{min} R_{max}$
$P_{t_{min} R_{min}}$	t_{min}	R_{min}	$\frac{M}{\pi E t_{min} R_{min}^3}$	$2\pi\rho t_{min} R_{min}$

5.3.2 *Case* ②

The minimum attainable radius R, corresponding to minimum m, is defined by the admissible stress, as shown in Fig. 5.4. The Pareto-optimal set in the design variable domain is given by the two expressions $t = t_{min}$ and $R = R_{max}$. The full analytical expressions of the Pareto-optimal sets, both in the design variable domain and in the objective function domain, are reported in Table 5.2.

5.3.3 *Case* ③

The minimum attainable radius R, corresponding to minimum mass m, is defined by the admissible stress and buckling, as shown in Fig. 5.5.

The full analytical expressions of the Pareto-optimal sets, both in the design variable domain and in the objective function domain, are reported in Table 5.3.

5.3.4 *Case* ④

The radius R to be adopted is R_{max}, the minimum m is defined by the admissible stress, as shown in Fig. 5.6.

The full analytical expressions of the Pareto-optimal sets, both in the design variable domain and in the objective function domain, are reported in Table 5.4.

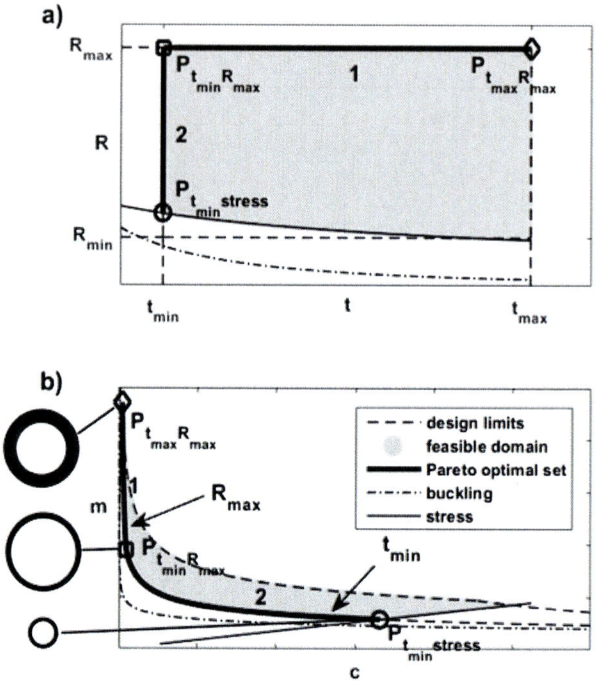

Fig. 5.4 Case ② of Fig. 5.2, Pareto-optimal sets defined, respectively, in the design variable domain (**a**) and in the objective function domain (**b**). Analytical expressions of Pareto-optimal sets in Table 5.2

Table 5.2 Case ② of Fig. 5.2. Analytical expressions of the Pareto-optimal sets in both the design variables domain and objective functions domain

Analytical expressions

Pareto optimal set in the design variable domain	Pareto optimal design in the objective functions domain	Boundaries (Fig. 5.4)	
		Pareto optimal subset	Points
$R = R_{max}$	$m = \frac{2\rho M}{E R_{max}^2 c}$	1	$\left[P_{t_{max}R_{max}}, P_{t_{min}R_{max}}\right]$
$t = t_{min}$	$m = 2\rho \sqrt[3]{\frac{\pi^2 M t_{min}^2}{Ec}}$	2	$\left[P_{t_{min}R_{max}}, P_{t_{min}stress}\right]$

Coordinates of points

Point	t	R	c	m
$P_{t_{max}R_{max}}$	t_{max}	R_{max}	$\frac{M}{\pi E t_{max} R_{max}^3}$	$2\pi\rho t_{max} R_{max}$
$P_{t_{min}R_{max}}$	t_{min}	R_{max}	$\frac{M}{\pi E t_{min} R_{max}^3}$	$2\pi\rho t_{min} R_{max}$
$P_{t_{min}stress}$	t_{min}	$\sqrt{\frac{M}{\pi \sigma_{adm} t_{min}}}$	$\frac{1}{E}\sqrt{\frac{\pi t_{min} \sigma_{adm}^3}{M}}$	$2\rho\sqrt{\frac{\pi M t_{min}}{\sigma_{adm}}}$

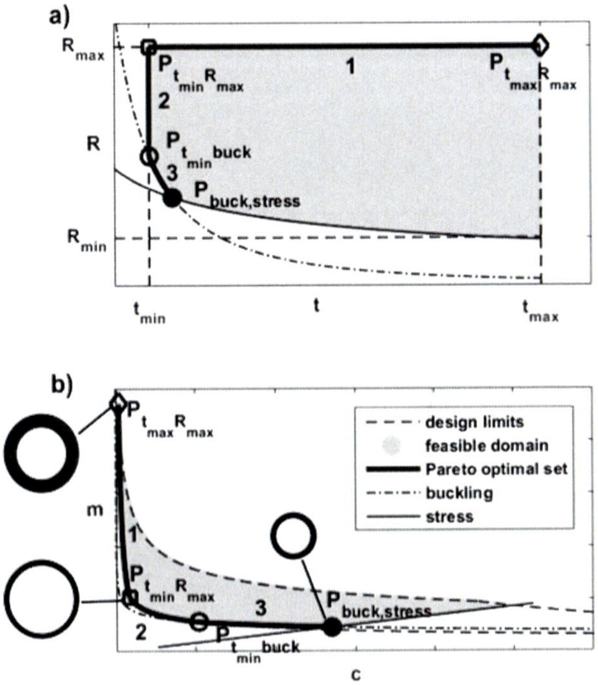

Fig. 5.5 Case ③ of Fig. 5.2, Pareto-optimal sets defined, respectively, in the design variable domain (**a**) and in the objective function domain (**b**). Analytical expressions of Pareto-optimal sets in Table 5.3

Table 5.3 Case ③ of Fig. 5.2. Analytical expressions of the Pareto-optimal sets in both the design variables domain and objective functions domain

Analytical expressions

Pareto optimal set in the design variable domain	Pareto optimal design in the objective functions domain	Boundaries (Fig. 5.5)	
		Pareto optimal subset	Points
$R = R_{max}$	$m = \frac{2\rho M}{E R_{max}^2 c}$	1	$\left[P_{t_{max}R_{max}}, P_{t_{min}R_{max}}\right]$
$t = t_{min}$	$m = 2\rho\sqrt[3]{\frac{\pi^2 M t_{min}^2}{Ec}}$	2	$\left[P_{t_{min}R_{max}}, P_{t_{min}buck}\right]$
$R = \frac{M}{k_{cr} E t_{min}^2}$	$m = 2\rho\sqrt[5]{\frac{\pi^4 M^3}{k_{cr}^2 E^3 c}}$	3	$\left[P_{t_{min}buck}, P_{buck,stress}\right]$

Coordinates of points

Point	t	R	c	m
$P_{t_{max}R_{max}}$	t_{max}	R_{max}	$\frac{M}{\pi E t_{max} R_{max}^3}$	$2\pi\rho t_{max} R_{max}$
$P_{t_{min}R_{max}}$	t_{min}	R_{max}	$\frac{M}{\pi E t_{min} R_{max}^3}$	$2\pi\rho t_{min} R_{max}$
$P_{t_{min}buck}$	t_{min}	$\frac{M}{k_{cr} E t_{min}^2}$	$\frac{k_{cr}^3 E^2 t_{min}^5}{\pi M^2}$	$2\pi\rho \frac{M}{k_{cr} E t_{min}}$
$P_{buck,stress}$	$\sqrt[3]{\frac{\pi M \sigma_{adm}}{k_{cr}^2 E^2}}$	$\sqrt[3]{\frac{k_{cr} E M}{\pi^2 \sigma_{adm}^2}}$	$\sqrt[3]{\frac{\pi^2 \sigma_{adm}^5}{k_{cr} E^4 M}}$	$2\rho\sqrt[3]{\frac{\pi^2 M^2}{k_{cr} E \sigma_{adm}}}$

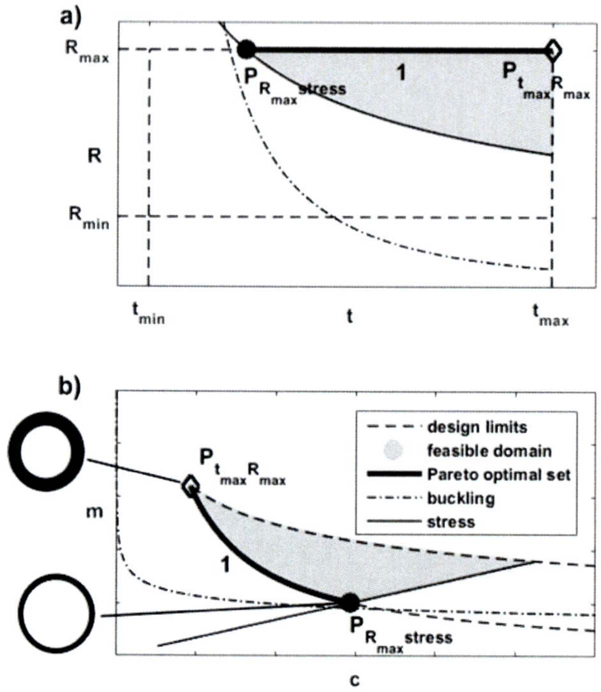

Fig. 5.6 Case ④ of Fig. 5.2, Pareto-optimal sets defined, respectively, in the design variable domain (**a**) and in the objective function domain (**b**). Analytical expressions of Pareto-optimal sets in Table 5.4

Table 5.4 Case ④ of Fig. 5.2. Analytical expressions of the Pareto-optimal sets in both the design variables domain and objective functions domain

Analytical expressions

Pareto optimal set in the design variable domain	Pareto optimal design in the objective functions domain	Boundaries (Fig. 5.6)	
		Pareto optimal subset	Points
$R = R_{max}$	$m = \frac{2\rho M}{E R_{max}^2 c}$	1	$\left[P_{t_{max} R_{max}}, P_{R_{max} stress}\right]$

Coordinates of points

Point	t	R	c	m
$P_{t_{max} R_{max}}$	t_{max}	R_{max}	$\frac{M}{\pi E t_{max} R_{max}^3}$	$2\pi\rho t_{max} R_{max}$
$P_{R_{max} stress}$	$\frac{M}{\pi R_{max}^2 \sigma_{adm}}$	R_{max}	$\frac{\sigma_{adm}}{E R_{max}}$	$2\rho \frac{M}{R_{max}\sigma_{adm}}$

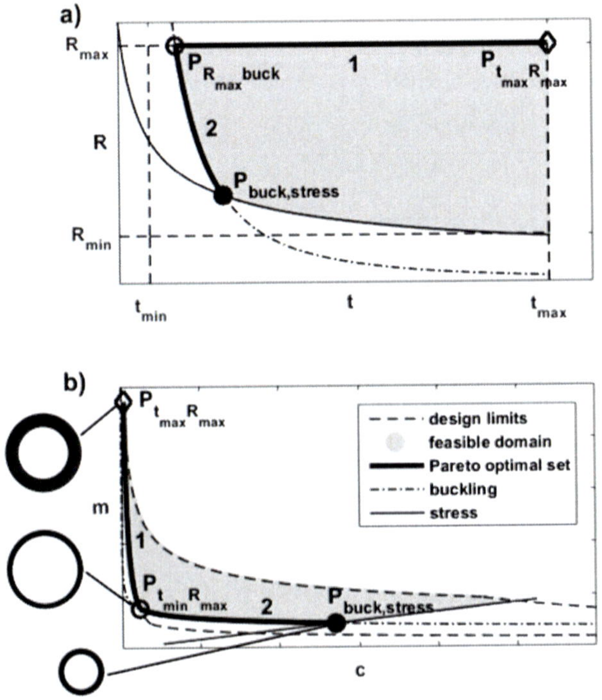

Fig. 5.7 Case ⑤ of Fig. 5.2, Pareto-optimal sets defined, respectively, in the design variable domain (**a**) and in the objective function domain (**b**). Analytical expressions of Pareto-optimal sets in Table 5.5

5.3.5 *Case* ⑤

The minimum radius R, corresponding to minimum m, is defined by the admissible stress and buckling, as shown in Fig. 5.7.

The full analytical expressions of the Pareto-optimal sets, both in the design variable domain and in the objective function domain, are reported in Table 5.5.

5.3.6 *Case* ⑥

In this case, we see from Fig. 5.2 that no solution within the bounds $t_{min} \leq t \leq t_{max}$ and $R_{min} \leq R \leq R_{max}$ is feasible since buckling and stress constraints are not satisfied.

Table 5.5 Case ⑤ of Fig. 5.2. Analytical expressions of the Pareto-optimal sets in both the design variables domain and objective functions domain

Analytical expressions

Pareto optimal set in the design variable domain	Pareto optimal design in the objective functions domain	Boundaries (Fig. 5.7)	
		Pareto optimal subset	Points
$R = R_{max}$	$m = \frac{2\rho M}{E R_{max}^2 c}$	1	$[P_{t_{max} R_{max}}, P_{R_{max} buck}]$
$R = \frac{M}{k_{cr} E t_{min}^2}$	$m = 2\rho \sqrt[5]{\frac{\pi^4 M^3}{k_{cr}^2 E^3 c}}$	2	$[P_{R_{max} buck}, P_{buck,stress}]$

Coordinates of points

Point	t	R	c	m
$P_{t_{max} R_{max}}$	t_{max}	R_{max}	$\frac{M}{\pi E t_{max} R_{max}^3}$	$2\pi\rho t_{max} R_{max}$
$P_{R_{max} buck}$	$\sqrt{\frac{M}{k_{cr} E R_{max}}}$	R_{max}	$\frac{1}{\pi}\sqrt{\frac{M k_{cr}}{E R_{max}^5}}$	$2\pi\rho\sqrt{\frac{M R_{max}}{k_{cr} E}}$
$P_{buck,stress}$	$\sqrt[3]{\frac{\pi M \sigma_{adm}}{k_{cr}^2 E^2}}$	$\sqrt[3]{\frac{k_{cr} E M}{\pi^2 \sigma_{adm}^2}}$	$\sqrt[3]{\frac{\pi^2 \sigma_{adm}^5}{k_{cr} E^4 M}}$	$2\rho\sqrt[3]{\frac{\pi^2 M^2}{k_{cr} E \sigma_{adm}}}$

5.4 Comparative Lightweight Design of Two Thin-Walled Tubes Made from Two Different Materials, Respectively

We will compare two Pareto-optimal sets referring, respectively, to two thin-walled tubes, one made from material A and one made from material B. Actually, as explained in Chap. 1, a correct comparison between different solutions to an engineering problem should be performed by comparing optimal solutions only [143]. By inspection of Fig. 5.2, we see that the Pareto-optimal sets can be divided into subsets, which refer to

- $R = R_{max}$ (subsets 1 in Fig. 5.2)
- $t = t_{min}$ (subsets 2 in Fig. 5.2)
- active constraint on buckling (subsets 3 in Fig. 5.2)

In the next subsections, we will compare the lightweight design of two thin-walled tubes made, respectively, from two different materials, by addressing the Pareto-optimal subsets 1,2 and 3 in Fig. 5.2. At the end of this section, we will focus on how obtaining minimum m when constraints are active.

5.4.1 Lightweight Design Referring to Pareto-Optimal Subset 1, $R = R_{max}$

Let us consider the Pareto-optimal subsets 1 in Fig. 5.2, which have expression $R = R_{max}$ (and t is variable). This case refers to the maximum exploitation of the available room, represented by $R = R_{max}$.

The Pareto-optimal subset 1 in the objective function domain is given by

$$m = \frac{2\rho M}{E R_{max}^2 c}$$

this equation is reported in Tables 5.1 to 5.5 where the proper lower and upper bounds on m and c are listed.

For any given value of c, the ratio between the mass per unit length of the thin-walled tube made from material A (m_A), and the mass per unit length of the thin-walled tube made from material B (m_B) can be computed as

$$\frac{m_A}{m_B} = \frac{\rho_A E_B}{\rho_B E_A} \tag{5.18}$$

Let us consider aluminium (A) and steel (B) ($\rho_A = 2800\,\text{kg/m}^3$, $\rho_B = 7800\,\text{kg/m}^3$, $E_A = 70000\,\text{MPa}$, $E_B = 210000\,\text{MPa}$, $\nu_A = \nu_B = 0.3$), Eq. 5.18 returns 1.08. When the maximum available room is saturated, steel can provide the same stiffness of aluminium with about 8% less mass. This allows steel to have a role in lightweight design.

5.4.2 *Lightweight Design Referring to Pareto-Optimal Subset 2, $t = t_{min}$*

Let us consider the case in which the minimum thickness $t = t_{min}$ gives the best compromise between minimum m and minimum c. From a practical point of view, this case means that there is some (technological) constraint that limits the minimum value of the thickness.

The Pareto-optimal subset 2 in the objective function domain is given by

$$m = 2\rho \sqrt[3]{\frac{\pi^2 M t_{min}^2}{Ec}}$$

This equation is reported in Tables 5.1, 5.2 and 5.5 where the proper lower and upper bounds on m and c are listed.

For any given value of c, with $t = t_{min}$, the ratio between the mass per unit length of the thin-walled tube made from material A (m_A), and the mass per unit length of the thin-walled tube made from material B (m_B) can be computed as

$$\frac{m_A}{m_B} = \frac{\rho_A}{\rho_B} \sqrt[3]{\frac{E_B}{E_A}} \tag{5.19}$$

Let us consider again aluminium (A) and steel (B), Eq. 5.19 returns 0.52. For the same compliance c, the mass of the aluminium tube is about half of the mass of the steel tube.

5.4.3 *Lightweight Design Referring to Pareto-Optimal Subset 3, Active Constraint on Buckling*

In this case, the buckling constraint is active and the Pareto-optimal set is described by

$$m = 2\rho \sqrt[5]{\frac{\pi^4 M^3}{k_{cr}^2 E^3 c}}$$

This equation refers to subset 3 in Tables 5.3 and 5.5 where the proper lower and upper bounds on m and c are listed.

For any given value of c, the ratio between the mass per unit length of the thin-walled tube made from material A (m_A), and the mass per unit length of the thin-walled tube made from material B (m_B) can be computed as

$$\frac{m_A}{m_B} = \frac{\rho_A}{\rho_B} \sqrt[5]{\frac{E_B^3}{E_A^3}} \tag{5.20}$$

By considering aluminium (A) and steel (B), Eq. 5.20 returns 0.69. This means that, when the buckling constraint is active, given the compliance c, the mass per unit length of the aluminium tube is about 30% less than the mass per unit length of the steel tube. Since only density and young modulus are considered in this equation, this result is only slightly dependent on the actual chemical composition (and thus strength) of the considered aluminium or steel alloy.

5.4.4 *Minimum Mass*

In the design of a thin-walled tube, the knowledge of the minimum theoretically attainable mass (per unit length of the tube) by changing design variables within prescribed constraints is a valuable information. The minimum mass depends on the active constraint(s). The minimum mass can be obtained when

- $t = t_{min}$ and $R = R_{min}$ (case ①, Fig. 5.2)
- the stress constraint is active (cases ② to ⑤, Fig. 5.2)

The first case represents the minimum absolute mass. In this case, the minimum mass reads (Table 5.1)

$$m_{min} = 2\pi\rho R_{min} t_{min} \tag{5.21}$$

and only material density has to be considered when comparing two materials.

In the second case, the minimum mass is given by the intersection between the stress constraints and one of the other constraints. If the other active constraint is $t = t_{min}$ (case ②), the minimum mass per unit length reads (Table 5.2)

$$m_{min} = 2\rho\sqrt{\frac{\pi M t_{min}}{\sigma_{adm}}} \tag{5.22}$$

and, when two materials (A and B) are considered, the ratio between the two minimum masses per unit length $m_{min,A}$ and $m_{min,B}$ can be computed as

$$\frac{m_{min,A}}{m_{min,B}} = \frac{\rho_A}{\rho_B}\sqrt{\frac{\sigma_{adm,B}}{\sigma_{adm,A}}} \tag{5.23}$$

In case the bucking constraint is active (cases ③ and ⑤), the minimum mass per unit length reads (Tables 5.3 and 5.5)

$$2\rho\sqrt[3]{\frac{\pi^2 M^2}{k_{cr} E \sigma_{adm}}} \tag{5.24}$$

and the ratio between the masses per unit length of the thin-walled tube made from two different materials can be computed as

$$\frac{m_{min,A}}{m_{min,B}} = \frac{\rho_A}{\rho_B}\sqrt[3]{\frac{E_B \sigma_{adm,B}}{E_A \sigma_{adm,A}}} \tag{5.25}$$

Finally, if the other active constraint is $R = R_{max}$, (case ④), the minimum mass per unit length reads (Table 5.4)

$$2\rho\frac{M}{R_{max}\sigma_{adm}} \tag{5.26}$$

and the ratio between the masses per unit length of the thin-walled tube made from two different materials can be computed as

$$\frac{m_{min,A}}{m_{min,B}} = \frac{\rho_A \sigma_{adm,B}}{\rho_B \sigma_{adm,A}} \tag{5.27}$$

5.4.5 Remarks

Summarising the analyses performed in the previous subsections, we give some hints for concept designers. Given $R_{min} \leq R \leq R_{max}$ and $t_{min} \leq t \leq t_{max}$, one Pareto-optimal solution is immediately defined, namely, the one referring to R_{max}, t_{max} which is the solution showing the minimum possible compliance. Other candidate Pareto-optimal solutions are the ones referring to R_{max}, t_{min}, R_{min}, t_{min}. If such solutions are feasible (i.e. they satisfy both the buckling and the stress constraints) they could be good options for, respectively, a good compromise between minimum m and c, and the best lightweight design (minimum m). Always the solution with minimum attainable radius and maximum attainable thickness must be discarded, which is not intuitive.

In general, to have minimum mass, high strength material should be employed, and this leads in general to higher compliance. This may appear trivial, but, by resorting to MOO we can state that such conclusion is ultimately proved.

5.5 Summary

In this chapter, the problem of the bending of thin-walled circular tubes has been addressed by employing MOO theory. The mass per unit axial length of the thin-walled tube and the bending rotation per unit axial length have been minimised under safety (admissible stress) and stability (buckling) constraints. Additionally, upper and lower bounds were given for exernal radius and thickness of the tube, respectively. The Pareto-optimal sets both in the objective function domain and in the design variable domain have been derived analytically. The proposed formulae can be used by the designer for the preliminary design of the structure. If the desired performance of the structure is given, the designer can immediately find the optimal parameters that satisfy the request. Even if the model used is elementary, the rigorous optimal design of a thin-walled tube turned out to be a rather complex task, in that a simple and straightforward formula for getting a solution does not exist. Nonetheless, the proposed approach makes easy, straightforward and rigorous the optimal design (sizing) of any thin-walled circular tube.

For the minimum mass/deflection design of a thin-walled tube, it has been demonstrated that:

- the Pareto-optimal solution is always given by the active constraints,
- when active, the buckling constraint is always part of the Pareto-optimal set,
- the volume constraint (maximum attainable radius of the tube) is always part of the Pareto-optimal set,
- the stress constraint, when active, is never part of the Pareto-optimal set, except for the tube with the minimum attainable mass.

As expected, such considerations are perfectly consistent with the outcomes of Chap. 4 related to general beams under bending.

An optimally sized thin-walled tube has a maximum attainable radius and maximum attainable thickness, which guarantees the minimum possible compliance. Another candidate Pareto-optimal solution is the one referring to maximum attainable radius and minimum attainable thickness which guarantees a good compromise between minimum mass and minimum compliance. Another candidate Pareto-optimal solution is the one referring to minimum attainable radius and minimum attainable thickness which guarantees lightweight design. If such two last candidate Pareto-optimal solutions are feasible (i.e. they satisfy both the buckling and the stress constraints) they could be immediately selected as initial concept design solutions. Always the solution with minimum attainable radius and maximum attainable thickness must be discarded, which is not intuitive.

Analytical formulae have been used to compare the lightweight performance of optimally sized thin-walled tubes made from different materials. Referring to aluminium and steel, if saturating the available room is the design driver, steel performs better than aluminium in terms of lightweight, at any given stiffness (i.e. compliance). In all other cases, the opposite occurs.

Focusing exclusively on minimum mass per unit length of the thin-walled tube, we gave analytical formulae for selecting materials, under the condition that buckling and admissible stress constraints are satisfied.

The results of the MOO analyses demonstrated that the use of steel can be a better choice with respect to aluminium for effective lightweight construction of circular tubes. This conclusion depends on the limit states which were considered in the analyses. If the admissible fatigue stress is considered as safety in the MOO, the conclusion could be different. Moreover, the presence of welds in the circular tubes could produce a completely different behaviour because of local effects not included in the analytical model. Future analyses could focus on the optimal design of tube geometry and local effects.

Appendix 1: Constraint Activity in the Pareto-Optimal Set

The necessary condition for for lightweight design of a thin-walled tube subject to bending is defined by Eq. 3.14. By applying this equation, we obtain

$$\begin{aligned} \frac{4M\rho}{\pi E t^2 r^5} \cdot (R - R_{max})^2 \cdot (R - R_{min})^2 \cdot (t - t_{max})^2 \cdot \\ \cdot (t - t_{min})^2 \cdot (M - ERk_{cr}t^2)^2 \cdot \\ \cdot (M - R^2 \pi \sigma_{adm} t)^2 = 0 \end{aligned} \quad (5.28)$$

Equation 5.28 represents a necessary condition, i.e. it contains the Pareto-optimal set plus other non-optimal solution. In the next subsections, the non-optimal solutions will be identified and removed.

A1.1 Available Room Constraint—Proof that $P_{t_{max},R_{min}}$ is Always Dominated

Let us consider the available room constraint defined in Eq. 5.3. The outer room constraint is related to R_{max}. The inner room constraint refers to $R_{min} - 2t_{max}$. In the design variables domain, the room constraints are described by a rectangle with vertices $P_{t_{min},R_{min}}$, $P_{t_{min},R_{max}}$, $P_{t_{max},R_{min}}$ and $P_{t_{max},R_{max}}$ (see Fig. 5.8).

$P_{t_{min},R_{min}}$ and $P_{t_{max},R_{max}}$ represent, respectively, the solution with minimum mass and the solution with minimum compliance. According to the definition of the Pareto-optimal solution (Definition 1.11 in Sect. 3.1), such two points, if feasible, are the extreme points of the Pareto-optimal set.

Let us consider such two points $P_{t_{min},R_{max}}$ and $P_{t_{max},R_{min}}$ and

$$R = R_{max} \qquad t = t_{min} \tag{5.29}$$

$$R_{min} = \frac{R_{max}}{k_R} \qquad t_{max} = k_t t \qquad k_R,\ k_t > 1 \tag{5.30}$$

$$\overline{c} = \frac{M}{\pi E R^3 t} \tag{5.31}$$

$$\overline{m} = 2\pi \rho R t \tag{5.32}$$

$$c' = \frac{M}{\pi E R^3 k_t t} \tag{5.33}$$

$$m' = 2\pi \rho \frac{R}{k_R} t \tag{5.34}$$

If $c' = \overline{c}$, by equating Eqs. 5.31 and 5.33, the following relationship between k_R and k_t can be obtained.

$$k_R^3 = k_t \tag{5.35}$$

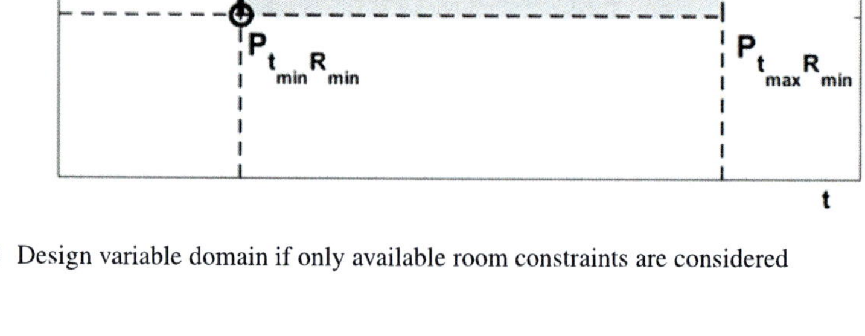

Fig. 5.8 Design variable domain if only available room constraints are considered

By replacing Eq. 5.35 into Eq. 5.30 and considering Eqs. 5.32 and 5.34, we obtain

$$m' = k_R^2 \overline{m} \qquad \Longrightarrow \qquad m' > \overline{m} \tag{5.36}$$

Conversely, if $m' = \overline{m}$ the relationship between k_R and k_t reads

$$k_R = k_t \tag{5.37}$$

By replacing Eq. 5.37 into Eq. 5.30 and considering Eqs. 5.31 and 5.33, we obtain

$$c' = k_R^2 \overline{c} \qquad \Longrightarrow \qquad c' > \overline{c} \tag{5.38}$$

From the definition of Pareto-optimal solution (Definition 1.11 in Sect. 3.1), Eqs. 5.36 and 5.38 imply that the point (c', m') cannot dominate the point $(\overline{c}, \overline{m})$. The corresponding point in the design variable domain $P_{t_{max}, R_{min}}$ cannot dominate the point $P_{t_{min}, R_{max}}$. But, from Eq. 5.13, one and only one of the two points must be part of the Pareto-optimal set. The only point that can be part of the Pareto-optimal set is $P_{t_{min}, R_{max}}$, while $P_{t_{max}, R_{min}}$ is not part of such set. For continuity, excluding $P_{t_{max}, R_{min}}$ from the Pareto-optimal set implies that also the boundaries $R = R_{min}$ and $t = t_{max}$ do not belong to the Pareto-optimal set. So in Eq. 5.28 the expressions $t = t_{max}$ and $R = R_{min}$ can be dropped.

Only the two expressions $R - R_{max} = 0$ and $t - t_{min} = 0$ remain in Eq. 5.28 to describe the available room constraint.

A1.2 Buckling and Structural Safety Constraints

If a sufficiently large design domain is considered, buckling and structural safety constraints become active. For any value of t, the two constraints (Eqs. 5.5 and 5.7) can be rewritten as

$$R \geq \sqrt{\frac{M}{\pi t \sigma_{adm}}} \tag{5.39}$$

$$R \leq \frac{M}{E k_{cr} t^2} \tag{5.40}$$

The two constraints are continuous and, from Eqs 5.39 and 5.40, it can be observed that for $R \to 0$ the constraint on the admissible stress is active, while for $R \to \infty$, the buckling constraint is active. Therefore, a switching point exists where the active constraints switches from the maximum admissible stress to the buckling limit. This point can be computed by equating eq 5.39 to Eq. 5.40 and reads

$$t = \sqrt[3]{\frac{\pi M \sigma_{adm}}{E^2 k_{cr}^2}} \qquad R = \sqrt[3]{\frac{M E k_{cr}^2}{\pi^2 \sigma_{adm}^2}} \tag{5.41}$$

By substituting Eq. 5.39 into the expressions of the mass and of the compliance, the following equation representing the maximum stress constraint in the objective functions domain can be obtained

$$m = \frac{2\rho E M c}{\sigma_{adm}^2} \tag{5.42}$$

Equation 5.42 is a straight line with positive angular coefficient, so the stress constraint is not part of the Pareto-optimal set.

The intersection between stress and buckling constraints is the extreme point of the Pareto-optimal set and is the point with the minimum value of mass.

The only active constraint in the Pareto-optimal set can be the buckling constraint which is active up to the maximum acceptable stress level. Considering only the buckling constraint, the L-matrix of Eq. 3.7 reads

$$\mathbf{L} = \begin{bmatrix} 2\pi R\, rho & -\frac{M}{\pi E R^3 t^2} & -2ERk_{cr}t \\ 2\pi t\, rho & -\frac{3M}{\pi E R^4 t} & -Ek_{cr}t^2 \\ 0 & 0 & M - ERk_{cr}t^2 \end{bmatrix} \tag{5.43}$$

By applying Eq. 5.11, the analytical expression of the Fritz John necessary condition when the buckling constraint is active can be derived.

$$R = \sqrt[3]{\frac{MEk_{cr}^2}{\pi^2\sigma_{adm}^2}} \tag{5.44}$$

This expression is coincident with the expression of the buckling itself (Eq. 5.40). By substituting Eq. 5.44 into the expressions of the mass and of the compliance, the following equation representing the buckling constraint in the objective functions domain can be obtained

$$m = \sqrt[5]{\frac{2\pi^4\rho M^3}{E^3 k_{cr}^2 c}} \tag{5.45}$$

It can be easily checked that this function is decreasing and convex for any value of c. Therefore, the points belonging to the buckling constraint condition belong to the Pareto-optimal set [90].

From Eqs. 5.42 and 5.45, it can be observed that for $c \to 0$ the admissible stress condition violates the buckling condition and for $c \to \infty$ the buckling condition violates the admissible stress condition. Thus, the buckling constraint condition is limited for high values of compliance by the stress constraint condition. The limit point corresponds to the switch between the two conditions reads

$$c = \sqrt[3]{\frac{\pi^2\sigma_{adm}^5}{E^4 k_{cr} M}} m = \sqrt[3]{\frac{2\pi^2 M^2 \rho}{E\sigma_{adm} k_{cr}}} \tag{5.46}$$

Chapter 6
Optimal Design of a Beam Subject to Bending: A Basic Application

In the framework of lightweight design, thin-walled structures have a primary role thanks to the high ratio between load carrying capabilities and mass. In order to maximise structural efficiency, i.e. minimise the mass of a general structure and maximise its stiffness at the same time, a rigorous optimisation approach is required.

Multi-objective minimisation approaches were followed by [106] and [160] for minimising the mass and deflection of members in bending. In the papers, optimised profiles of thin-walled open cross sections were computed by means of numerical methods. Optimisation of frame structures was studied in [30], where also buckling constraints have been considered.

The cited studies provided significant contributions in the field of MOO of beam structures, however, especially in the preliminary design stage, the possibility of relying on analytical solutions offers the designer a broader view and a profound understanding of the physical problem he/she is facing.

In this chapter, a basic study on a uniform beam under bending load is performed. A simple and straightforward MOO approach is followed, i.e. both the beam mass and deflection are minimised at the same time. Such a simple approach is based on the analytical method for finding the Pareto-optimal solutions described in Chap. 3. A number of beams with different cross sections are considered. The method is quite general and can, in principle, be applied to any (simple) multi-objective minimisation problem. The method guarantees that the analytical solution that is found provides necessary conditions for Pareto optimality.

The possibility of having analytical solutions could be a helpful tool for the designer since provides a broader view of the problem and a deeper understanding of its solution.

In this chapter, analytical solutions to the problem of minimising the mass and deflection of a uniformly bended beam are provided. The formalism and the mathematical approach adopted are the same as of Chap. 4, however, in this chapter, a more detailed analysis referring to particular beam cross sections is presented. Three

F. M. Ballo et al., *Optimal Lightweight Construction Principles*,
https://doi.org/10.1007/978-3-030-60835-4_6

beams with different cross sections are optimised under the same loading condition and the obtained solutions are compared. Dedicated buckling formulations are employed for each section in order to exploit their maximum performances. The computed analytical solutions are validated by applying a numerical procedure.

6.1 Uniformly Bended Beam

In this section, the problem of the optimal design of a uniformly bended beam of length L (Fig. 6.1) is formulated. The beam is optimised in order to minimise both mass and deflection, while subjected to constraints on the maximum admissible stress and buckling. The optimisation problem is solved for three different cross sections of the beam, namely, the hollow square, the I-shaped and the hollow rectangular cross section (Fig. 6.2). For each cross section, two design variables are considered in the optimisation process. For the hollow square cross section, the section width b and thickness s are considered as design variables. For all the other cross sections, the design variables are the cross section width b and height h. In the case of I-shaped and hollow rectangular sections, also the thickness s is required to fully describe the sections. In these cases, the thickness is given as a function of the other two variables in order to avoid local buckling.

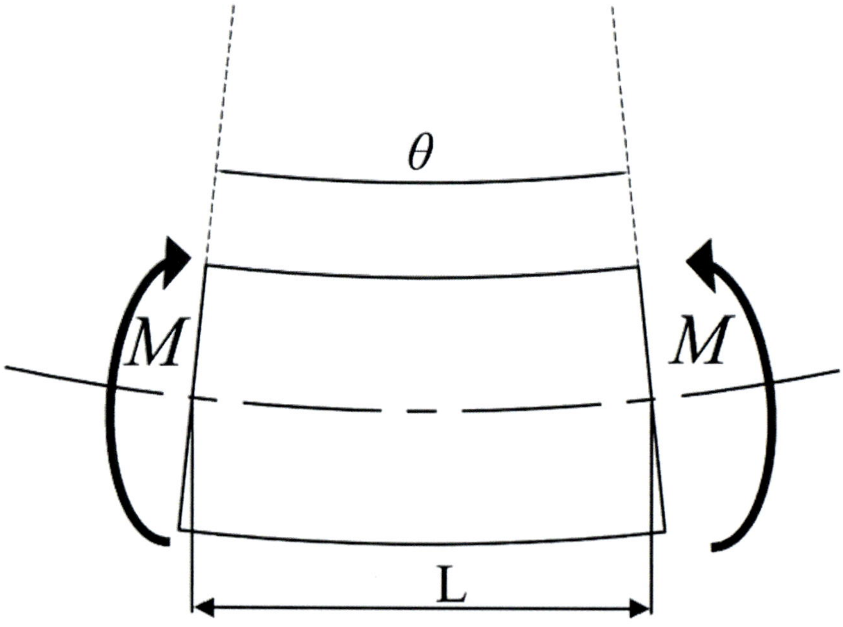

Fig. 6.1 Uniformly bended beam

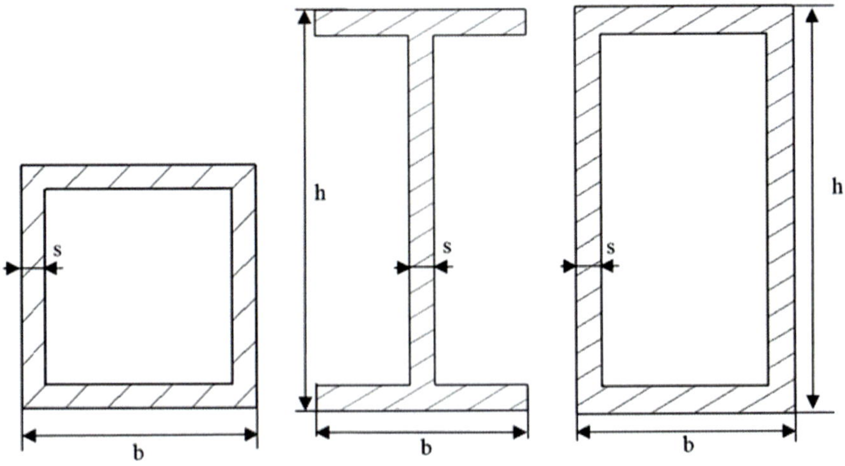

Fig. 6.2 Different cross sections considered

Actually when thin-walled cross sections are employed, failures by buckling can occur. Buckling phenomena depend on many parameters (cross-sectional geometry, material...) but mainly on the wall thickness [181]. As the thickness of the cross section is decreased, the structure gets more exposed to local buckling phenomena. That's why in our design approach the cross-sectional thickness was related to the other dimensions b and h through a linear relation, i.e. as a larger cross section is employed, the thickness s is proportionally increased.

Being x_1 and x_2 the design variables, the multi-objective structural optimisation problem can be formulated as

$$min \begin{bmatrix} \overline{\theta}(x_1, x_2) \\ \overline{m}(x_1, x_2) \end{bmatrix} \quad s.t.$$

$$\begin{aligned} &\sigma_{max}\,(x_1, x_2) \le \frac{\sigma_y}{\eta} \\ &M\,(x_1, x_2) \le M_{cr}\,(x_1, x_2) \\ &\zeta\,(x_1, x_2) \le cost \\ &x_{1min} \le x_1 \le x_{1max} \\ &x_{2min} \le x_2 \le x_{2max} \end{aligned} \tag{6.1}$$

where $\overline{m}$ is the mass per unit of length, $\overline{\theta}$ is the rotation of the beam end per unit of length, σ_y and η are the yield stress at the safety coefficient, respectively, and M_{cr} is the critical buckling load [69]. The constraint $\zeta\,(x_1, x_2) \le cost$ represents the local buckling of the cross section members. This constraint depends on the cross section types and limits the aspect ratio $\zeta\,(x_1, x_2)$ of the beam cross section members in order to prevent them to collapse by local buckling [69].

The problem stated in Eq. 6.1 has two objective functions $f_j(\mathbf{x})$ and two design variables. In this case, the number of design variables n_{dv} equals the number of objective functions n_{of} and Eq. 3.13 applies.

6.1.1 Hollow Square Cross Section

The problem of Eq. 6.1 is formulated for the hollow square beam.

In this case, the considered design variables are b and s. For the existence of such section, in addition to the obvious conditions $b > 0$ and $s > 0$, also the condition

$$b > 2s \tag{6.2}$$

must be satisfied (see Fig. 6.2). Under these conditions, the beam normalised rotation is given by

$$\frac{\theta}{L} = \frac{M}{EJ_{xx}} \tag{6.3}$$

where E is the material Young's modulus and J_{xx} the cross-sectional moment of inertia around the x-axis. The beam rotation per unit of length can be expressed as function of the design variables as

$$\frac{\theta}{L} = \frac{12M}{E\left(b^4 - (b-2s)^4\right)} \tag{6.4}$$

The mass per unit of length of the beam is

$$\frac{m}{L} = \rho\left(b^2 - (b-2s)^2\right) = 4\rho s\,(b-s) \tag{6.5}$$

with ρ the material density. The maximum stress in the beam reads

$$\sigma_{max} = \frac{6Mb}{b^4 - (b-2s)^4} \tag{6.6}$$

The (lateral-torsional) buckling constraint defines an upper limit on the sustainable bending moment of the beam. The critical bending moment of a hollow squared uniformly bended beam can be expressed as [69]

$$M_{cr} = \frac{\pi}{L}\sqrt{GJ_t EJ_{yy}} \tag{6.7}$$

where J_t is the torsional moment of inertia and J_{yy} is the moment of inertia around the weak axis of the cross section. Additionally, a constraint on local buckling of the cross section is required. In fact, in case of reduced cross-sectional thickness (i.e. thin-walled sections) local instabilities can occur [181]. The constraint on local buckling is

$$\frac{b-3s}{s} \leq k\epsilon \tag{6.8}$$

where k and ϵ are constants that depend on material and cross section.

The multi-objective minimum compliance problem for the hollow square cantilever beam therefore reads

$$min \begin{bmatrix} \dfrac{12M}{E\left(b^4-(b-2s)^4\right)} \\ 4\rho s\,(b-s) \end{bmatrix} \quad s.t.$$

$$\begin{aligned}
&\frac{6Mb}{b^4-(b-2s)^4} \leq \frac{\sigma_y}{\eta} \\
&M \leq \frac{\pi}{2L}\sqrt{GJ_tEJ_{yy}} \\
&\frac{b-3s}{s} \leq k\epsilon \\
&b_{min} \leq b \leq b_{max} \\
&s_{min} \leq s \leq s_{max}
\end{aligned} \tag{6.9}$$

The values of all the constant parameters are reported in Table 6.1.

The two objective functions are continuous and differentiable. The Hessian matrix **H** of each objective function can be computed as

Table 6.1 Structural parameters—hollow square beam

Notation	Description	Value	Unit
ρ	Material density	7800	$\frac{kg}{m^3}$
L	Beam length	8	m
E	Material modulus of elasticity	$210 \cdot 10^9$	Pa
G	Material modulus of tangential elasticity	$80.769 \cdot 10^9$	Pa
σ_y	Material yield stress	$260 \cdot 10^6$	Pa
η	Safety factor	1.5	–
b_{min}	Lower bound for b	0.15	m
b_{max}	Upper bound for b	0.6	m
s_{min}	Lower bound for s	0.001	m
s_{max}	Upper bound for s	0.05	m
M	Bending moment	50000	Nm
ϵ	Material parameter [69]	$\sqrt{\frac{235e6}{\sigma_y}}$	–
k	Local buckling coefficient [69]	42	–

$$\mathbf{H} = \begin{bmatrix} \dfrac{\partial^2 f}{\partial b^2} & \dfrac{\partial^2 f}{\partial b \partial s} \\ \dfrac{\partial^2 f}{\partial s \partial b} & \dfrac{\partial^2 f}{\partial s^2} \end{bmatrix} \tag{6.10}$$

It is easy to verify that, both mass and beam rotation have continuous second derivatives, thus the Hessian matrix is symmetric (Schwarz' theorem). A symmetric real matrix is positive definite if the determinants of the matrix and of all its principal minors are positive [137]. For a $2x2$ symmetric real matrix, as $\mathbf{H}$, this reduces to require that one of the main diagonal terms and its determinant are positive.

For the beam rotation, the diagonal term given by the second derivative with respect to the thickness reads

$$\frac{\partial^2 \overline{\theta}}{\partial s^2} = \frac{9M\,(b-2s)^2}{E\left(\dfrac{b^4-(b-2s)^4}{8}\right)^2} + \frac{3M\,(b-2s)^6}{E\left(\dfrac{b^4-(b-2s)^4}{8}\right)^3} \tag{6.11}$$

and is always positive since it is the sum of positive terms. The determinant of the Hessian matrix of the rotation reads

$$det\,(\mathbf{H}) = \frac{135Mb^2\,(b-2s)^2}{4E^2s^4\,(b-s)\left(b^2-2bs+2s^2\right)\left(b^3-3b^2s+4bs^2-2s^3\right)^3} \tag{6.12}$$

The numerator and denominator of Eq. 6.12 are always positive when the condition given by Eq. 6.2 holds. The beam rotation objective function is therefore convex (in the considered design space).

The Hessian matrix of the mass objective function reads

$$\mathbf{H} = \begin{bmatrix} 0 & 4\rho \\ 4\rho & -8\rho \end{bmatrix} \tag{6.13}$$

which has a positive and a negative eigenvalue and therefore the mass objective function is not convex. However, it can be observed that the condition $\nabla \overline{m} > \mathbf{0}$ gives

$$\begin{cases} s > 0 \\ b > 2s \end{cases} \tag{6.14}$$

which means that for any feasible value of b and s (see Eq. 6.2) the mass objective function is strictly monotonically increasing and no critical points are present. Therefore, this function is pseudo-convex for any feasible value of b and s [12].

Accordingly, the Fritz John necessary condition introduced in Chap. 3 can be applied to the current problem. Since $n_{dv} = n_{of} = 2$ Eq. 3.13 can be employed for solving the optimisation problem. The gradient of the objective functions reads

$$\nabla \mathbf{f} = \begin{bmatrix} \frac{\partial \overline{\theta}}{\partial b} & \frac{\partial \overline{m}}{\partial b} \\ \frac{\partial \overline{\theta}}{\partial s} & \frac{\partial \overline{m}}{\partial s} \end{bmatrix} = \begin{bmatrix} -\frac{48M}{E}\frac{b^3-(b-2s)^3}{\left(b^4-(b-2s)^4\right)^2} & 4\rho s \\ -\frac{48M}{E}\frac{(b-2s)^3}{\left(b^4-(b-2s)^4\right)^2} & 4\rho\,(b-2s) \end{bmatrix} \tag{6.15}$$

By imposing $det\,(\nabla \mathbf{f}) = 0$, we obtain

$$-\frac{12\rho M b\,(b-2s)}{Es\,(b-s)\left(b^2-2bs+2s^2\right)^2} = 0 \tag{6.16}$$

which has solution for $b = 0$, for $b = 2s$, for $b \to \infty$ and for $s \to -\infty$.

From a mathematical point of view, the solution $b = 2s$ is not a minimum and cannot belong to the Pareto-optimal set. In fact, from Eq. 6.14, the mass is neither convex nor pseudo-convex for that subset and the region $b = 2s$ represents a saddle. From a physical point of view, the solution $b = 2s$ violates the condition of Eq. 6.2 to have a hollow square section.

The other solutions, being outside of the feasible set, have to be excluded.

The Pareto-optimal set is therefore given by the active constraints only, see Eq. 3.13 for details.

By solving the buckling equation, we obtain

$$\frac{s}{L} \geq \frac{1}{2}\left(\frac{b}{L} - \sqrt[4]{\left(\frac{b}{L}\right)^4 - \frac{6\sqrt{2}M}{\pi L^3\sqrt{EG}}}\right) \tag{6.17}$$

which, considering the values of Table 6.1, is less restrictive than Eq. 6.2. Therefore, it can be assumed that the buckling constraint is never active for this problem and can be removed [164].

The local buckling constraint is a straight line in the design variables domain and reads

$$\frac{s}{L} \geq \frac{1}{k\epsilon + 3}\frac{b}{L} \tag{6.18}$$

The stress constraint reads

$$\frac{s}{L} \geq \frac{1}{2}\frac{b}{L} - \frac{1}{2}\sqrt[4]{\left(\frac{b}{L}\right)^4 - \frac{6M\eta}{\sigma_y L^3}\frac{b}{L}} \tag{6.19}$$

The feasible set of solutions, i.e. the values of the design parameters that satisfy all the design constraints [164], is shown in Fig. 6.3.

In the feasible set (defined by Eq. 6.2), the expression of the beam rotation of Eq. 6.4 is monotonically decreasing both with b and s, while the mass is monotonically increasing with b and s. This means that, if we consider only the bounds on b and s, the Pareto-optimal solution will be given either by the combination $b = b_{max}$ and $s = s_{min}$ or by $b = b_{min}$ and $s = s_{max}$.

Now, by substituting $b = b_{min}$ and $b = b_{max}$ in the objective functions of Eq. 6.9, the following relations are obtained

$$\overline{\theta}\left(\overline{m}\right)\Big|_{b=b_{min}} = \frac{12\rho^2 M}{\overline{m}E\left(2\rho b_{min}^2 - \overline{m}\right)} \tag{6.20}$$

$$\overline{\theta}\left(\overline{m}\right)\Big|_{b=b_{max}} = \frac{12\rho^2 M}{\overline{m}E\left(2\rho b_{max}^2 - \overline{m}\right)} \tag{6.21}$$

By inspecting Eqs. 6.20 and 6.21, it is clear that the function of Eq. 6.21 is always lower than the one of Eq. 6.20, meaning that all the solutions that belong to Eq. 6.20 do not satisfy Definition 1.11 of Pareto optimality. Therefore, the Pareto-optimal set is given by $b = b_{max}$ and $s = s_{min}$.

If a sufficiently large design space is considered, the solution $s = s_{min}$ violates the constraints on local buckling and maximum stress, that therefore are active constraints. The intersection point $\hat{b}_1$ between the stress constraint and the local buckling reads

$$\frac{\hat{b}_1}{L} = \frac{1}{2L}\sqrt[3]{\frac{6M\eta\left(k\epsilon + 3\right)^4}{\sigma_y\left(k^3\epsilon^3 + 6k^2\epsilon^2 + 13k\epsilon + 10\right)}} \tag{6.22}$$

For $b \leq \hat{b}_1$, the stress equation is more binding than the local buckling. This means that for $b \leq \hat{b}$ the stress constraint is active, then from $b = \hat{b}$ the active constraint switches from the maximum stress to the local buckling.

The constraint expression in Eq. 6.19 written as an equality constraint (i.e. active constraint) gives the equation of the maximum stress in the objective functions domain. In fact, Eq. 6.19 provides a relation between b and s. If we substitute Eq. 6.19 in the objective functions given by Eqs. 6.4 and 6.5 the dependency on s is dropped and $\overline{\theta}$ and $\overline{m}$ can be expressed as functions of b only. Finally, a general relation $b = g^{-1}\left(\overline{\theta}\right)$ is obtained which, once substituted into the mass objective function, provides the following expression

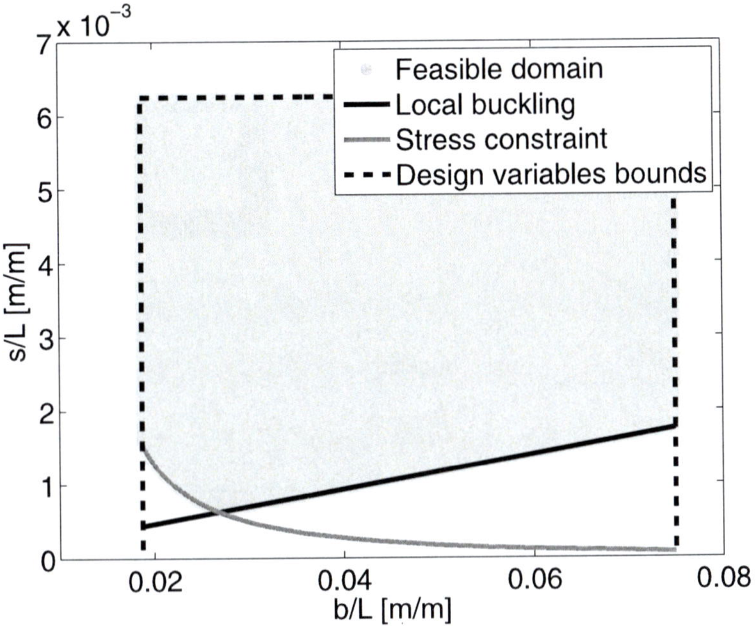

Fig. 6.3 Feasible set in design variables domain—hollow square section. Data in Table 6.1

$$\overline{m}\left(\overline{\theta}\right)\Big|_{\sigma=\frac{\sigma_y}{\eta}} = \rho\left(\frac{4\sigma_y^2}{E^2\eta^2\overline{\theta}^2} - \sqrt{\frac{2}{E\eta\overline{\theta}}\left(\frac{8\sigma_y^4}{E^3\eta^3\overline{\theta}^3} - 6M\eta\right)}\right) \tag{6.23}$$

which represents the stress constraint in the objective functions domain.

In the same way, the analytical expressions of the other design constraints in the objective functions domain are obtained. The expression of the local buckling constraint reads

$$\overline{m}\left(\overline{\theta}\right)\Big|_{buckling} = 2\rho\left(k\epsilon+2\right)\sqrt{\frac{6M}{E\overline{\theta}\left(k^3\epsilon^3+6k^2\epsilon^2+13k\epsilon+10\right)}} \tag{6.24}$$

The feasible set of solutions is therefore mapped into the objective functions domain and is shown in Fig. 6.4.

According to the Fritz John necessary condition, the Pareto-optimal set is given by the composition of the active constraints, i.e. the maximum stress, local buck-

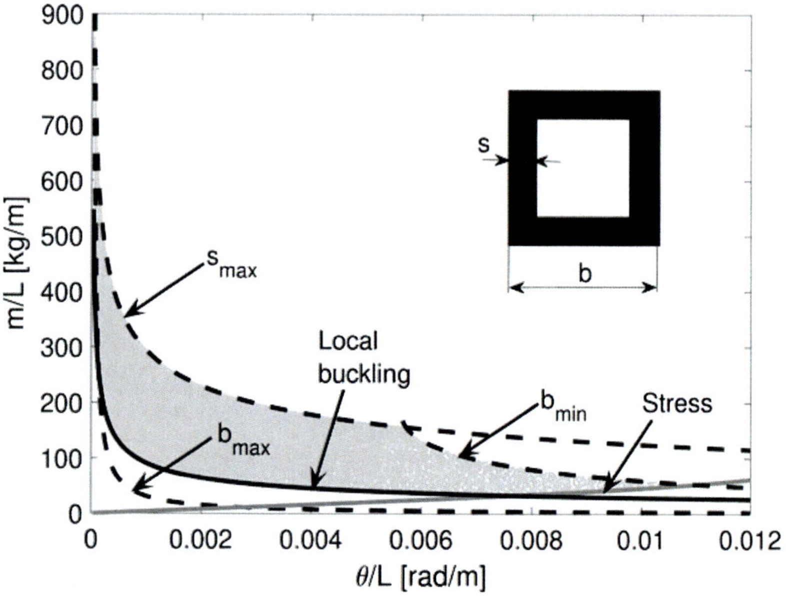

Fig. 6.4 Feasible set in objective functions domain—hollow square section. Data in Table 6.1

ling and the equation $b = b_{max}$. Equation $b = b_{max}$ and the local buckling are both monotonically decreasing in the objective functions domain and therefore belong to the Pareto-optimal set. For the constraint $b = b_{max}$, the derivative of the inverse of Eq. 6.21 is required

$$\frac{d\overline{m}}{d\overline{\theta}} = \left(\frac{d\overline{\theta}}{d\overline{m}}\right)^{-1} = -\frac{6\rho M}{\sqrt{E\overline{\theta}^3\left(E\overline{\theta}b_{max}^4 - 12M\right)}} \tag{6.25}$$

which is always negative. The constraint on local buckling (Eq. 6.24) is proportional to $1/\sqrt{\overline{\theta}}$ and therefore is monotonically decreasing in the objective functions domain. The derivative of the stress constraint (Eq. 6.23) with respect to $\overline{\theta}$ can be computed by applying the chain rule. In fact by substituting Eq. 6.19 in the expressions of $\overline{\theta}$ and $\overline{m}$, the dependency on s is dropped and the derivative reads

$$\frac{d\overline{m}}{d\overline{\theta}} = \frac{\partial\overline{m}}{\partial b}\frac{\partial b}{\partial\overline{\theta}} = \frac{\partial\overline{m}}{\partial b}\left(\frac{\partial\overline{\theta}}{\partial b}\right)^{-1} \tag{6.26}$$

It is possible to verify (analytical passages are not reported for sake of space) that Eq. 6.26 is always positive, meaning that the stress constraint is increasing with $\overline{\theta}$. So, the maximum stress does not satisfy Definition 1.11 of Pareto optimality and

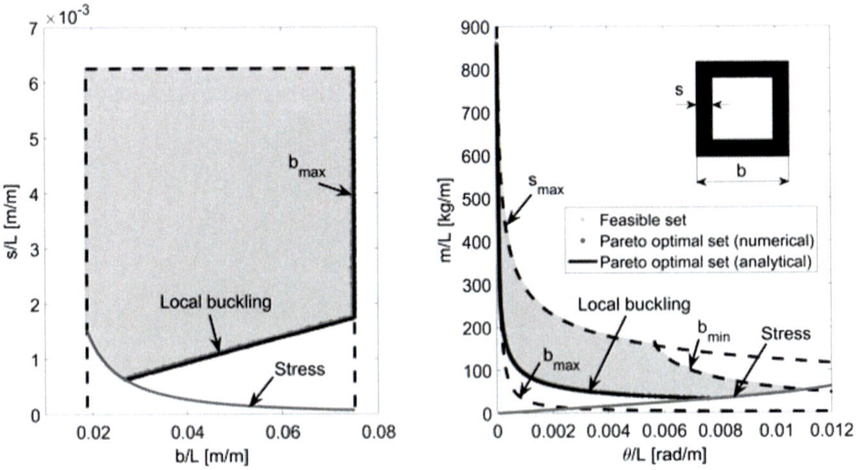

Fig. 6.5 Pareto-optimal set in the design variables domain (left) and objective functions domain (right) for a beam with hollow square cross section. Data in Table 6.1

therefore is only limiting the constraint on local buckling as shown in Figs. 6.3 and 6.4.

Summarising, from the previous analysis, the analytical Pareto-optimal set in the design variables domain is:

- $b = (k\epsilon + 3)\, s$ from $\hat{b}_1$ to b_{max}
- $b = b_{max}$ from $\hat{s}_1 = \dfrac{\hat{b}_1}{k\epsilon + 3}$ to s_{max}

The obtained solution is reported in Fig. 6.5 both in the design variables and objective functions domain. A numerical validation is also reported.

6.1.2 I-Shaped Cross Section

In this section, the I-shaped beam is optimised with respect to the two design variables b and h (Fig. 6.2). The thickness s was considered as a function of the design variables in order to prevent local buckling on the cross section. A linear relation is derived by considering the existing IPE and HE standardised cross sections [67]. All the existing IPE cross sections are included in the interpolation, while for the HE only the A-series is included. Figure 6.6 shows the results obtained from the linear interpolation.

The interpolated expression of the cross-sectional thickness reads

$$s = p_1 + p_2 b + p_3 h \tag{6.27}$$

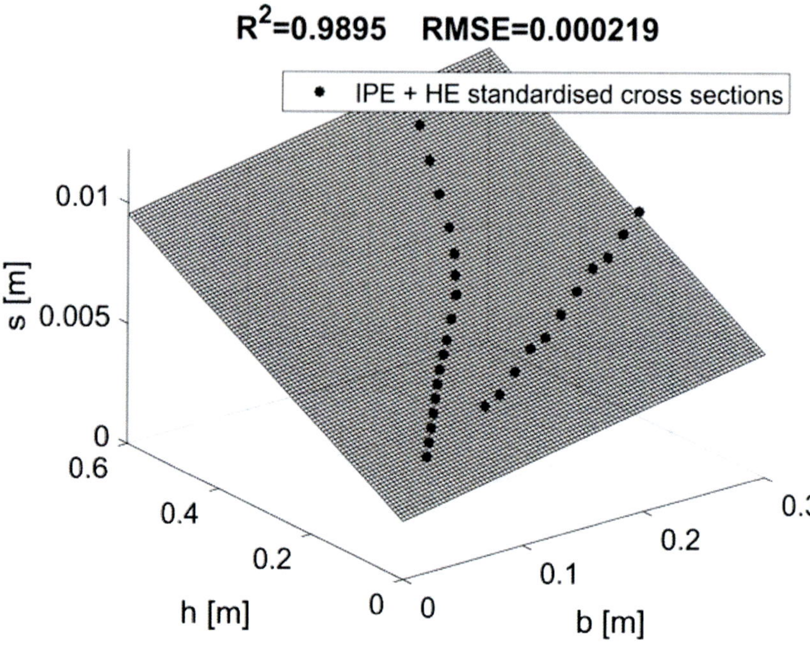

Fig. 6.6 Linear interpolation of thickness s—I-shaped cross section

The values of the coefficients are reported in Table 6.2, the cross-sectional dimensions are expressed in meters.

By considering the formulae for the I-shaped cross section, the MOO problem reads

$$\begin{aligned}
& min \begin{bmatrix} \dfrac{12M}{E(2bs^3+s(h-2s)^3+6bs(h-s)^2)} \\ \rho(2bs+s(h-2s)) \end{bmatrix} \quad s.t. \\
& \frac{6Mh}{2bs^3+s(h-2s)^3+6bs(h-s)^2} \leq \frac{\sigma_y}{\eta} \\
& M \leq \frac{\pi}{2L}\sqrt{GJ_t EJ_{yy}}\sqrt{1+\frac{EJ_\omega}{GJ_t}\frac{1}{4}\left(\frac{\pi}{L}\right)^2} \\
& \frac{h-3s}{s} \leq k_w\epsilon \\
& \frac{b}{2s} \leq k_f\epsilon \\
& b_{min} \leq b \leq b_{max} \\
& h_{min} \leq h \leq h_{max}
\end{aligned} \tag{6.28}$$

where J_ω is the warping moment of inertia of the cross section. Constraints on fourth and fifth lines of Eq. 6.28 represent the limit for the local buckling of the web and the flanges, respectively [69]. The other structural parameters are listed in Table 6.2. By

Table 6.2 Structural parameters–I-shaped beam

Notation	Description	Value	Unit
ρ	Material density	7800	$\frac{kg}{m^3}$
L	Beam length	8	m
E	Material modulus of elasticity	$210 \cdot 10^9$	Pa
G	Material modulus of tangential elasticity	$80.769 \cdot 10^9$	Pa
σ_y	Material yield stress	$260 \cdot 10^6$	Pa
b_{min}	Lower bound for b	0.01	m
b_{max}	Upper bound for b	0.15	m
h_{min}	Lower bound for h	0.01	m
h_{max}	Upper bound for h	0.6	m
p_1		0.002419	–
p_2		0.00904	–
p_3		0.01186	–
c_1		$-4.6526 \cdot 10^6$	–
c_2		$1.328 \cdot 10^5$	–
c_3		−982.5	–
c_4		−3.68	–
c_5		0.0619	–
M	Bending moment	50000	Nm
ϵ	Material parameter [69]	$\sqrt{\frac{235e6}{\sigma_y}}$	–
k_w	Local buckling coefficient for web [69]	124	–
k_f	Local buckling coefficient for flange [69]	15	–

replacing Eq. 6.27 in Eq. 6.28, the optimisation problem can be expressed as function of the two design variables b and h.

The solution of the unconstrained problem can be obtained by imposing $det(\nabla \mathbf{f}) = 0$ As shown in Fig. 6.8, the solution of the unconstrained problem is feasible for certain values of b and h and intersects $b = b_{min}$ and $h = h_{max}$.

Actually, the presence of an unconstrained solution may appear in contrast with what stated at the beginning of Sect. 4.2. The explanation stands in Eq. 6.27. In fact, interpolation 6.27, introduced to account for typical aspect ratios deriving from the standards, is implicitly adding a further constraint to the problem (Eq. 6.27), which is not directly related to structural requirements. That's why, in the following of the chapter, we will refer to that solution as *unconstrained solution* anyway.

If a sufficiently large design space is considered, the unconstrained solution is bounded by the constraints equations. Equation 3.13 can be applied to solve the problem.

The stress constraint is provided by the following equation:

$$\frac{6Mh}{bh^3 - (h-2s)^3(b-2s)} - \frac{\sigma_y}{\eta} = 0 \tag{6.29}$$

by solving Eq. 6.29, a relation between b and h is obtained.

An explicit analytical solution cannot be found for the buckling equation, therefore an approximated expression is derived by interpolating the numerical results with a fourth-degree polynomial function (Fig. 6.7):

$$\frac{h}{L} \simeq c_1 \left(\frac{b}{L}\right)^4 + c_2 \left(\frac{b}{L}\right)^3 + c_3 \left(\frac{b}{L}\right)^3 + c_4 \left(\frac{b}{L}\right) + c_5 \tag{6.30}$$

The chosen expression is able to approximate the changes in the curvature of the numerical solution in the design domain. The interpolation error tends to increase for low values of b.

By comparing the expressions of the stress and buckling constraints (or comparing Figs. 6.7 and 6.8), one realises that the buckling curve is less restrictive than the stress one for this particular problem. The buckling constraint can be therefore removed from the problem. The same condition holds also for the constraints on local buckling, which are removed from the optimisation problem.

The feasible set of solutions is shown in Figs. 6.8 and 6.9 in the design variables and objective functions domain, respectively.

Figure 6.8 shows that the stress constraint is limiting the minimum value of h. The unconstrained solution is thus limited by the stress constraint and cannot reach the condition $h = h_{min}$. According to [127], this part of the unconstrained solution belongs to the Pareto-optimal set.

The upper bound on h is an active constraint since is limiting the unconstrained solution from above. These two constraints are also convex and monotonically decreasing in the objective functions domain. Figure 6.9 shows this situation. The upper bound on h (h_{max}) between b_{max} and the unconstrained solution belongs to the Pareto-optimal set.

The stress equation limits the unconstrained solution from below. This constraint presents a minimum in the objective function domain and is increasing for high values of deflection (this situation can be noticed in Fig. 6.9). The portion of this curve from the intersection with the unconstrained solution to the minimum of the stress constraint belongs to the Pareto-optimal set.

The complete Pareto-optimal set is therefore composed by three regions:

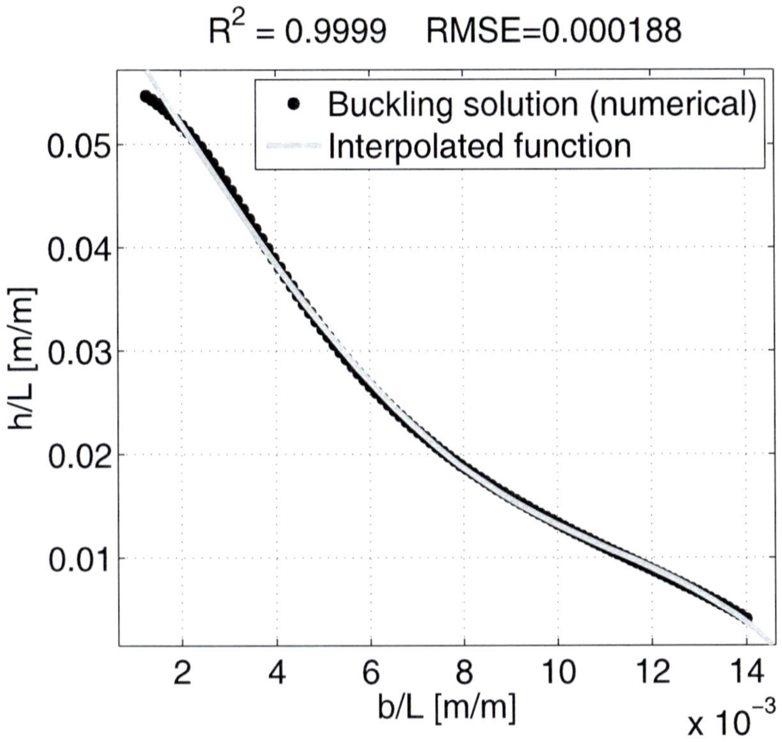

Fig. 6.7 Interpolated buckling expression—I-shaped cross section

- the portion of the stress equation (Eq. 6.29) from the minimum of $\left.\frac{d\overline{m}}{d\overline{\theta}}\right|_{\sigma=\frac{\sigma_Y}{\eta}}$ to the intersection with the unconstrained solution;
- the unconstrained solution from this point to h_{max};
- $h = h_{max}$ from this point to b_{max}.

The Pareto-optimal set for the I-shaped cantilever beam is shown in Fig. 6.10 together with a numerical validation.

6.1.3 Hollow Rectangular Cross Section

In this section, the hollow rectangular beam is optimised. The design variables are the dimensions b and h of Fig. 6.2, the wall thickness s is kept as a linear function of the design variables in order to avoid local buckling. The relation between the wall

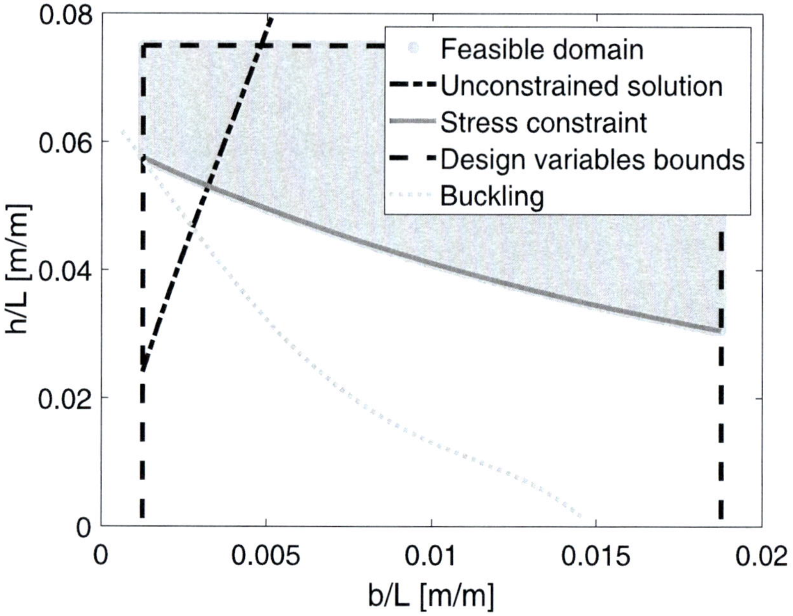

Fig. 6.8 Feasible set in design variables domain—I-shaped cross section. Data in Table 6.2

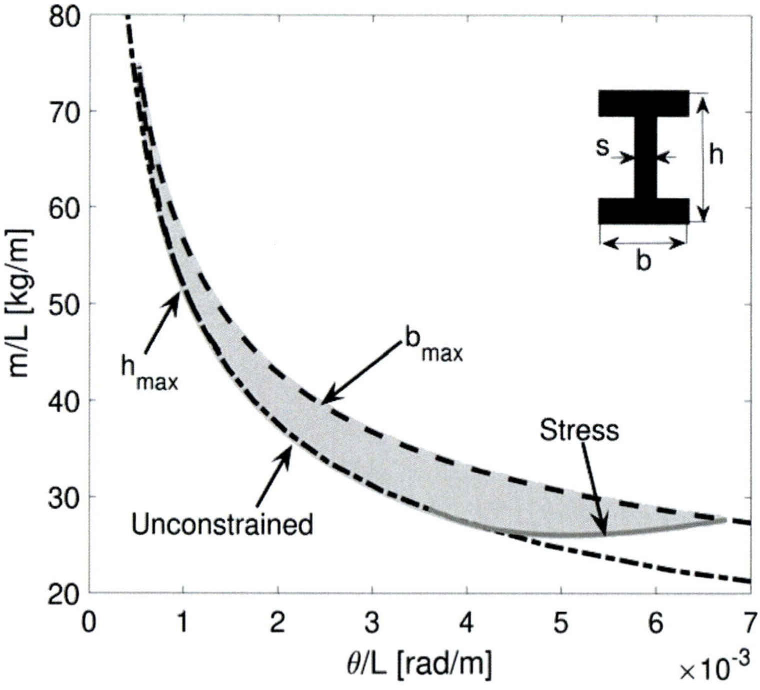

Fig. 6.9 Feasible set in objective functions domain—I-shaped cross section. Data in Table 6.2

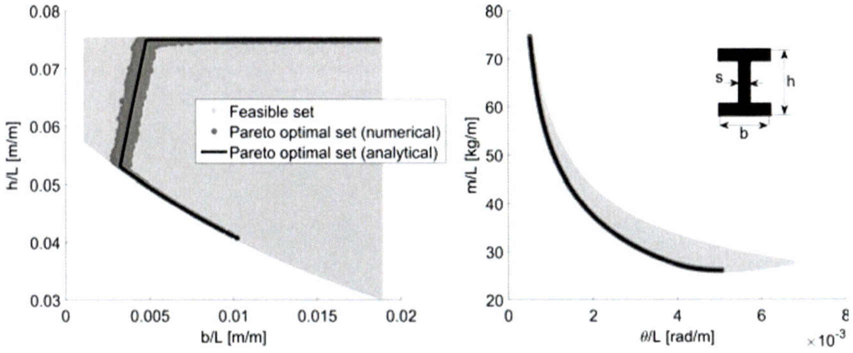

Fig. 6.10 Pareto optimal set in the design variables domain (left) and objective functions domain (right) for a I-shaped beam. Data in Table 6.2

Table 6.3 Structural parameters—cantilever hollow rectangular cross section

Notation	Description	Value	Unit
ρ	Material density	7800	$\frac{kg}{m^3}$
L	Beam length	8	m
E	Material modulus of elasticity	$210 \cdot 10^9$	Pa
G	Material modulus of tangential elasticity	$80.769 \cdot 10^9$	Pa
σ_y	Material yield stress	$260 \cdot 10^6$	Pa
b_{min}	Lower bound for b	0.01	m
b_{max}	Upper bound for b	0.15	m
h_{min}	Lower bound for h	0.01	m
h_{max}	Upper bound for h	0.6	m
p_1		$6.561 \cdot 10^{-4}$	–
p_2		0.0104	–
p_3		0.01312	–
c_1		$-9.298 \cdot 10^8$	–
c_2		$4.072 \cdot 10^7$	–
c_3		$-6.894 \cdot 10^5$	–
c_4		$5.700 \cdot 10^3$	–
c_5		-25.03	–
c_6		0.0645	–
M	Bending moment	50000	Nm
ϵ	Material parameter [69]	$\sqrt{\frac{235e6}{\sigma_y}}$	–
k_w	Local buckling coefficient for web [69]	124	–
k_f	Local buckling coefficient for flange [69]	42	–

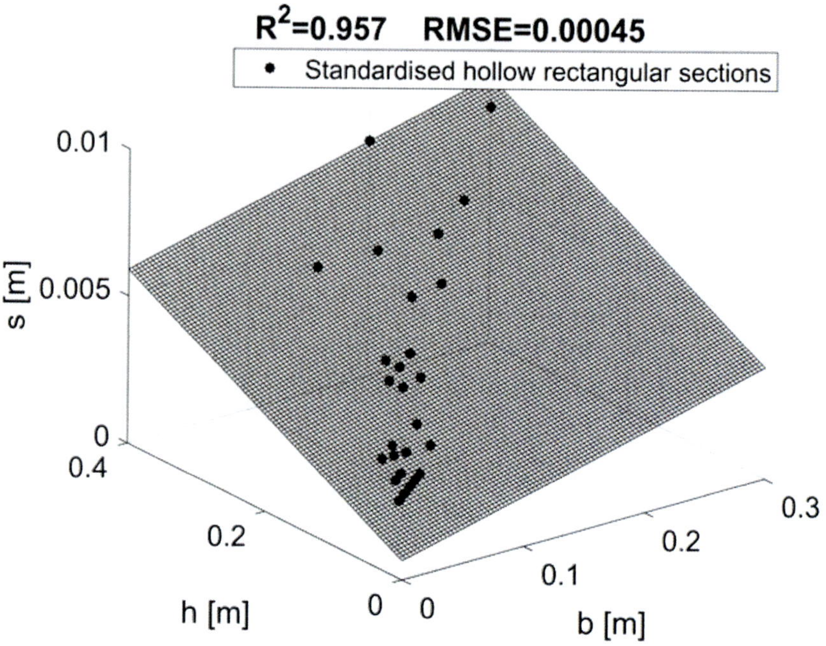

Fig. 6.11 Linear interpolation of thickness s—hollow rectangular section

thickness and the design variables is obtained by a linear fitting on the standardised hollow rectangular beams [68]. In the standard, for a given combination of b and h, several values of the thickness are available. Considering that both objective functions as defined in Eq. 6.32 are monotonically decreasing with s, only the cross sections with the lowest available thickness for each combination of b and h are included in the interpolation. The thickness s is described by the following linear function of b and h

$$s = p_1 + p_2 b + p_3 h \tag{6.31}$$

The coefficients p_1, p_2 and p_3 are obtained by minimising the mean square error and are reported in Table 6.3. Figure 6.11 shows the results of the linear interpolation.

The optimisation problem now reads

$$
\begin{aligned}
& min \left[\begin{array}{c} \dfrac{12M}{E(bh^3-(h-2s)^3(b-2s))} \\ \rho(bh-(b-2s)(h-2s)) \end{array} \right] \quad s.t. \\
& \frac{6Mh}{bh^3-(h-2s)^3(b-2s)} \le \frac{\sigma_y}{\eta} \\
& M \le \frac{\pi}{L}\sqrt{GJ_t EJ_{yy}} \\
& \frac{h-3s}{s} \le k_w\epsilon \\
& \frac{b-3s}{s} \le k_f\epsilon \\
& b_{min} \le b \le b_{max} \\
& h_{min} \le h \le h_{max}
\end{aligned}
\tag{6.32}
$$

As for the I-shaped beam, constraints on local buckling both of the two webs (vertical members) and the flanges (horizontal members) were included [69]. The complete list of all the model parameters is provided in Table 6.3. By substituting Eq. 6.31 in Eq. 6.32, the optimisation problem can be expressed as a function of the two design variables.

The solution of the unconstrained problem is feasible and intersects $b = b_{min}$ and $h = h_{max}$ as shown in Fig. 6.13. If a sufficiently large design space is considered, the effect of the constraints has to be taken into account. Equation 3.13 can be applied to solve the problem.

The stress constraint is given by

$$
\frac{6Mh}{bh^3-(h-2s)^3(b-2s)} - \frac{\sigma_y}{\eta} = 0 \tag{6.33}
$$

As for the I-shaped cross section, the buckling equation is approximated from numerical results [181] (Fig. 6.12) and reads

$$
\begin{aligned}
\frac{h}{L} \simeq & c_1\left(\frac{b}{L}\right)^5 + c_2\left(\frac{b}{L}\right)^4 + c_3\left(\frac{b}{L}\right)^3 + \\
& c_4\left(\frac{b}{L}\right)^3 + c_5\left(\frac{b}{L}\right) + c_6
\end{aligned}
\tag{6.34}
$$

The adopted polynomial expression seems to fit well the numerical data (Fig. 6.12), with a root mean square error smaller than 0.3 mm.

The feasible set, considering constraints, is shown in Figs. 6.13 and 6.14 both in the design variables and objective functions domain.

As for the I-shaped beam, the stress constraint is more restrictive than the constraints on lateral-torsional buckling and local buckling (which then are removed from the problem). For the given values of the design variables, the unconstrained solution is feasible and is bounded by the stress constraint from below, and by h_{max} from above. The stress constraint presents a minimum in the objective functions

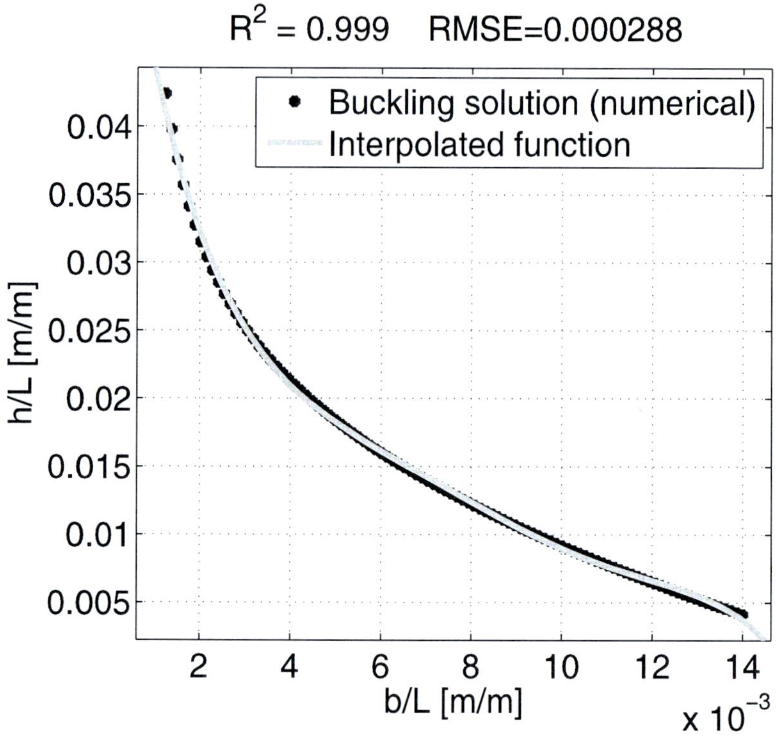

Fig. 6.12 Interpolated buckling expression—hollow rectangular section

domain. The Pareto-optimal set comprises this constraint from the intersection with the unconstrained solution to this minimum.

The Pareto-optimal set is given by the composition of the three active constraints as follows:

- the portion of the stress equation (Eq. 6.33) from the minimum of $\left.\frac{d\overline{m}}{d\overline{\theta}}\right|_{\sigma=\frac{\sigma_Y}{\eta}}$ to the intersection with the unconstrained solution;
- the unconstrained solution from this point to the intersection with $h = h_{max}$;
- $h = h_{max}$ from this point to b_{max}.

The obtained Pareto-optimal set was validated numerically and is shown in Fig. 6.15.

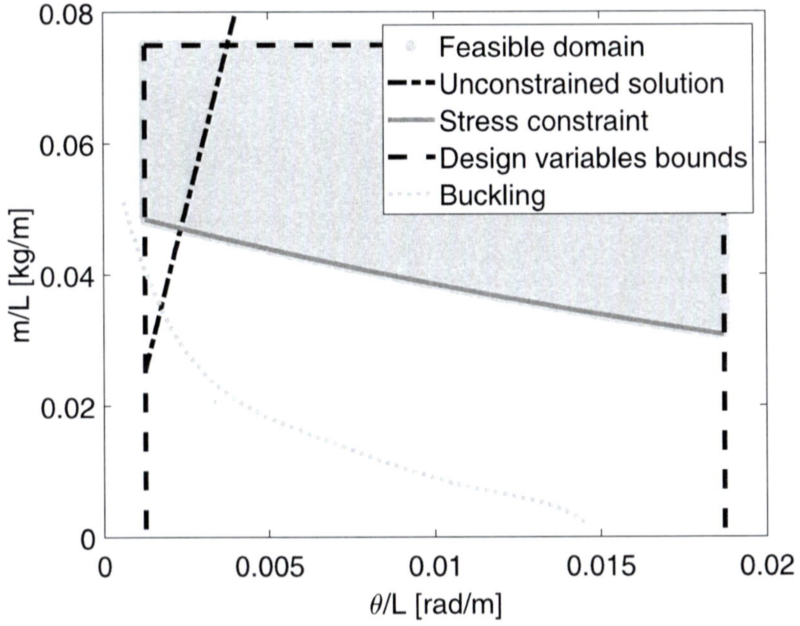

Fig. 6.13 Feasible set in design variables domain—hollow rectangular section. Data in Table 6.3

6.1.4 Comparison of Optimal Solutions

In this section, the obtained optimal solutions for the considered cross sections are compared. The three Pareto-optimal sets are reported in Fig. 6.16.

Figure 6.16 shows that the I-shaped beam provides the best structural efficiency. In fact, for a prescribed value of beam deflection, the I-shaped beam is able to reach the lowest mass. The hollow squared section on the other hand has the highest mass for a prescribed value of rotation. The results are consistent with the ones reported in [129] and based on a numerical solution.

6.2 Summary

In this chapter, the optimal design of uniformly bended beams has been analysed from a MOO perspective. A bi-objective (mass and deflection) minimisation problem was solved referring to a uniformly bended beam. Structural integrity and elastic stability were addressed and introduced as constraints in the minimisation process. Pareto-optimal solutions for a hollow squared, I-shaped and hollow rectangular beams were obtained and compared.

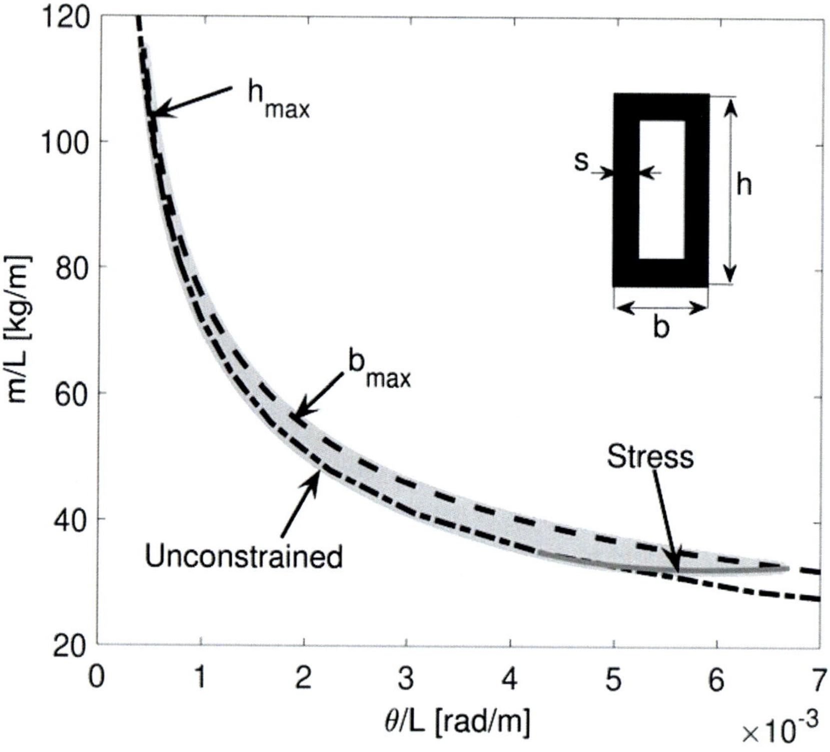

Fig. 6.14 Feasible set in objective functions domain—hollow rectangular section. Data in Table 6.3

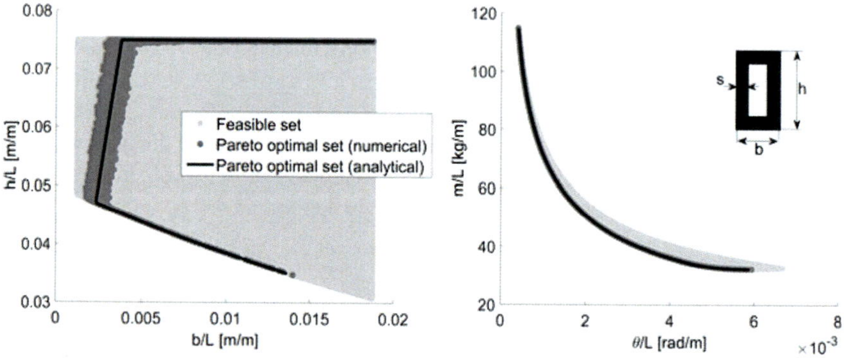

Fig. 6.15 Pareto-optimal set in the design variables domain (left) and objective functions domain (right) for a hollow rectangular beam. Data in Table 6.3

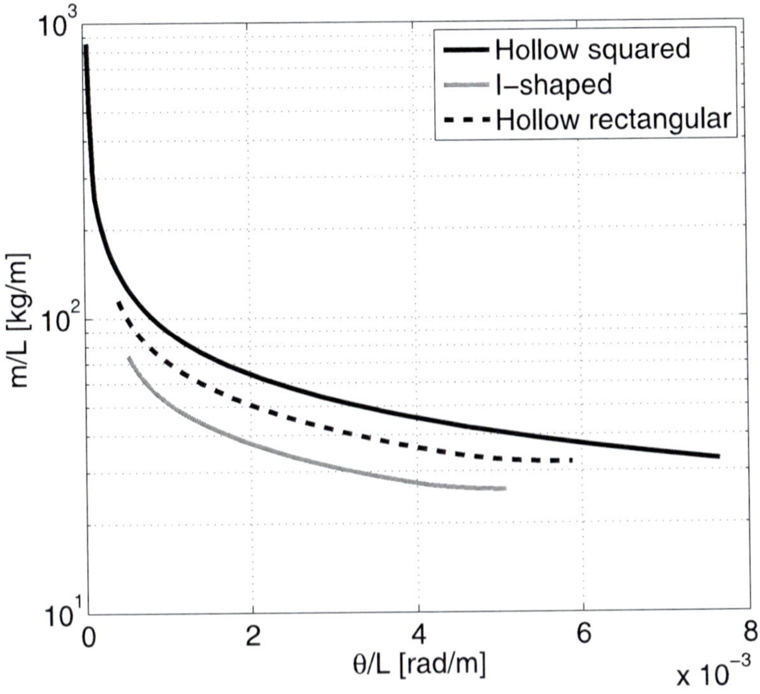

Fig. 6.16 Comparison between the obtained Pareto-optimal sets in the objective functions domain

As demonstrated also in Chap. 4, when active, the buckling constraint defines a portion of the Pareto-optimal set (see the case of the hollow square cross section). In Chap. 4, the same (simplified) formula for the buckling limit was employed for all the studied cross sections. In this chapter, as a specific study on three particular cross sections was conducted, more accurate buckling limits have been employed.

Another portion of the Pareto-optimal set is defined by the constraint on the admissible volume (i.e. the maximum height of the cross section). This outcome is consistent with the one of Chap. 4.

Finally, results show that the I-shaped beam exhibits the best structural performance for this particular problem in which bending is considered.

Chapter 7
Bending of Lightweight Inflated Circular Tubes—Optimal Design

Inflated tubes, and in general air inflated structures, have unique properties and advantages that can be exploited by designers. Such structures, with their very high load carrying to mass ratio, are well suited for lightweight design. Also, they present interesting properties. In fact, they can be rapidly and autonomously deployed, can be quickly moved and easily transported and, in case, can be rigidified [42, 134, 242]. These structures are utilised in many engineering applications such as aerospace, naval engineering and structural engineering, both for military and civil applications [42].

The mathematical modelling of inflated tubes has been discussed in a number of papers. The first attempts to build a mathematical model of these structures dates back to 1963 [54] where a analytical model describing the bending behaviour of inflated tubes is presented. In this work, however, the Eulero–Bernulli beam formulation was used and the effect of the internal pressure was not correctly included. In 1966, Fichter [78] effectively included the inflating pressure effects in the bending formulation of the tube by using an energy approach.

Later experiments [42, 134, 139, 140, 179, 225, 230, 231, 235, 262] confirmed that the internal pressure has a primary role on the tube deformation. On the basis of such observations, different models for the bending of inflated tubes have been realised. In [124], a quite comprehensive model of the inflated tube is presented. The tube is modelled as a shell and both the structural stiffness of the tube and the stiffening effect of the internal pressure are considered.

In [133], it is observed that, if the inflating pressure is high enough, the structural stiffness of fabric tubes can be neglected. In this case the tube is considered as a membrane. Different membrane models can be found [116, 133, 225, 235]. In [225], a membrane tube model for highly inflated tubes based on Timoshenko's theory is presented and experimentally validated. The agreement between the model and the experimental data is quite good. In the paper, it is observed that the equations have to be written in the deformed configuration.

F. M. Ballo et al., *Optimal Lightweight Construction Principles*,
https://doi.org/10.1007/978-3-030-60835-4_7

Inflated tube instability is a rather discussed problem. Given the membrane-like behaviour of the inflated tube, the main instability is related to the so-called wrinkling instability [139, 179, 231, 235, 243, 262]. Such instability happens locally where the traction stress due to inflation is reduced to almost zero by the applied load. The instability can happen for compressive load and in the compressed part of a bended tube. The wrinkling stability equation has been derived firstly in [243]. Similar expressions can be found in other papers [139, 179, 231, 235, 262], where different empiric coefficients have been considered.

In this chapter, starting from published bending and stability formulae for an inflated tube, the problem of its design is analytically solved in the framework of the Pareto-optimal theory [21, 86, 143, 152, 164]. The design objective is the minimization of both mass and deflection of the inflated tube subject to bending, under the constraint of available room, maximum inflating pressure, wrinkling stability and structural safety.

7.1 Equations Describing the Bending of an Inflated Circular Tube

The bending of an internally inflated thin-walled circular tube is shown in Fig. 7.1.

The tube has length l, external radius r and wall thickness t. The tube is inflated with an internal pressure p and a load F is applied to its free end. The other end of the tube is clamped. The pressure is supposed to be high enough to consider small the shell effects of the tube which is considered as a membrane [229]. Additionally, the material is supposed to be sufficiently stiff in tension to neglect changes in the geometrical dimensions and sufficiently soft in compression to neglect compressive stresses when contracted.

The applied force and the tube length are given. The designer can choose the best combination of radius, thickness and internal pressure in order to minimise deflection

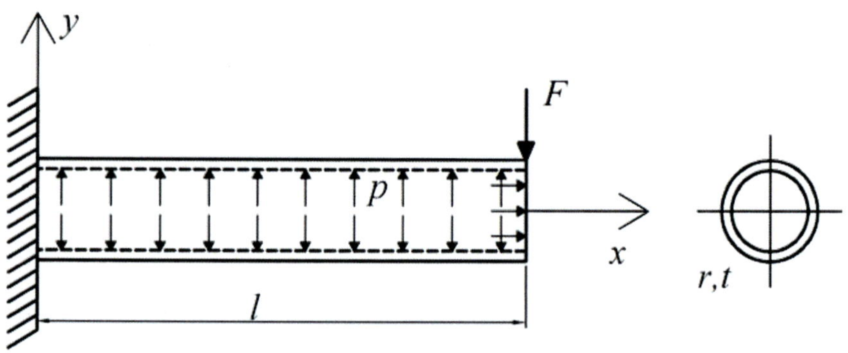

Fig. 7.1 Internally inflated circular cantilever beam subjected to bending moment

and mass of the tube. Referring to the optimisation problem, radius, thickness and internal pressure are the design variables, while mass and deflection are the objective functions (to be minimised).

For the bending deformation of the tube, the following equation, developed in [225] for highly inflated fabric tubes, is considered:

$$v\left(x\right)=F\frac{K^2G^2S^2-P^2}{KGS\Omega P}\frac{\sinh\left(\Omega\left(l-x\right)\right)-\sinh\left(\Omega l\right)}{\cosh\left(\Omega l\right)}+\frac{KGS}{P^2}Fx \tag{7.1}$$

where G is the shear modulus of the material, K is the shear coefficient ($K=0.5$, [57, 225]), S is the surface of the section

$$S=2\pi rt \tag{7.2}$$

P is the traction force due to the pressure and has expression

$$P=\pi r^2p \tag{7.3}$$

and Ω is given by

$$\Omega=\frac{P}{\sqrt{KEIGS}} \tag{7.4}$$

where E is the elastic modulus of the material and I the moment of inertia of the section. I can be computed as

$$I=\pi tr^3 \tag{7.5}$$

If the displacement at the free end is considered, i.e. $x=l$, the tip deflection reads

$$v=F\frac{P^2-K^2G^2S^2}{KGS\Omega P^2}\tanh\left(\Omega l\right)+\frac{KGS}{P^2}Fl \tag{7.6}$$

The mass of the tube reads

$$m=\rho Sl \tag{7.7}$$

being ρ the material density.

The load carrying capacity of an inflated tube in bending is limited by the wrinkling instability. This instability happens when the membrane representing the tube is no more in traction. The wrinkling condition on bending moment (M_w) reads

$$M_w\leq\alpha\pi pr^3 \tag{7.8}$$

where α is an empirical coefficient to reduce the theoretical wrinkling moment (M_{wt}) to match experimental data ($\alpha<1$) [243]. Different values of α can be found in the literature. In this chapter a value of 0.75 is considered [235].

When inflated, the tube is subjected to a axial and a circumferential stress. The stress normal to the membrane is null. For a thin-walled tube, axial (σ_a) and circumferential (σ_c) stresses read

$$\sigma_a = \frac{p\,r}{2\,t} \tag{7.9}$$

$$\sigma_c = \frac{p\,r}{t} \tag{7.10}$$

Before the wrinkling condition is reached, the axial stress on the most stretched fibres of the cylinder grows but remains lower than the circumferential stress. After wrinkling, the axial stress on these fibres can exceed the circumferential stress. For sake of simplicity, and since we are focused on the pre-wrinkling condition of the tube, we consider as structural limit a circumferential stress lower than the admissible stress of the material

$$\sigma^* = \sigma_c = \frac{pR}{t} \leq \frac{\sigma_{lim}}{\eta} = \sigma_{adm} \tag{7.11}$$

where σ_{lim} is the limit stress of the material and η is a safety coefficient. σ^* represents the Guest-Tresca equivalent stress.

7.2 Statement of the Multi-objective Optimisation Problem

The problem of the mass-deflection optimisation of an inflated circular tube subjected to bending can be stated in mathematical form as follows:

Given

r	nominal radius	[m]
t	thickness	[m]
p	internal pressure	[p]
F	applied force (see Fig.7.1)	[N]
σ_{adm}	admissible stress of the material	[MPa]
E	material modulus of elasticity (Young's modulus)	[MPa]
G	material shear modulus	[MPa]
ρ	material density	[kg/m^3]

and defining

$m = m\,(r, t)$	beam mass (Eq. 7.7)	[kg]
$v = v\,(r, t, p)$	deflection of the inflated tube (Eq. 7.6)	[m]
$\sigma^* = \sigma^*\,(r, t, p)$	equivalent stress (Eq. 7.11)	[MPa]
$M_w = M_w\,(r, p)$	maximum moment for wrinkling stability (Eq. 7.8)	[Nm]

find r, t and p such that

$$
\begin{aligned}
r_{min} &\leq r \leq rmax \\
t_{min} &\leq t \leq tmax \\
p_{min} &\leq p \leq p_{max}
\end{aligned}
\tag{7.12}
$$

and such that

$$
\min \begin{pmatrix} m(r,t) \\ v(r,t,p) \end{pmatrix} = \min \begin{pmatrix} \rho\pi rtl \\ F\frac{P^2-K^2G^2S^2}{KGS\Omega P^2}\tanh(\Omega l) + \frac{KGS}{P^2}Fl \end{pmatrix} \tag{7.13}
$$

subject to

$$
\sigma^* = \frac{pr}{t} \leq \sigma_{adm} \tag{7.14}
$$

$$
Fl \leq M_w = \alpha\pi pr^3 \tag{7.15}
$$

In this problem, the objective functions to be minimised are the mass m and the deflection v (Eq. 7.13) under the constraints of allowable room (Eq. 7.12), structural safety (Eq. 7.14) and elastic stability (Eq. 7.15).

7.2.1 Monotonicity Analysis and Problem Reduction

Before solving the problem, a *Monotonicity analysis* [164] has to be performed to check proper boundedness of the problem, identify the active constraints and eventually simplify the problem itself.

Let us consider the design variable p. Only the deflection objective function v (Eq. 7.6) is function of p. The mass m (Eq. 7.7) does not depend on p, if we assume, at a first approximation, the mass of the air negligible with respect to that of the tube. Moreover, v is monotonically decreasing with p [225]. For any combination of the other two design variables, the best solution is obtained by the maximum allowable value of p. p is bounded from above by Eq. 7.11 (structural safety) and by the third condition of Eq. 7.12 ($p \leq p_{max}$). According to the first monotonicity principle [164], in a well constrained objective function, every decreasing variable is bounded above by at least one active constraint, so the inflation pressure can be *optimised out* as

$$
p = \min\left(p_{max}, \sigma_{adm}\frac{t}{r}\right) \tag{7.16}
$$

Based on Eq. 7.16, the optimisation problem can be divided into three cases, namely,

- case 1: the maximum pressure constraint is always more binding than the structural safety constraint, i.e.

$$
p_{max} \leq \sigma_{adm}\frac{t}{r} \quad \forall t, r \tag{7.17}
$$

- case 2: the structural safety constraint is always more binding than the maximum pressure constraint, i.e.

$$p_{max} \geq \sigma_{adm} \frac{t}{r} \quad \forall t, r \tag{7.18}$$

- case 3: depending on the combinations of r and t, p can be limited by the maximum pressure or by the structural safety condition. The expression of p can switch between the two conditions of Eq. 7.16.

It can be observed that the three resulting optimisation problems are function of two design variables (r ant t) with two objective functions (m and v). The solutions can be obtained by considering Eq. 3.13.

7.2.2 Solution of the Optimisation Problem: Case 1

In this case, the maximum inflation pressure of the tube depends on technological limits on the maximum available pressure of the inflating fluid. The structural limit of the tube is never reached. The inflating pressure of the tube is the same for any combination of the two design variables r and t.

By substituting $p = p_{max}$ in Eqs. 7.13 and 7.15 and by remembering Eqs. 7.3 and 7.4, the optimisation problem stated in Sect. 7.2 can be rewritten as

Given the quantities defined in Sect. 7.2, and $p = p_{max}$, find r and t such that

$$\begin{aligned} r_{min} &\leq r \leq r_{max} \\ t_{min} &\leq t \leq t_{max} \end{aligned} \tag{7.19}$$

and such that

$$\min \begin{pmatrix} m(r,t) \\ v(r,t,p) \end{pmatrix} = \min \begin{pmatrix} \rho \pi r t l \\ F \frac{\sqrt{E}(p_{max}^2 r^2 - 4G^2K^2t^2)T}{\pi\sqrt{2GK}p_{max}^2 r^3} + \frac{2FGKlt}{\pi p_{max}^2 r^3} \end{pmatrix} \tag{7.20}$$

with $T = \tanh\left(\frac{l p_{max}}{\sqrt{2EKGt}}\right)$
subject to

$$r \geq \sqrt[3]{\frac{Fl}{\alpha \pi p_{max}}} \tag{7.21}$$

The new optimisation problem (Eq. 7.19 to Eq. 7.21) is constrained with $n_{dv} = n_{of}$. For the solution, Eq. 3.13 can be applied. This equation states that the expression containing the Pareto-optimal set is given by the product of the solution of the unconstrained problem times the expressions of the active constraints.

Let us start with the computation of the Pareto-optimal set for the unconstrained problem, i.e. the problem of Eq. 7.13 where all constraints (Eqs. 7.19 and 7.21) are removed from the formulation. The unconstrained solution can be derived by applying Eq. 3.14 and reads

$$det\left(\nabla \mathbf{f}\right) = det\left(\begin{bmatrix} \frac{\partial m}{\partial r} & \frac{\partial v}{\partial r} \\ \frac{\partial m}{\partial t} & \frac{\partial v}{\partial t} \end{bmatrix}\right) = 0 \tag{7.22}$$

which leads to the equation

$$\frac{Fl\rho\left(lp_{max}^3 r^2 T^2 - 20G^2K^2 l p_{max} t^2 + 20\sqrt{2EG^5K^5}t^3 T + G^2K^2 l p_{max} t^2 T^2 - \sqrt{2EGK}\, p_{max}^2 r^2 t T\right)}{GKp_{max}^3 r^3 t} = 0 \tag{7.23}$$

with solutions $r^3 t \to \infty$ or for $r = \overline{r}$, where $\overline{r}$ reads

$$\overline{r} = \pm \frac{2GKt\sqrt{l^2 p_{max}^2 (5-6T^2+T^4) + 10EGKt^2T^2 + 2\sqrt{2EGK}\, l p_{max} t \left(3T^3 - 5T\right)}}{p_{max}\left(lp_{max}T^2 - lp_{max}\sqrt{2EGK}tT\right)} \tag{7.24}$$

For usual values of the parameters, $\overline{r}$ has imaginary values, while $r^3 t \to \infty$ does not belong to the set of the positive and finite numbers. Both solutions have no physical meaning [90, 164] and must be discarded.

The unconstrained problem does not have any acceptable physical solution, therefore, the Pareto-optimal set is given only by the active constraints.

Referring to the stability constraint (Eq. 7.21), it can be easily removed from the problem formulation. In fact, when $p = p_{max}$, this constraint is a lower constant bound on r and it can be included into the available room constraint by defining a new minimum radius as

$$r_{min}^* = \max\left(r_{min}, \sqrt[3]{\frac{Fl}{\alpha\pi p_{max}}}\right) \tag{7.25}$$

the available room constraint (Eq. 7.19) can be rewritten as

$$\begin{aligned} r_{min}^* &\leq r \leq r_{max} \\ t_{min} &\leq t \leq t_{max} \end{aligned} \tag{7.26}$$

Given Eqs. 7.25 and 7.26, only the available room constraint remains in the problem formulation. Being the Pareto-optimal set given only by the active constraints, they have to be identified.

If t is fixed to its extreme value ($t = t_{lim}$ with $t_{lim} = t_{min}$ or $t_{lim} = t_{max}$), the available room constraint on the thickness in the objective function domain reads

$$v\left(m\right)|_{t=t_{lim}} = \frac{\sqrt{2E}Fl\rho t_{lim}\left(M^2 p_{max}^2 - 16\pi^2 G^2 K^2 l^2 \rho^2 t_{lim}^2\right)A}{\sqrt{GK}p_{max}^3 m^3} + \frac{16FGKl^4\rho^3\pi^2 t_{lim}}{m^3 p_{max}} \tag{7.27}$$

with $A = \tanh\left(\frac{lp_{max}}{\sqrt{2EGKt_{lim}}}\right)$. Equation 7.27 is convex and decreasing in the objective functions domain. Also, Eq. 7.27 is monotonically increasing as t_{lim} grows. Therefore, the constraint $t = t_{min}$, when active, is part of the Pareto-optimal set, while $t = t_{max}$ has to be discarded.

Similarly, if r is fixed to its extreme value ($r = r_{lim}$ with $r_{lim} = r_{min}^*$ or $r_{lim} = r_{max}$), the available room constraint on the radius in the objective functions domain reads

$$v\left(m\right)|_{r=r_{max}} = \frac{\sqrt{E}F\left(\pi^2\rho^2 l^2 p_{max}^2 r_{lim}^4 - G^2K^2m^2\right)B}{\sqrt{2GK}l^2 p_{max}^3 \rho^2\pi^3 r_{lim}^5} + \frac{FGKm}{\pi^2\rho p_{max}^2 r_{lim}^4} \tag{7.28}$$

with $B = \tanh\left(\frac{\sqrt{2}\pi\rho l^2 p_{max} r_{lim}}{\sqrt{EGKm}}\right)$. Equation 7.28 decreases monotonically as r_{lim} grows, therefore, only $r = r_{max}$ can be part of the Pareto-optimal set, while $r = r_{min}^*$ has to be discarded.

Equation 7.28 is the sum of a straight line with positive coefficient and a concave decreasing non-linear function. If the limits for $m \rightarrow 0$ and $m \rightarrow \infty$ are computed (expressions are reported in Appendix 7.3), the straight line prevails for small values of m, while for large values of m, the non-linear part prevails and the constraint equation decreases asymptotically to 0. Therefore, Eq. 7.28 has at least one maximum. Let us call $\hat{m}$ the value of the mass at maximum v. Therefore, for $m \in \left[0, \hat{m}\right]$ the constraint is never part of the Pareto-optimal set. For $m \in \left[\hat{m}, \infty\right]$ the constraint can be part of the Pareto-optimal set. If the constraint is part or not of the Pareto-optimal set depends on the values of the parameters.

In Fig. 7.2 the Pareto-optimal sets for three different values of maximum pressure are reported in the design variables and objective functions domains (data in Table 7.1).

The minimum mass point is given by the design variables combination r_{min}^* and t_{min} and, for the properties of Eq. 7.27, is part of the Pareto-optimal set. This point is also the point with maximum deflection. The solution t_{min} and r_{max} is always part of the Pareto-optimal set for the properties of Eq. 7.27. If the mass of this solution is greater than $\hat{m}$, the portion of the constraint $r = r_{max}$ from this point up to t_{max} belongs to the Pareto-optimal set as for the case depicted in Fig. 7.2 on the left. In this case the point at maximum mass ($r = r_{max}$,$t = t_{max}$) is part of the Pareto-optimal set.

The point at maximum mass is given by r_{max} and t_{max} and, given the properties of Eq. 7.28, can be part of the Pareto-optimal set. If the point at maximum mass is part of the Pareto-optimal set, it is also the point at minimum deflection.

In case this point is not part of the Pareto-optimal set (as in the right case of Fig. 7.2), the constraint $r = r_{max}$ is not Pareto-optimal and the Pareto-optimal set is given only by the constraint $t = t_{min}$. In this case the point at minimum compliance is given by the solution t_{min} and r_{max}.

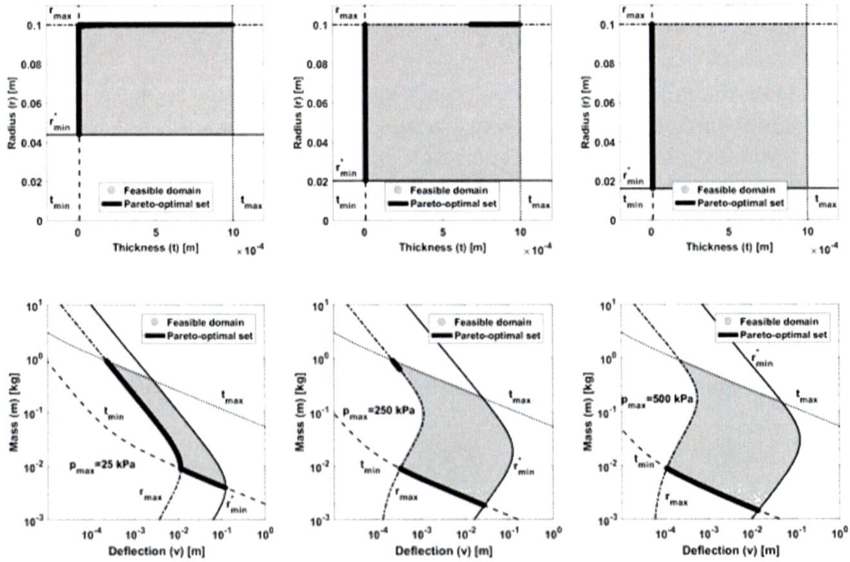

Fig. 7.2 Pareto-optimal sets for an inflated tube subject to bending for different levels of inflating pressure. Top: design variables domain. Bottom: objective functions domain. From left to right, inflating pressures of 25 kPa, 250 kPa and 500 kPa. Data in Table 7.1

Table 7.1 Inflated tube data

Parameter	Symbol	Value
Minimum radius	r_{min}	0.01 m
Maximum radius	r_{max}	0.1 m
Minimum thickness	t_{min}	10^{-5} m
Maximum thickness	t_{max}	10^{-3} m
Length	l	1 m
Material modulus of elasticity	E	2500 MPa
Material shear modulus	G	930 MPa
Material density	ρ	1420 kgm^{-3}
Applied force	F	5 N

Finally, if the point at maximum mass is part of the Pareto-optimal set and the mass of the solution t_{min} and r_{max} is lower than $\hat{m}$, the Pareto-optimal set is disconnected because of the non-monotonic shape of the constraint equation $r = r_{max}$ in the objective functions domain (Eq. 7.28). This last scenario is depicted in the central case of Fig. 7.2.

7.2.3 Solution of the Optimisation Problem: Case 2

In this case the inflating pressure is limited by the resistance of the material. The inflating pressure depends on the actual combination of the two design variables r and t. From Eq. 7.14, the admissible pressure p_{adm} can be computed as

$$p_{adm} = \sigma_{adm} \frac{t}{r} \tag{7.29}$$

By substituting Eq. 7.29 in Eqs. 7.13 and 7.15 and by remembering Eqs. 7.3 and 7.4, the optimisation problem of Sect. 7.2 can be rewritten as

Given the quantities defined in Sect. 7.2, and $p = p_{adm} = \sigma_{adm} \frac{t}{r}$, find r and t such that

$$\begin{aligned} r_{min} &\leq r \leq rmax \\ t_{min} &\leq t \leq tmax \end{aligned} \tag{7.30}$$

and such that

$$\min \begin{pmatrix} m(r,t) \\ v(r,t,p) \end{pmatrix} = \min \begin{pmatrix} \rho \pi r t l \\ F \frac{\sqrt{2E}(\sigma_{adm}^2 - 4G^2K^2)T}{2\pi \sqrt{GK\sigma_{adm}^3} t} + \frac{2FGKl}{\pi \sigma_{adm}^2 r t} \end{pmatrix} \tag{7.31}$$

with $T = \tanh\left(\frac{l\sigma_{adm}}{\sqrt{2EGKr}}\right)$
subject to

$$Fl \leq \pi \alpha \sigma_{adm} r^2 t \tag{7.32}$$

By applying Eq. 3.14, the unconstrained solutions of the problem can be found by solving the following equation:

$$det\,(\nabla \mathbf{f}) = \frac{Fl\rho\left(\sigma_{adm}^2 - 4G^2K^2\right)\left(l\sigma_{adm}\left(T^2 - 1\right) + \sqrt{2EGK}rT\right)}{GK\sigma_{adm} r t} = 0 \tag{7.33}$$

Equation 7.33 has solution for $rt \to \infty$ and for $r = \bar{r}$ with

$$\bar{r} = \left\{ r \in [r_{min}, r_{max}] : l\sigma_{adm}\left(T^2 - 1\right) + \sqrt{2EGK}rT = 0 \right\} \tag{7.34}$$

If $r = \bar{r}$, t can assume any value. The solution $rt \to \infty$ does not belong to the set of the positive finite numbers and must be discarded [164].

Being Eq. 3.14 a necessary condition, we have to check if the solution $r = \bar{r}$ is actually part of the Pareto-optimal set. For r equal to a constant value $\hat{r}$ and any t, the expression $v = v(m)$ reads

$$v(m)\,|_{r=\hat{r}} = \frac{\sqrt{2E}FL\rho\hat{r}\left(\sigma_{adm}^2 - 4G^2K^2\right)\tanh\left(\frac{l\sigma_{adm}}{\sqrt{2EGK\hat{r}}}\right)}{\sqrt{GK}\sigma_{adm}m} + \frac{4FGKl^2\rho}{\sigma_{adm}^2 m} \tag{7.35}$$

Equation 7.35, for reasonable values of the parameters $4G^2K^2 > \sigma_{adm}$, is decreasing with $\hat{r}$. This implies that the non-dominated solution of $v(m)\,|_{r=\hat{r}}$ is given by the upper bound of r, i.e. when $\hat{r} = r_{max}$. Therefore, $r = \overline{r}$, if feasible, is not a Pareto-optimal solution, unless the very particular case of $\overline{r} = r_{max}$. For these considerations, the available room constraint $r = r_{max}$, when active, is part of the Pareto-optimal set.

If t is fixed to an extreme value t_{lim}, the expression of the deformation as function of mass reads

$$v(m)\,|_{t=t_{lim}} = \frac{\sqrt{2E}F\left(\sigma_{adm}^2 - 4G^2K^2\right)\tanh\left(\frac{\sqrt{2}\pi l^2\rho\sigma_{adm}t_{lim}}{\sqrt{EGK}m}\right)}{2\pi\sqrt{GK}\sigma_{adm}^3 t_{lim}} + \frac{4FGKl^2\rho}{\sigma_{adm}^2 m} \tag{7.36}$$

Equation 7.36 is decreasing in the objective functions domain and, for reasonable valued of G, K and σ_{adm} is increasing as t_{lim} grows. The constraint $t = t_{min}$ is therefore part of the Pareto-optimal set, while the constraint $t = t_{max}$ has to be discarded.

Finally, the expression of the wrinkling stability (Eq. 7.32) in the objective functions domain reads

$$v(m)\,|_{FL=M_w} \leq \frac{\sqrt{2E}F^2l^3\rho^2\left(\sigma_{adm}^2 - 4G^2K^2\right)\tanh\left(\frac{\sqrt{2}\alpha\sigma_{adm}^2 m}{4\sqrt{EGKFl\rho}}\right)}{\sqrt{GK}\sigma_{adm}^4\alpha m^2} + \frac{4FGKl^2\rho}{\sigma_{adm}^2 m} \tag{7.37}$$

From Eq. 7.37 the feasible domain is below the wrinkling stability condition, therefore, this condition is not part of the Pareto-optimal set.

For the limit conditions of $m \to 0$ and $m \to \infty$, the following relationships can be derived (for sake of clarity, the analytical expressions of the limits are reported in 7.3).

$$m \to 0 \Rightarrow v(m)\,|_{FL=M_w} < v(m)\,|_{r=r_{max}} < v(m)\,|_{t=t_{min}} \tag{7.38}$$

$$m \to \infty \Rightarrow v(m)\,|_{FL=M_w} > v(m)\,|_{r=r_{max}} > v(m)\,|_{t=t_{min}} \tag{7.39}$$

Given the continuity of Eqs. 7.35, 7.36 and 7.37 and the conditions of Eqs. 7.38 and 7.39, the following considerations can be derived.

- For $m \to 0$ the wrinkling condition is not satisfied. The point at minimum mass, given by the design variables r_{min} and t_{min} is the first point that can be excluded by the stability condition.

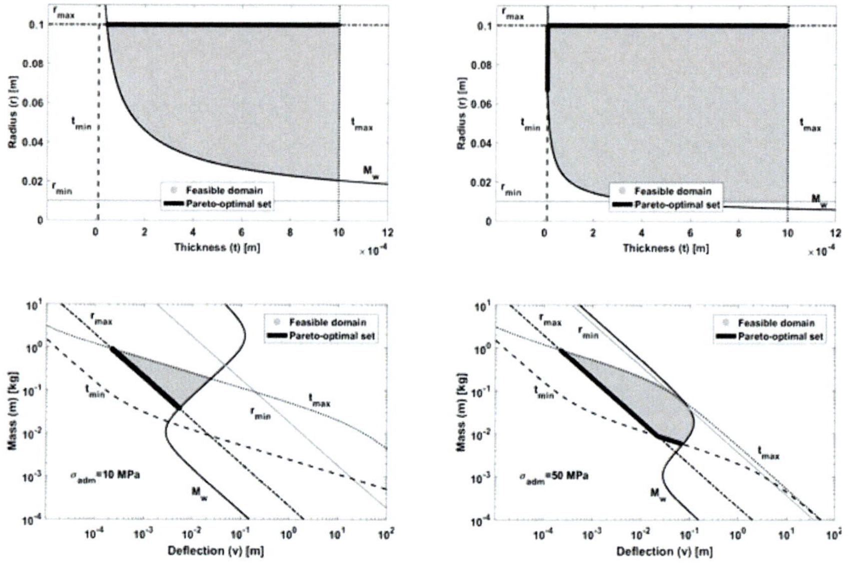

Fig. 7.3 Pareto-optimal sets for an inflated tube subject to bending for different levels of maximum admissible stress. Top: design variables domain. Bottom: objective functions domain. Left: $\sigma_{adm} = 10\,\text{MPa}$. Right: $\sigma_{adm} = 50\,\text{MPa}$. Data in Table 7.1

- The wrinkling stability condition intersects both of the available room constraints. The intersection point is the point at minimum mass and maximum deflection.
- Since also the two available room constraints intersect each other (at the point with design variables r_{max} and t_{min}), depending on the parameter values, the wrinkling stability constraint can intersect firstly one or the other constraint. The point at minimum mass, and maximum deflection, is the intersection with higher mass (and lower deflection). The other intersection is not feasible.
- The point at maximum mass, and minimum deflection, is given by the design variables r_{max} and t_{max}. If this point is not feasible, i.e. does not comply with the wrinkling stability constraint, no other point is feasible.

In Fig. 7.3 the Pareto-optimal sets in the design variables and objective functions domains are depicted for different values of the admissible stress (data in Table 7.1).

7.2.4 Solution of the Optimisation Problem: Case 3

This case is the most general case in which the inflating pressure is limited either by the resistance of the material or by the available inflating pressure. If both material resistance and maximum inflating pressure are reached at the same time, the following relationship holds

$$r = \frac{\sigma_{adm}}{p_{max}} t \tag{7.40}$$

This relationship divides the design variable space into two parts. The part of the space above the straight line represented by Eq. 7.40 defines the design variable combinations for which the admissible stress is the limiting parameter. The part of the space below the straight line represents the design variable combinations for which the maximum inflating pressure is the limiting parameter.

The Pareto-optimal set for this case is therefore given by the union of the two Pareto-optimal sets computed for the two subsets of design variables. The expression of the frontier between the two subsets in the objective functions domain reads

$$v(m)\Big|_{\substack{p = p_{max} \\ \sigma = \sigma_{adm}}} \leq \frac{\sqrt{\rho l E} F \left(\sigma_{adm}^2 - 4G^2K^2\right) T}{\sqrt{\pi G K \sigma_{adm}^5 p_{max} m}} + \frac{4FGKl^2\rho}{\sigma_{adm}^2 m} \tag{7.41}$$

where $T = \tanh\left(\frac{\sqrt{\pi\rho\sigma_{adm} p_{max} l^3}}{\sqrt{EGKm}}\right)$.

In Fig. 7.4 the Pareto-optimal set for a maximum pressure of 250 kPa and an admissible stress of 50 MPa is shown (data in Table 7.1). Equation 7.40 and 7.41 represent the borders in the design variables and objective functions domains between the two optimisation problems considered in Sects. 7.2.2 and 7.2.3. For larger values of mass and lower values of deflection, the optimal solutions are obtained by considering the limit value of pressure p_{max}. Conversely, optimal solutions with higher values of deflections and lower values of mass are obtained when the constraint $\sigma = \sigma_{adm}$ is active. The point at minimum mass and maximum deflection is limited by the wrinkling stability constraint.

7.3 Summary

The present chapter has dealt with the optimisation of an inflated tube under bending. The deflection of the tube under the hypothesis of internal pressure high enough to consider the tube as a membrane has been considered. Mass and deflection have been minimised under the constraints of available room, technological limitation on the maximum inflating pressure, wrinkling stability and structural safety.

Tube radius and thickness and internal inflating pressure have been considered as design variables. A monotonicity analysis has shown that the inflating pressure has to be chosen as high as possible, considering the limits on the maximum inflating pressure that can be reached and on the structural safety. Also, it has been proved that the optimal solution, according to the Pareto-optimal theory, is obtained when one of these two constraints is active.

Firstly, the two cases of maximum inflating pressure for technological limitation or of structural safety of the tube have been considered separately. In both cases,

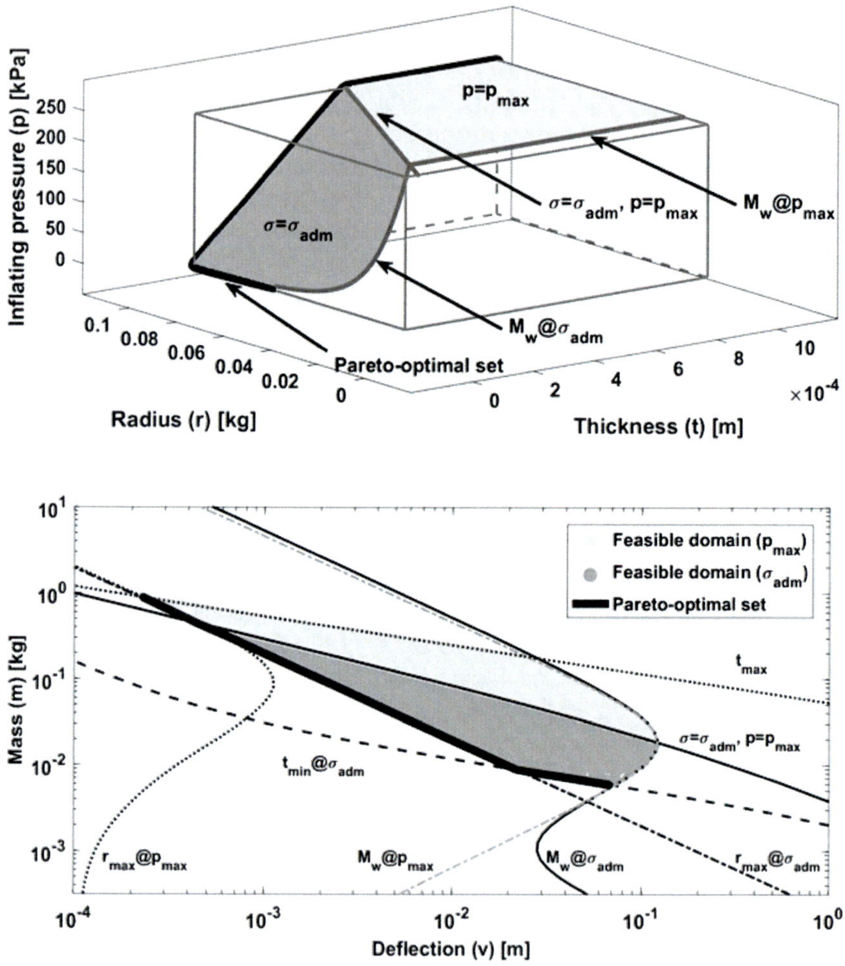

Fig. 7.4 Pareto-optimal sets for an inflated tube subject to bending. Top: design variables domain. Bottom: objective functions domain. Maximum pressure 250 kPa. Admissible stress 50 MPa. Data in Table 7.1

the optimal solutions lay exclusively on the maximum allowed radius or minimum allowed thickness. Solutions laying on the wrinkling stability condition are not a best compromise design, but wrinkling stability is a limiting parameter. In fact, when this condition is reached, only one design variable combination belongs to the Pareto-optimal set and represents the point at minimum mass and maximum deflection.

Finally, the most general case in which one or the other constraints on the inflating pressure is reached depending on the actual values of radius and thickness has been considered. In this case, the Pareto-optimal set is given by the combination of the two previously studied cases. In particular, the maximum pressure limitation leads to solutions characterised by high mass and small deflection. Conversely, structural safety limitation leads to a solution characterised by larger deflection and lower mass.

Appendix: Limit Expressions of the Constraints

Limit Expressions of the Constraints for Case 1

The limit expressions for the constraint of Eq. 7.28 read

For $m \to 0$

$$v|_{r=r_{max}} \sim \frac{FGKm}{\pi^2 \rho p_{max}^2 r_{max}^2} \tag{7.42}$$

For $m \to \infty$

$$v|_{r=r_{max}} \to -\frac{FGKm}{\rho\pi^2 r_{max}^4} + \frac{FGKm}{\rho\pi^2 r_{max}^4} \to 0 \tag{7.43}$$

Limit Expressions of the Constraints for Case 2

The limit expressions of the functions of Eqs. 7.38 and 7.39 read

For $m \to 0$

$$v|_{t=t_{min}} \sim \frac{4FGKl^2\rho}{\sigma_{adm}^2 m} \tag{7.44}$$

$$v|_{r=r_{max}} = \frac{\sqrt{2E}FL\rho r_{max}\left(\sigma_{adm}^2 - 4G^2K^2\right)\tanh\left(\frac{l\sigma_{adm}}{\sqrt{2EGK}r_{max}}\right)}{\sqrt{GK}\sigma_{adm} m} + \frac{4FGKl^2\rho}{\sigma_{adm}^2 m} \tag{7.45}$$

$$v|_{FL=M_w} \sim \frac{Fl^2\rho}{GKm} \tag{7.46}$$

For $m \to \infty$

$$v|_{t=t_{min}} \sim \frac{Fl^2\rho}{GKm} \tag{7.47}$$

$$v|_{r=r_{max}} = \frac{\sqrt{2E}FL\rho r_{max}\left(\sigma_{adm}^2 - 4G^2K^2\right)\tanh\left(\frac{l\sigma_{adm}}{\sqrt{2EGK}r_{max}}\right)}{\sqrt{GK}\sigma_{adm} m} + \frac{4FGKl^2\rho}{\sigma_{adm}^2 m} \tag{7.48}$$

$$v|_{FL=M_w} \sim \frac{4FGKl^2\rho}{\sigma_{adm}^2 m} \tag{7.49}$$

where $\sigma_{adm}^2 - 4G^2K^2 < 0$ for reasonable values of the parameters.

Chapter 8
Torsion of Lightweight Circular Tubes—Optimal Design

Thin-walled structures have a high ratio between load carrying capabilities and mass and play a primary role in lightweight design. In order to improve the structural efficiency, i.e. stiffness and mass ratio, a rigorous optimisation approach is required.

Multi-Objective Optimisation (MOO) can be effectively applied to structural design [21, 99, 143, 152] with particular reference to mass minimization while maximising the structural stiffness (or, equivalently, minimising the compliance of the structure). In such problems, mass and compliance are considered objective functions (to be minimised). The design variables are the geometrical dimensions of the structural members. Constraints on safety (i.e. admissible stress), elastic stability and maximum available room have to be satisfied.

The solution of the optimisation process in the Pareto-optimal framework is composed of a set of optimal solutions (the so-called Pareto-optimal set). Those solutions represent the best compromise in terms of both minimum mass and compliance. The designer can then choose the final structure configuration among these optimal solutions [164].

In [106, 160] multi-objective optimisation theory was employed for mass and deflection minimization of structural members in bending. Optimised profiles of thin-walled open cross sections were obtained by means of numerical methods. Shape optimisation of bars under torsion was addressed by Wang [234]. The shape and rounding of the corners of polygonal bars was optimised by means of numerical methods with the aim to maximise the torsional stiffness for a prescribed target of mass. The topology of the cross section of thin-walled beams under torsion was optimised in [111]. A multi-objective optimisation approach was followed for maximising the torsional stiffness and minimising the distortion of the thin-walled cross section. The optimal solution was derived numerically by a weighted sum method. Gobbi and Mastinu [86] presented a method for the optimal design of composite material tubular helical springs. In the paper, Multi-Objective Programming (MOP) was adopted. Both theoretical studies and experimental activities were conducted.

F. M. Ballo et al., *Optimal Lightweight Construction Principles*,
https://doi.org/10.1007/978-3-030-60835-4_8

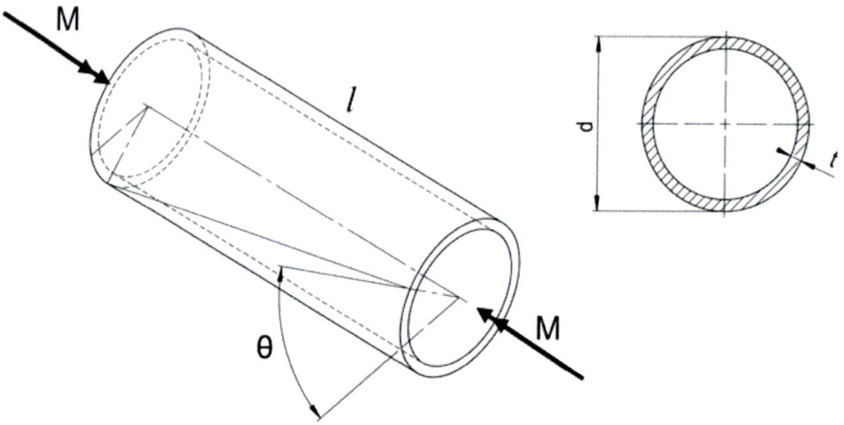

Fig. 8.1 Thin-walled circular tube subject to torsion

The solution of (simple) multi-objective optimisation problems can be derived analytically for a number of engineering problems (see, for example, [20, 90, 92, 143] or the four-bars plane truss problem in [11]). When available, analytical formulae can be very useful for designers since they provide a broader view of the problem and may guide the designers at the conceptual design stage.

Referring to beams, analytical formulae for designing *optimal* beams under torsion or bending are presented in several papers. In [10], a set of design formulae for beams of arbitrary cross section under torsion or bending is provided to compare different materials and shapes.

In this chapter, the optimal design of a thin-walled tube under torsion is discussed. A rigorous multi-objective optimisation approach is followed. The mass of the tube is minimised together with its compliance. Analytical formulae providing the cross-sectional dimensions of the optimised tube are derived. These formulae may be a useful tool for designers, who can choose an optimised cross section without any further iteration.

8.1 Equations for a Thin-Walled Tube Under Torsion

The torsion of a thin-walled circular tube is shown in Fig. 8.1.

The tube has length l, external diameter d, wall thickness t and is subject to a torsional moment M. The design problem refers to mass and compliance minimization. For the sake of generality, mass and compliance are divided by the tube length, the objective functions of the problem are therefore the mass per unit of length m and the compliance per unit of length c of the tube.

The *design variables* are the tube diameter d and its thickness t.

The analytical expressions of the objective functions read

$$m = \frac{m_t}{l} = \pi \rho d t \tag{8.1}$$

$$c = \frac{\theta}{l} = \frac{4M}{G\pi d^3 t} \tag{8.2}$$

where ρ is the density of the material and G is the shear modulus, while m_t and θ are the overall mass of the tube and the relative rotation of the two end sections of the tube (Fig. 8.1), respectively. The design variables d and t are limited by maximum and minimum attainable values [164]

$$d_{min} \leq d \leq d_{max} \tag{8.3}$$

$$t_{min} \leq t \leq t_{max} \tag{8.4}$$

The upper bound of the diameter d_{max} can be interpreted as a constraint on the available room, while the lower bound on the wall thickness t_{min} as a technological constraint. The other bounds d_{min} and t_{max} can assume, theoretically, any value. In practice, the thin-walled condition $\frac{t}{d} \leq \frac{1}{20}$ (and therefore the validity of the mathematical model employed) should always be checked after a solution is obtained.

The structural safety is related to the maximum stress occurring in the tube that introduces a constraint in the optimisation problem

$$\tau_{max} \leq \frac{\tau_y}{\eta} \tag{8.5}$$

The left-hand side of Eq. 8.5 represents the maximum shear stress acting in the cross section, that can be expressed as [250]

$$\tau_{max} = \frac{2M}{\pi d^2 t} \tag{8.6}$$

The right-hand side of Eq. 8.5 represents the admissible shear stress, given by the yielding shear stress of the material τ_y divided by the safety coefficient η.

When thin-walled cross sections are employed, the structure gets more exposed to failure by local buckling. Local buckling phenomena depend on many parameters (cross-sectional geometry, material, etc.) but mainly on the wall thickness. To avoid buckling failure, an additional constraint on the maximum admissible torsional moment is introduced

$$M \leq \frac{M_{cr}}{\eta} \tag{8.7}$$

where M_{cr} is the critical buckling moment and η the safety coefficient.

The critical moment M_{cr} depends on the material properties and on the geometry of the cross section. Many analytical expressions of this quantity can be found in the literature. The first work on the topic was conducted by Schwerin in 1924 [198].

Other expressions of the critical torsional moment of a thin-walled tube can be found in [250] or in Donnell's [63] and Lundquist's [135] works.

For sake of simplicity, in this analysis, Lundquist's relation has been considered.

$$M_{cr} = \pi \frac{d^2}{2} t K_s E \left(\frac{d}{2t} \right)^{\alpha} \tag{8.8}$$

where E is the material elastic modulus, α is a constant equal to -1.35 and K_s can be related to the geometry of the tube as [58]

$$K_s = B \left(\frac{2l}{d} \right)^{\beta} \tag{8.9}$$

with B and β numerical constants equal to 1.27 and -0.46, respectively.

8.2 Optimal Design of a Thin-Walled Tube Subject to Torsion

In this section the optimal design problem of the tube is formulated by following a multi-objective optimisation (MOO [143, 152]) approach.

The problem is formulated as follows:

Find

$$min \begin{bmatrix} c(t,d) = \dfrac{4M}{\pi G t d^3} \\ m(t,d) = \pi \rho t d \end{bmatrix} \quad such\ that$$

$$\begin{aligned} \tau_{max} &= \frac{2M}{\pi d^2 t} \le \frac{\tau_y}{\eta} \\ M &\le \frac{M_{cr}}{\eta} = \pi \frac{d^2}{2\eta} t K_s E \left(\frac{d}{2t} \right)^{\alpha} \\ K_s &= B \left(\frac{2l}{d} \right)^{\beta} \end{aligned} \tag{8.10}$$

$$\begin{aligned} t_{min} &\le t \le t_{max} \\ d_{min} &\le d \le d_{max} \end{aligned}$$

The solution of problem (5.9) can be obtained by applying the theory described in Chap. 3. Being the number of design variables equal to the number of objective functions Eq. 3.13 applies. The solution is therefore given either by the solution of the unconstrained problem or by the active constraint(s).

The buckling constraint provides the following relation:

$$d \geq \left(\frac{2^{\alpha+1} M \eta}{\pi B E \left(2l\right)^{\beta}} \right)^{\frac{1}{2+\alpha-\beta}} t^{\frac{\alpha-1}{2+\alpha-\beta}} \tag{8.11}$$

while the stress constraint gives

$$d \geq \sqrt{\frac{2M\eta}{\pi t \tau_y}} \tag{8.12}$$

Let us consider the unconstrained problem which reads

$$min \begin{bmatrix} c(t,d) = \dfrac{4M}{\pi G t d^3} \\ m(t,d) = \pi \rho t d \end{bmatrix} \tag{8.13}$$

The solution of the unconstrained problem in Eq. 8.13 is given by Eq. 8.14.

$$det \left(\begin{bmatrix} \frac{\partial c}{\partial d} & \frac{\partial m}{\partial d} \\ \frac{\partial c}{\partial t} & \frac{\partial m}{\partial t} \end{bmatrix} \right) = det \left(\begin{bmatrix} -\frac{12M}{\pi G d^4 t} & \pi \rho t \\ -\frac{-4M}{\pi G d^3 t^2} & \pi \rho d \end{bmatrix} \right) = 0 \tag{8.14}$$

which leads to

$$-\frac{8M\rho}{Gd^3 t} = 0 \tag{8.15}$$

Equation 8.15 has solution for $d^3 t \to \infty$. Such solution, not belonging to the set of the finite positive numbers has no physical meaning and has to be discarded [164].

The Pareto-optimal set is therefore given by the combination of active constraints (Eq. 3.13). By applying Eq. 3.13 the following result is obtained:

$$\begin{aligned} -\left(d - \left(\frac{2^{\alpha+1} M \eta}{\pi B E \left(2l\right)^{\beta}} \right)^{\frac{1}{2+\alpha-\beta}} t^{\frac{\alpha-1}{2+\alpha-\beta}} \right) \left(d - \sqrt{\frac{2M\eta}{\pi t \tau_y}} \right) \cdot \\ (d - d_{max})(t - t_{max})(d - d_{min})(t - t_{min}) \frac{8M\rho}{Gd^3 t} = 0 \end{aligned} \tag{8.16}$$

Equation 8.16 gives a necessary condition for the Pareto-optimal solutions of the problem. The Pareto-optimal solution is, in general, a subset of the solution given by Eq. 8.16.

In order to extract Pareto-optimal solutions the intersections among the active constraints have to be studied. Depending on the relative values of the parameters, different scenarios are possible. These scenarios are analysed in the following section.

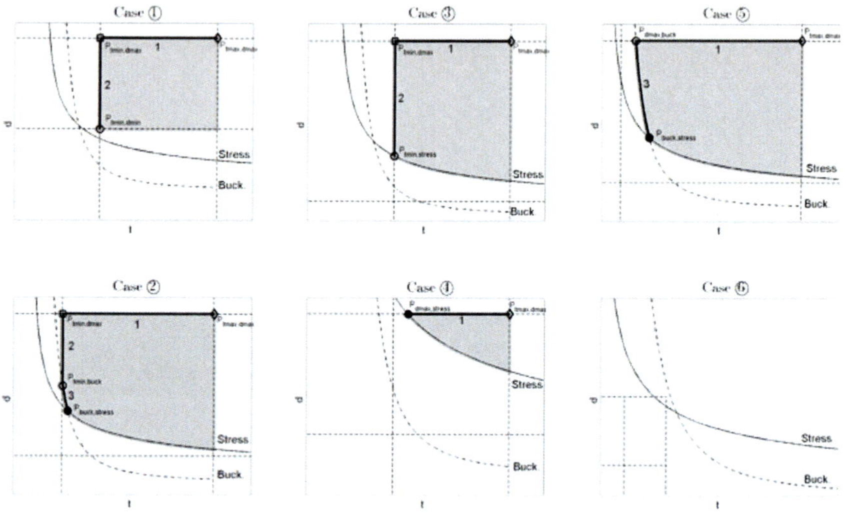

Fig. 8.2 Possible scenarios in the design variables domain. The grey area is the set of feasible solutions, the black lines are the Pareto-optimal sets. Intersection points are marked with diamond, square and circle

8.3 Sizing of Thin-Walled Tubes with Constraints on Available Room, on Minimum Thickness, on Buckling and Admissible Stress

Figures 8.2 and 8.3 show the number of possible scenarios in the design variables domain. The grey area represents the feasible set of solutions (i.e. solutions that satisfy the design constraints), while the black lines are the Pareto-optimal solutions.

Case ①

The lower bounds d_{min} and t_{min} of the design variables prevent both static and buckling failures. Buckling and stress constraints are not active.

By inspecting the expressions of the objective functions in Eq. 5.9, one can observe that the compliance objective function is monotonically decreasing with d and t, while the mass objective function is monotonically increasing with d and t. This means that the Pareto-optimal solution (i.e. the set of solutions that minimise the mass and compliance at the same time) lies on the borders of the design domain and is given either by the combination $d = d_{max}$ and $t = t_{min}$ or $d = d_{min}$ and $t = t_{max}$.

By substituting $d = d_{min}$ and $d = d_{max}$ in the objective functions, the respective expression in the objective functions domain (i.e. the m, c domain) can be computed

$$m\left(c\right)\Bigg|_{d=d_{min}} = \frac{4M\rho}{Gd_{min}^{2}c} \tag{8.17}$$

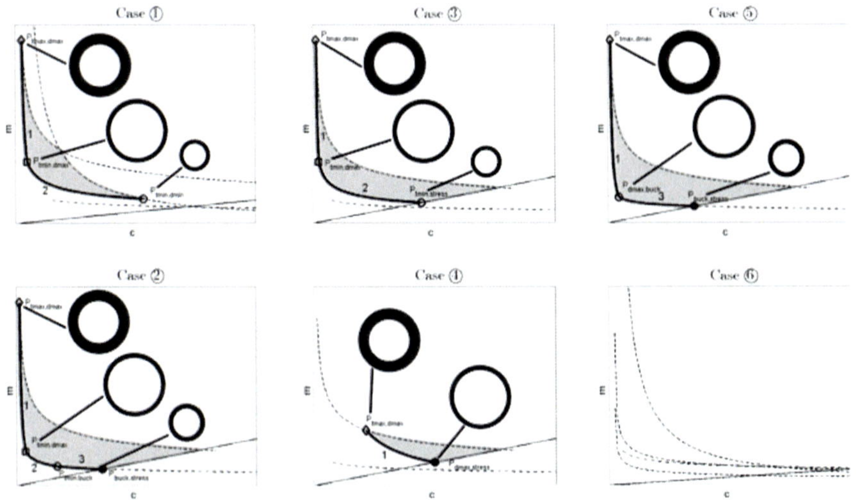

Fig. 8.3 Possible scenarios in the objective functions domain. The grey area is the set of feasible solutions, the black lines are the Pareto-optimal sets. Intersection points are marked with diamond, square and circle

$$m\left(c\right)\Bigg|_{d=d_{max}} = \frac{4M\rho}{Gd_{max}^{2}c} \tag{8.18}$$

By comparing Eqs. 8.17 and 8.18, the solution for $d = d_{max}$ (Eq. 8.18) is always lower than Eq. 8.17 and therefore $t = t_{min}$ and $d = d_{max}$ are subsets of the Pareto-optimal set (see Figs. 8.2 and 8.3).

The expression of Pareto-optimal solutions for Case ① are reported in Table 8.1 in the design variables and objective functions domain.

Case ②

This is the most general case since both the buckling constraint and stress constraint are active and part of the Pareto-optimal set.

From Case ① we have demonstrated that solutions $t = t_{min}$ and $d = d_{max}$ are Pareto-optimal. If a sufficiently large design space is considered, the solutions $t = t_{min}$ can violate the constraints on buckling and admissible stress. The intersection point $P_{buck,stress}$ between the buckling and stress constraint reads

$$\hat{t} = \left(\frac{\left(\pi BE\right)^{2}\left(2l\right)^{2\beta}}{2^{\alpha+\beta}\left(M\eta\right)^{\beta-\alpha}\left(\pi\tau_{y}\right)^{2+\alpha-\beta}}\right)^{\frac{1}{3\alpha-\beta}} \tag{8.19}$$

for $t \geq \hat{t}$ the stress equation is more binding than the buckling equation. This means that for $t \leq \hat{t}$ the buckling constraint is active, then the active constraint switches from the buckling equation to the admissible stress.

Table 8.1 Case ① of Figs. 8.2 and 8.3. Analytical expressions of the Pareto-optimal sets in both the design variables domain and objective functions domain

Analytical expressions

Pareto optimal set in the design variables domain	Pareto optimal design in the objective functions domain	Boundaries (Figs. 8.2 and 8.3)	
		Pareto optimal subset	Points
$d = d_{max}$	$m = \frac{4\rho M}{G d_{max}^2 c}$	1	$\left[P_{t_{max} d_{max}}, P_{t_{min} d_{max}}\right]$
$t = t_{min}$	$m = \rho \sqrt[3]{\frac{4\pi^2 M t_{min}^2}{Gc}}$	2	$\left[P_{t_{min} d_{max}}, P_{t_{min} d_{min}}\right]$

Coordinates of points

		$P_{t_{max} d_{max}}$	$P_{t_{min} d_{max}}$	$P_{t_{min} d_{min}}$
Design variables	t	t_{max}	t_{min}	t_{min}
	d	d_{max}	d_{max}	d_{min}
Objective functions	c	$\frac{4M}{\pi G t_{max} d_{max}^3}$	$\frac{4M}{\pi G t_{min} d_{max}^3}$	$\frac{4M}{\pi G t_{min} d_{min}^3}$
	m	$\pi \rho t_{max} d_{max}$	$\pi \rho t_{min} d_{max}$	$\pi \rho t_{min} d_{min}$

By substituting the stress constraint of Eq. 8.12 (where the $\geq$ is replaced by $=$) in the objective functions expressions, a relation between m and c when the stress constraint is active can be obtained and reads

$$m(c)\Big|_{\tau=\frac{\tau_y}{\eta}} = \frac{G M \eta^2 \rho}{\tau_y^2} c \tag{8.20}$$

which is a straight line (monotonically increasing) in the objective functions domain and therefore does not belong to the Pareto-optimal set.

With the same procedure the buckling constraint in the objective functions domain can be derived and reads

$$m(c)\Big|_{buck.} = \rho\pi \left(\frac{2^{\alpha+1} M\eta}{\pi B E (2l)^{\beta}}\right)^{\frac{1}{2+\alpha-\beta}} \left(\frac{4M}{G\pi \left(\frac{2^{\alpha+1} M\eta}{\pi B E (2l)^{\beta}}\right)^{\frac{3}{2+\alpha-\beta}}}\right)^{\frac{2\alpha-\beta+1}{4\alpha-\beta-1}} \frac{1}{c^{\frac{2\alpha-\beta+1}{4\alpha-\beta-1}}} \tag{8.21}$$

which is monotonically decreasing in the objective functions domain and therefore belongs to the Pareto-optimal set.

The expression of Pareto-optimal solutions for Case ② is reported in Table 8.2 in the design variables and objective functions domain.

Case ③

The minimum attainable diameter d is defined by the admissible stress.

Table 8.2 Case ② of Figs. 8.2 and 8.3. Analytical expressions of the Pareto optimal sets in both the design variables domain and objective functions domain

Analytical expressions

Pareto optimal set in the design variables domain	Pareto optimal design in the objective functions domain	Boundaries (Figs. 8.2 and 8.3) Pareto optimal subset	Boundaries (Figs. 8.2 and 8.3) Points
$d = d_{max}$	$m = \frac{4\rho M}{G d_{max}^2 c}$	1	$\left[P_{t_{max} d_{max}}, P_{t_{min} d_{max}}\right]$
$t = t_{min}$	$m = \rho \sqrt[3]{\frac{4\pi^2 M t_{min}^2}{Gc}}$	2	$\left[P_{t_{min} d_{max}}, P_{t_{min} d_{min}}\right]$
$d = \left(\frac{2^{\alpha+1} M\eta}{\pi BE (2l)^\beta}\right)^{\frac{1}{2+\alpha-\beta}} t^{\frac{\alpha-1}{2+\alpha-\beta}}$	$m = \rho\pi \left(\frac{2^{\alpha+1} M\eta}{\pi BE (2l)^\beta}\right)^{\frac{1}{2+\alpha-\beta}} \left(\frac{4M}{G\pi \left(\frac{2^{\alpha+1} M\eta}{\pi BE (2l)^\beta}\right)^{\frac{3}{2+\alpha-\beta}}}\right)^{\frac{2\alpha-\beta+1}{4\alpha-\beta-1}} c^{-\frac{2\alpha-\beta+1}{4\alpha-\beta-1}}$	3	$\left[P_{t_{min} buck}, P_{buck,stress}\right]$

Coordinates of points

		$P_{t_{max} d_{max}}$	$P_{t_{min} d_{max}}$	$P_{buck,stress}$	$P_{t_{min},buck}$
Design variables	t	t_{max}	t_{min}	$\left(\frac{(\pi BE)^2 (2l)^{2\beta}}{2^{\alpha+\beta} (M\eta)^{\beta-\alpha} (\pi\tau_y)^{2+\alpha-\beta}}\right)^{\frac{1}{3\alpha-\beta}}$	t_{min}
	d	d_{max}	d_{max}	$\left(\frac{2^{\alpha+1} M\eta}{\pi BE (2L)^\beta}\right)^{\frac{1}{2+\alpha-\beta}} \left(\frac{(\pi BE)^2 (2l)^{2\beta}}{2^{\alpha+\beta} (M\eta)^{\beta-\alpha} (\pi\tau_y)^{2+\alpha-\beta}}\right)^{\frac{\alpha-1}{(3\alpha-\beta)(2+\alpha-\beta)}}$	$\left(\frac{2^{\alpha+1} M\eta}{\pi BE (2L)^\beta}\right)^{\frac{1}{2+\alpha-\beta}} t_{min}^{\frac{\alpha-1}{2+\alpha-\beta}}$
Objective functions	c	$\frac{4M}{\pi G t_{max} d_{max}^3}$	$\frac{4M}{\pi G t_{min} d_{max}^3}$	$\frac{4M}{G\pi \left(\frac{(BE\pi)^2 (M\eta)^{\alpha-\beta} (\pi\tau_y)^{\beta-\alpha-2} (2l)^{2\beta}}{2^{\alpha+\beta}}\right)^{\frac{4\alpha-\beta-1}{(3\alpha-\beta)(\alpha-\beta+2)}} \left(\frac{2^{\alpha+1} M\eta}{BE\pi (2l)^\beta}\right)^{\frac{3}{\alpha-\beta+2}}}$	$\frac{4M}{G\pi t_{min}^{\frac{4\alpha-\beta-1}{\alpha-\beta+2}} \left(\frac{2^{\alpha+1} M\eta}{BE\pi (2l)^\beta}\right)^{\frac{3}{\alpha-\beta+2}}}$
	m	$\pi\rho t_{max} d_{max}$	$\pi\rho t_{min} d_{max}$	$\pi\rho \left(\frac{(BE\pi)^2 (M\eta)^{\alpha-\beta} (\pi\tau_y)^{\beta-\alpha-2} (2l)^{2\beta}}{2^{\alpha+\beta}}\right)^{\frac{2\alpha-\beta+1}{(3\alpha-\beta)(\alpha-\beta+2)}} \left(\frac{2^{\alpha+1} M\eta}{BE\pi (2l)^\beta}\right)^{\frac{1}{\alpha-\beta+2}}$	$\pi\rho t_{min}^{\frac{2\alpha-\beta+1}{\alpha-\beta+2}} \left(\frac{2^{\alpha+1} M\eta}{BE\pi (2l)^\beta}\right)^{\frac{1}{\alpha-\beta+2}}$

Table 8.3 Case ③ of Figs. 8.2 and 8.3. Analytical expressions of the Pareto-optimal sets in both the design variables domain and objective functions domain

Analytical expressions

Pareto optimal set in the design variables domain	Pareto optimal design in the objective functions domain	Boundaries (Figs. 8.2 and 8.3)	
		Pareto optimal subset	Points
$d = d_{max}$	$m = \frac{4\rho M}{G d_{max}^2 c}$	1	$\left[P_{t_{max} d_{max}}, P_{t_{min} d_{max}}\right]$
$t = t_{min}$	$m = \rho \sqrt[3]{\frac{4\pi^2 M t_{min}^2}{Gc}}$	2	$\left[P_{t_{min} d_{max}}, P_{t_{min}, stress}\right]$

Coordinates of points

		$P_{t_{max} d_{max}}$	$P_{t_{min} d_{max}}$	$P_{t_{min} stress}$
Design variables	t	t_{max}	t_{min}	t_{min}
	d	d_{max}	d_{max}	$\sqrt{\frac{2M\eta}{\pi \tau_y t_{min}}}$
Objective functions	c	$\frac{4M}{\pi G t_{max} d_{max}^3}$	$\frac{4M}{\pi G t_{min} d_{max}^3}$	$\frac{1}{G}\sqrt{\frac{2\pi t_{min} \tau_y^3}{M\eta^3}}$
	m	$\pi\rho t_{max} d_{max}$	$\pi\rho t_{min} d_{max}$	$\rho\sqrt{\frac{2\pi M \eta t_{min}}{\tau_y}}$

The Pareto-optimal set is given by the solutions $t = t_{min}$ and $d = d_{max}$. The stress constraint, as demonstrated in Case ②, does not belong to the Pareto-optimal set and it defines the solution with the lowest mass as shown in Fig. 8.3.

The expression of Pareto-optimal solutions for Case ③ is reported in Table 8.3 in the design variables and objective functions domain.

Case ④

The solution is defined by the maximum diameter $d = d_{max}$, the stress constraint is limiting t from below and defines the point with the minimum mass as shown in Fig. 8.3.

The expression of Pareto-optimal solutions for Case ④ is reported in Table 8.4 in the design variables and objective functions domain.

Case ⑤

The minimum diameter d, corresponding to minimum mass is defined by the admissible stress and buckling, as shown in Figs. 8.2, 8.3.

The full analytical expressions of the Pareto-optimal sets, both in the design variable domain and in the objective function domain, are reported in Table 8.5.

Case ⑥

In this case, we see from Fig. 8.2 that no solution within the bounds $t_{min} \leq t \leq t_{max}$ and $d_{min} \leq d \leq d_{max}$ is feasible since buckling and stress constraints are not satisfied.

Table 8.4 Case ④ of Figs. 8.2 and 8.3. Analytical expressions of the Pareto optimal sets in both the design variables domain and objective functions domain

Analytical expressions

Pareto optimal set in the design variables domain	Pareto optimal design in the objective functions domain	Boundaries (Figs. 8.2 and 8.3)	
		Pareto optimal subset	Points
$d = d_{max}$	$m = \frac{4\rho M}{G d_{max}^2 c}$	1	$[P_{t_{max} d_{max}}, P_{d_{max}, stress}]$

Coordinates of points

		$P_{t_{max} d_{max}}$	$P_{t_{min} d_{max}}$	$P_{d_{max} stress}$
Design variables	t	t_{max}	t_{min}	$\frac{2M\eta}{\pi \tau_y d_{max}^2}$
	d	d_{max}	d_{max}	d_{max}
Objective functions	c	$\frac{4M}{\pi G t_{max} d_{max}^3}$	$\frac{4M}{\pi G t_{min} d_{max}^3}$	$\frac{2\tau_y}{G d_{max} \eta}$
	m	$\pi \rho t_{max} d_{max}$	$\pi \rho t_{min} d_{max}$	$\frac{2M\eta\rho}{d_{max} \tau_y}$

8.4 Comparison of Tubes Made from Different Materials

In this section a comparison between optimised tubes made from two different materials (material A and B) is performed. The comparison is made referring to Pareto-optimal solutions [143]. Considering Fig. 8.2, the Pareto-optimal sets can be divided into a number of subsets, depending on the considered case, namely,

- $d = d_{max}$ (subsets 1 in Fig. 8.2)
- $t = t_{min}$ (subsets 2 in Fig. 8.2)
- active constraint on buckling (subsets 3 in Fig. 8.2)

In the following, the single subsets will be compared for two thin-walled tubes made from different materials.

8.4.1 Comparison Referring to Pareto-Optimal Subset 1, $d = d_{max}$

If the Pareto-optimal subsets 1 of Fig. 8.2 are considered, the optimised tubes have $d = d_{max}$ for any value of t. This case represents the maximum exploitation of the available room.

The Pareto-optimal subset 1 in the objective functions domain reads

$$m = \frac{4\rho M}{G d_{max}^2 c} \tag{8.22}$$

Table 8.5 Case ⑤ of Figs. 8.2 and 8.3. Analytical expressions of the Pareto optimal sets in both the design variables domain and objective functions domain

Analytical expressions

Pareto optimal set in the design variables domain	Pareto optimal design in the objective functions domain	Boundaries (Figs. 8.2 and 8.3)	
		Pareto optimal subset	Points
$d = d_{max}$	$m = \frac{4\rho M}{G d_{max}^2 c}$	1	$\left[P_{t_{max} d_{max}}, P_{d_{max},buck}\right]$
$d = \left(\frac{2^{\alpha+1} M\eta}{\pi BE\,(2l)^{\beta}}\right)^{\frac{1}{2+\alpha-\beta}} t^{\frac{\alpha-1}{2+\alpha-\beta}}$	$m = \rho\pi\left(\frac{2^{\alpha+1} M\eta}{\pi BE\,(2l)^{\beta}}\right)^{\frac{1}{2+\alpha-\beta}} \left(\frac{4M}{G\pi\left(\frac{2^{\alpha+1} M\eta}{\pi BE\,(2l)^{\beta}}\right)^{\frac{3}{2+\alpha-\beta}}}\right)^{\frac{2\alpha-\beta+1}{4\alpha-\beta-1}} c^{-\frac{2\alpha-\beta+1}{4\alpha-\beta-1}}$	3	$\left[P_{d_{max},buck}, P_{buck,stress}\right]$

Coordinates of points

		$P_{t_{max} d_{max}}$	$P_{d_{max} buck}$	$P_{buck,stress}$
Design variables	t	t_{max}	$\left(\frac{\pi BE\,(2l)^{\beta}}{2^{\alpha+1} M\eta}\right)^{\frac{1}{\alpha-1}} d_{max}^{\frac{2+\alpha-\beta}{\alpha-1}}$	$\left(\frac{(\pi BE)^2\,(2l)^{2\beta}}{2^{\alpha+\beta}\,(M\eta)^{\beta-\alpha}\,(\pi\tau_y)^{2+\alpha-\beta}}\right)^{\frac{1}{3\alpha-\beta}}$
	d	d_{max}	d_{max}	$\left(\frac{2^{\alpha+1} M\eta}{\pi BE\,(2l)^{\beta}}\right)^{\frac{1}{2+\alpha-\beta}} \left(\frac{(\pi BE)^2\,(2l)^{2\beta}}{2^{\alpha+\beta}\,(M\eta)^{\beta-\alpha}\,(\pi\tau_y)^{2+\alpha-\beta}}\right)^{\frac{\alpha-1}{(3\alpha-\beta)(2+\alpha-\beta)}}$
Objective functions	c	$\frac{4M}{\pi G t_{max} d_{max}^3}$	$\frac{4M}{G\pi d_{max}^{\frac{4\alpha-\beta-1}{\alpha-1}} \left(\frac{BE\pi\,(2l)^{\beta}}{2^{\alpha+1} M\eta}\right)^{\frac{1}{\alpha-1}}}$	$\frac{4M}{G\pi\left(\frac{(BE\pi)^2\,(M\eta)^{\alpha-\beta}\,(\pi\tau_y)^{\beta-\alpha-2}\,(2l)^{2\beta}}{2^{\alpha+\beta}}\right)^{\frac{4\alpha-\beta-1}{(3\alpha-\beta)(\alpha-\beta+2)}} \left(\frac{2^{\alpha+1} M\eta}{BE\pi\,(2l)^{\beta}}\right)^{\frac{3}{\alpha-\beta+2}}}$
	m	$\pi\rho t_{max} d_{max}$	$\rho\pi d_{max}^{\frac{2\alpha-\beta+1}{\alpha-1}} \left(\frac{BE\pi\,(2l)^{\beta}}{2^{\alpha+1} M\eta}\right)^{\frac{1}{\alpha-1}}$	$\pi\rho\left(\frac{(BE\pi)^2\,(M\eta)^{\alpha-\beta}\,(\pi\tau_y)^{\beta-\alpha-2}\,(2l)^{2\beta}}{2^{\alpha+\beta}}\right)^{\frac{2\alpha-\beta+1}{(3\alpha-\beta)(\alpha-\beta+2)}} \left(\frac{2^{\alpha+1} M\eta}{BE\pi\,(2L)^{\beta}}\right)^{\frac{1}{\alpha-\beta+2}}$

If two tubes made from different materials (let's say material A and material B) are considered, the ratio between the mass per unit of length at a given stiffness of the two tubes can be written as

$$\frac{m_A}{m_B} = \frac{\rho_A E_B}{\rho_B E_A} \tag{8.23}$$

where G has been replaced by E through the relation $G = \dfrac{E}{2\,(1+\nu)}$

If we assume aluminium for material A and steel for material B ($\rho_A = 2800\,\mathrm{kg/m^3}$, $\rho_B = 7800\,\mathrm{kg/m^3}$, $E_A = 70\,\mathrm{GPa}$, $E_B = 210\,\mathrm{GPa}$, $\nu_A = \nu_B = 0.3$) Eq. 8.23 returns 1.077, meaning that when all the available room is exploited (i.e. $d = d_{max}$) steel allows to design a tube with about 7.7% less mass than the aluminium counterpart.

8.4.2 Comparison Referring to Pareto-Optimal Subset 2, $t = t_{min}$

This case can be interpreted as a technological constraint that limits the minimum manufacturing thickness of the tube. This constraint may depend on the material and technological manufacturing process.

The Pareto-optimal subset 2 in the objective functions domain reads

$$m = \rho \sqrt[3]{\frac{4\pi^2 M t_{min}^2}{Gc}} \tag{8.24}$$

By considering the two materials A and B the ratio $\dfrac{m_A}{m_B}$ has the following expression:

$$\frac{m_A}{m_B} = \frac{\rho_A}{\rho_B} \sqrt[3]{\frac{E_B}{E_A}} \tag{8.25}$$

If we consider again aluminium for material A and steel for material B Eq. 8.25 returns 0.52, thus making the mass of aluminium tube about one half of the steel one for the same compliance c.

8.4.3 Comparison Referring to Pareto-Optimal Subset 3, Active Constraint on Buckling

In this case the minimum allowable thickness is determined by the constraint on buckling which is active. The expression of the Pareto-optimal subset 3 in the objective functions domain is

$$m = \rho\pi \left(\frac{2^{\alpha+1}M\eta}{\pi BE\,(2l)^{\beta}}\right)^{\frac{1}{2+\alpha-\beta}} \left(\frac{4M}{G\pi\left(\frac{2^{\alpha+1}M\eta}{\pi BE\,(2l)^{\beta}}\right)^{\frac{3}{2+\alpha-\beta}}}\right)^{\frac{2\alpha-\beta+1}{4\alpha-\beta-1}} c^{-\frac{2\alpha-\beta+1}{4\alpha-\beta-1}} \tag{8.26}$$

Again by considering the relation between G and E, the ratio $\frac{m_A}{m_B}$ reads

$$\frac{m_A}{m_B} = \frac{\rho_A}{\rho_B}\left(\frac{E_B}{E_A}\right)^{0.545} \tag{8.27}$$

Substituting the values of aluminium (material A) and steel (material B) Eq. 8.27 gives 0.65. Therefore it turns out that, when the buckling constraint is active, the aluminium tube is about 35% lighter for a prescribed compliance.

8.5 Optimal Design of a Race Car Driveshaft

In this section, a practical example of an application of the derived formulae is presented. The analysed problem refers to the optimal design of the main driveshaft of a high-performance race car. The component is highlighted in the scheme of Fig. 8.4 and has the role of transmitting the drive torque M from the engine to the drive axle.

The shaft has a tubular shape, the design variables to be optimised are the tube diameter and its wall thickness, whereas the design objectives are

- minimization of the overall mass of the shaft.
- minimization of the deflection of the shaft when subject to the applied load.

The driveshaft is subjected to the following constraints. The available room for the driveshaft limits the maximum diameter to 90 mm ($d_{max} = 90$ mm) and constraints on structural safety and elastic stability have to be satisfied to avoid failures when the maximum drive torque is applied. The maximum torque M on the driveshaft is 3600 Nm, obtained by multiplying the maximum engine torque (800 Nm) by the first gear ratio (4.5). Additionally, a safety factor of 2.5 is considered in the design process to account for overloads on the driveline and durability requirements.

A C45 quenched and tempered steel is assumed as reference material for the shaft. All the parameters that are necessary for the optimisation are listed in Table 8.6.

By substituting the numerical values of Table 8.6 in the analytical expressions obtained in Sect. 8.3, one realises that Case ④ applies. The Pareto-optimal solution is therefore given by the constraint $d = d_{max}$, the solution with the minimum attainable mass is defined by the intersection between the stress constraint and $d = d_{max}$; the relative analytical expressions are reported in Table 8.4.

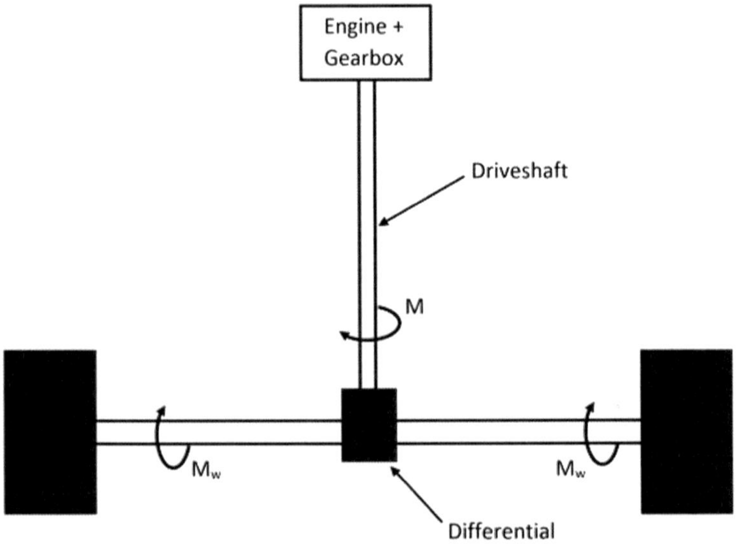

Fig. 8.4 General schematics of the driveline of a two-wheel drive vehicle

Table 8.6 Design of the main driveshaft of a race car—input data

Description	Notation	Value	Unit
Applied torque	M	3600	Nm
Material density	ρ	7800	$\frac{kg}{m^3}$
Material tangential modulus	G	80.77	GPa
Material yielding shear stress	τ_y	261	MPa
Safety coefficient	η	2.5	–
Lower bound on tube diameter	d_{min}	0.02	m
Upper bound on tube diameter	d_{max}	0.09	m
Lower bound on wall thickness	t_{min}	0.0005	m
Upper bound on wall thickness	t_{max}	0.005	m

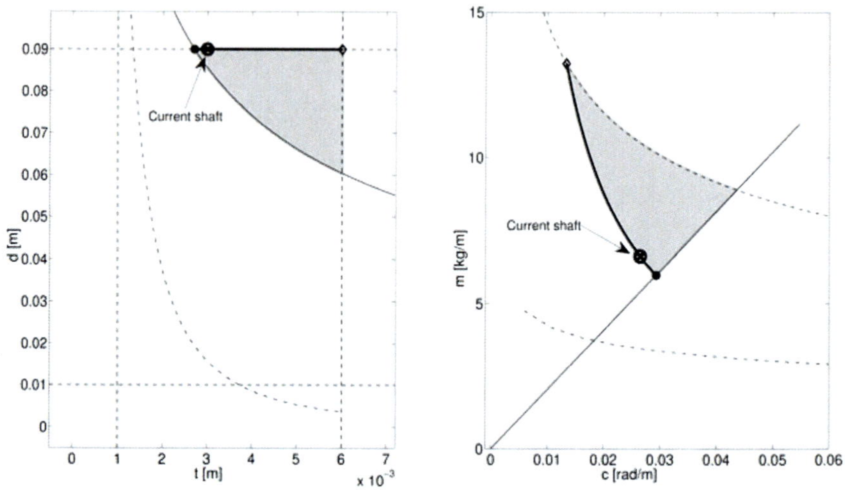

Fig. 8.5 Optimal design of the driveshaft of a race car—Pareto-optimal solution (black line) both in the design variables domain (left) and objective functions domain (right)

Figure 8.5 shows the set of feasible solutions both in terms of design variables (i.e. tube diameter and wall thickness) and objective functions (mass and compliance per unit of length). The (Pareto) optimal solutions are identified by the black line in the graphs of Fig. 8.5, the designer has to select his final design among this set of solutions. The solution with the lowest attainable mass is identified by the black dot of Fig. 8.5 and is given by the intersection point of the stress constraint with the constraint $d = d_{max}$. Such a solution is characterised by an outer diameter of 90 mm and a wall thickness of 2.71 mm, with a mass per unit of length of 5.98 $\frac{\text{kg}}{\text{m}}$; this solution exhibits also the highest deflection (0.0293 $\frac{\text{rad}}{\text{m}}$) when subject to the torsional moment.

On the other hand, the solution with the lowest compliance (marked with a diamond in Fig. 8.5) is given by a tube with the maximum admissible diameter (90 mm) and the maximum admissible wall thickness (5 mm); regarding the objective functions, this solution is the stiffest (with a deflection of 0.0132 $\frac{\text{rad}}{\text{m}}$) but the heaviest one (13.23 $\frac{\text{kg}}{\text{m}}$).

The driveshaft currently mounted on the car is also highlighted in the graphs of Fig. 8.5 and has an outer diameter of 90 mm and a wall thickness of 3 mm. As evidenced from Fig. 8.5, the currently adopted solution lies on the Pareto front, showing that the proposed approach for the design of thin walled tubes under torsion is in accordance with practical designs obtained by experienced specialists.

8.6 Summary

In the chapter, the analytical multi-objective optimisation for the lightweight design of a thin-walled tube subject to torsion has been dealt with. Tube mass and compliance have been minimised at the same time. Constraints on safety (i.e. admissible stress), elastic stability (buckling), available room (maximum diameter) and manufacturing constraints (minimum thickness) have been considered.

Analytical formulae of the Pareto-optimal set have been obtained for the considered design problem. The analytical expressions are derived both in the design variables (tube diameter and wall thickness) and in the objective functions (mass and compliance of the tube) domain. A comparison of optimised designs made from different materials has been included.

It has been demonstrated that

- the Pareto-optimal set of the thin-walled tube under torsion is given by the combination of the buckling limit, minimum thickness and maximum diameter.
- apart from the solution having maximum thickness and maximum diameter, all the other designs with $t = t_{max}$ are non-optimal and have to be discarded.
- the stress constraint is not part of the Pareto-optimal set, actually it has a role only in the definition of the minimum attainable mass.

The comparative lightweight design of tubes made from different materials showed that aluminium alloy allows an effective lightweight construction, but when the available room is saturated and proper stiffness is requested, steel allows to obtain a lighter structure by 8%.

A simple engineering problem related to the design of the main driveshaft of a race car has been solved by employing the obtained analytical expressions, showing the practical use of the method and its effectiveness in the definition of early stage optimal design solutions. From a direct comparison with the currently adopted driveshaft, it has been shown that the proposed method is in accordance with the solution obtained by experienced specialists.

Chapter 9
Stochastic Optimisation Applied to Multi-axial Bending of Lightweight Beams

The following chapter, adapted from [129], deals with the stochastic optimisation of complex mechanical systems like a beam subject to bending and torsion [88, 177, 178]. The adopted (and proper) theoretical basis for such an optimisation is multi-objective programming [152]. Multi-objective programming allows the designer to find a preferred solution by means of a proper selection among the set of optimal solutions called "Pareto-optimal set". The designer, by means of the addressed selection, defines a set of parameters (design variable set) which defines the system under consideration. The choice (selection) is made after the optimal combination of the system performances has been identified by the designer. This process is deterministic in nature, as multi-objective programming relies on a deterministic approach. In recent years many authors have applied multi-objective programming methods to design structural systems within a deterministic framework [79, 152]. However, almost every engineering design should be performed within a stochastic framework, in fact, actual systems are subject to variations and uncertainties that arise from a variety of sources, i.e. the manufacturing process, variability of the material properties, operating conditions, ... Not only the system parameters and external disturbances (loads), but also the system design variables can be—in general—considered as stochastic quantities. Usually, the "robustness" of a system depends not only on external disturbances (loads), but—mostly—on the random variation of the system parameters and/or design variables. For this reason an optimal design based on a deterministic approach may turn out to be not robust, leading to substantial performance deterioration or design constraints violation [65, 177, 178, 206, 207]. As observed in [226], optimal designs based on deterministic approach are often prone to be the most sensitive to parameters uncertainty. According to the knowledge of the authors only a limited number of papers exists referring to a stochastic multi-objective approach applied to the optimal and robust design of mechanical systems [87, 88, 144, 145, 154, 177]. One of the most popular approaches seems to be the physical programming [46, 47, 147] but it requires an accurate description of the designer's preferences. The approach adopted in this chapter requires, instead,

F. M. Ballo et al., *Optimal Lightweight Construction Principles*,
https://doi.org/10.1007/978-3-030-60835-4_9

only the specification of a tolerable level of risk, as in [38]. The method proposed in this chapter is based on previous papers of the authors [87, 88, 129], but it can also be seen as a substantial evolution of one of the stochastic optimisation methods proposed in [38]. Actually the approach (and method) introduced and exploited in this paper takes some of the theoretical concepts reported in [38] and, introducing stochastic constraints, makes the concepts usable for actual applications. The integration of a stochastic or probabilistic approach within the optimisation process is of basic importance since it combines the two most important tasks of the designer: to create the best possible designs and to estimate the tolerable level of risk for these optimal designs to fail.

The structural problem considered in this chapter refers to the definition of both the optimal shape and the dimensions of the cross section of a cantilever beam. The beam is loaded at the free end by a vertical force whose amplitude is subject to stochastic variations of given distribution. Since it is assumed that the applied load may not have a perfect vertical direction, the beam is also subject to a horizontal load of stochastic nature, with zero mean value. Moreover the limit strength and the elastic modulus of the material are subject to uncertainty. The objectives are to minimise the mass and the vertical deflection of the free end of the beam. A number of different beam cross sections and two different materials have been considered. The problem describes an actual situation of a cantilever beam designed to carry mostly vertical loads.

9.1 Multi-objective Stochastic Problem

A stochastic system is described by a mathematical model in which some random quantities are subject to uncertainty. These quantities may be parameters, whose values are not varied by the designer, external disturbances (loads) or design variables, whose expected (i.e. mean) value can be defined by the designer. The uncertainties (on the parameters and/or design variables and/or external disturbances) are transmitted to the objective functions and constraints so that they acquire a stochastic nature too.

We will denote by $\mathbf{x}$ the vector of the n design variables (deterministic and stochastic), by $\mathbf{c}$ the vector of the m random parameters and by $g_i(\mathbf{x}, \mathbf{c})$ the generic stochastic objective function whose expected value (mean value) and standard deviation are $\overline{g}_i(\mathbf{x}, \mathbf{c})$ and $\sigma_{gi}(\mathbf{x}, \mathbf{c})$, respectively.

When considering a stochastic system, the solution can be obtained, see [38], by transforming the original stochastic problem into an *equivalent deterministic problem*. This transformation can be carried out by using some statistical characteristics, such as expected value and variance, of the random variables involved. Among the many possible formulations of the equivalent deterministic problem the one adopted in the present paper is the K_β formulation (β-efficient problem) [38]

$$\begin{aligned} &given\ the\ probabilities\quad \beta_1, \beta_2, \ldots, \beta_k \\ &find\quad \min\left(u_1(\mathbf{x}, \mathbf{c}), u_2(\mathbf{x}, \mathbf{c}), \ldots, u_k(\mathbf{x}, \mathbf{c})\right) \\ &such\ that\quad Prob\left(g_i\left(\mathbf{x}, \mathbf{c} \leq u_i(\mathbf{x}, \mathbf{c})\right)\right) \geq \beta_i \quad i = 1, 2, \ldots, k \end{aligned} \tag{9.1}$$

The probabilities β_i (that describe the tolerable level of risk) are fixed and the values u_i of the objective functions (that guarantee that the risk is not superior to β_i) are minimised. There is therefore a probability of $1 - \beta_i$ that the performance of the optimal solutions will be worse than expected. This formulation seems to be the best for many engineering problems in which, in general, the levels of risk are fixed by standards or good design practice while little is known about the maximum obtainable performance. If the objective functions have normal distribution, the K_β-problem can be converted to the so-called β-efficient problem with normal distribution (K_α-formulation) [38]

$$\begin{aligned} &given\ the\ probabilities\quad \beta_1, \beta_2, \ldots, \beta_k \\ &find\quad \min\left(\overline{g}_1(\mathbf{x}, \mathbf{c}) + \alpha_1\sigma_{g1}(\mathbf{x}, \mathbf{c}), \overline{g}_2(\mathbf{x}, \mathbf{c}) + \alpha_2\sigma_{g2}(\mathbf{x}, \mathbf{c}), \ldots, \overline{g}_k(\mathbf{x}, \mathbf{c}) + \alpha_k\sigma_{gk}(\mathbf{x}, \mathbf{c})\right) \\ &with\quad \alpha_i = \Phi^{-1}(\beta_i) \quad i = 1, 2, \ldots, k \end{aligned} \tag{9.2}$$

where the function $\Phi^{-1}(\cdot)$ is the inverse of the standard normal distribution. Although the idea of minimising a weighted sum of the mean and the standard deviation of the objective functions is quite popular, see, for example, [87, 88, 144, 206], here the focus is on the choice of the weights α_i through the definition of the probabilities β_i. Due to the probabilistic nature of the design variables $\mathbf{x}$ and uncertain parameters $\mathbf{c}$, the design constraints do have a stochastic nature, that is constraints can be violated, even if the violation will be limited by a pre-defined tolerable risk β_i. In other terms there is a probability of $1 - \beta_i$ that the optimal solutions will not satisfy a constraint.

For sake of simplicity, and according to the K_α formulation, it is useful to assume that the constraints and the design variables have a normal distribution, so that the variability of constraints (and of the objective functions) can be estimated by computing the standard deviation of a multi-variables function, see [32]. If the generic constraint is $h(x_1, \ldots, x_n, c_1, \ldots c_m) < 0$ and if x_j, x_k and c_j, c_k with $j \neq k$ are independent stochastic processes, its mean and standard deviation can be written as

$$\mu_{h_i} = h_i\left(\mu_{x1}, \ldots, \mu_{xn}, \mu_{c1}, \ldots, \mu_{cm}\right) \quad \sigma_{h_i} \cong \sqrt{\sum_{j=1}^{n}\left(\frac{\partial h_i}{\partial x_j}\right)^2 \cdot \sigma_{xj}^2 + \sum_{j=1}^{m}\left(\frac{\partial h_i}{\partial c_j}\right) \cdot \sigma_{cj}^2} \tag{9.3}$$

where μ_{xj} and σ_{xj} are the mean and standard deviation of the variable x_j, and μ_{cj} and σ_{cj} are the mean and standard deviation of the parameter c_j.

In order to guarantee a constraint satisfaction probability β_i the following equations must hold

$$Prob\left(h_i(\mathbf{x}, \mathbf{c}) < 0\right) \geq \beta_i \rightarrow \Phi\left(\frac{0 - \mu_{h_i}}{\sigma_{h_i}}\right) \geq \beta_i \rightarrow \mu_{h_i} + \Phi^{-1}(\beta_i) \cdot \sigma_{h_i} \leq 0 \tag{9.4}$$

In recent years many authors have proposed numerical algorithms for the computation of the Pareto-optimal set within a deterministic framework, see [34, 75, 156, 247, 256]. The numerical procedure used for the solution of the problem presented in this paper is based on a Quasi-Monte Carlo Search [79, 88]. The working principle is simple:

- many points are chosen in the space of design variables through a low-discrepancy Sobol sequence [88],
- the points that do not comply with the design stochastic constraints are discarded,
- the expected (mean) values and standard deviations of the objective functions are computed at the selected points,
- the Pareto-optimal points are selected according to the definition (see Eq. 1.11).

The numerical procedure to find optimal solutions that have been applied in this chapter is actually more complex than explained above. It allows to refine the search through a multi-step procedure based on the *direct* algorithm [34, 79]. The key idea of the method is to subdivide the space of the variables in many sub-domains and to select the sub-domains which contain optimal solutions in order to direct the search only in these "promising" sub-domains.

The designer has to select the number n of subdivisions of the design space. The space of the design variables is therefore divided into $n^{\#dimension}$ hypercubes. A low-discrepancy (Sobol) sequence of design points is created covering each hypercube. The objective functions values are computed, the design points that do not comply with the design constraints are discarded, and then the Pareto-optimal points are selected. The algorithm searches for the hypercubes in which there is at least one Pareto-optimal point: these hypercubes are selected; the others are discarded. A new step of the algorithm will therefore carry out the same subdivision by considering only the selected hypercubes. The search becomes more and more accurate in the most promising areas of the design space. This multi-step Quasi-Monte Carlo Search gives a good approximation of the Pareto-optimal region. A standard algorithm for constrained optimisation is used to refine the search.

9.2 Formulation of the Design Problem

The design problem deals with the definition of the optimal shape and dimensions of the cross section of a cantilever beam represented in Fig. 9.1. The beam is loaded at the free end by means of a vertical force P_y whose amplitude is subject to stochastic variations with normal distribution applied on the centre of the cross section. Moreover a horizontal force P_x with zero mean value and normal distribution acts on the shear centre. This problem describes an actual situation of a cantilever beam designed to carry mainly vertical loads. The objectives are the mass reduction and the minimisation of vertical deflection of the free end of the beam. Two different materials are considered in the paper: steel and aluminium alloy. The limit strength and the elastic modulus of these materials are supposed to be subject to uncertainty. The

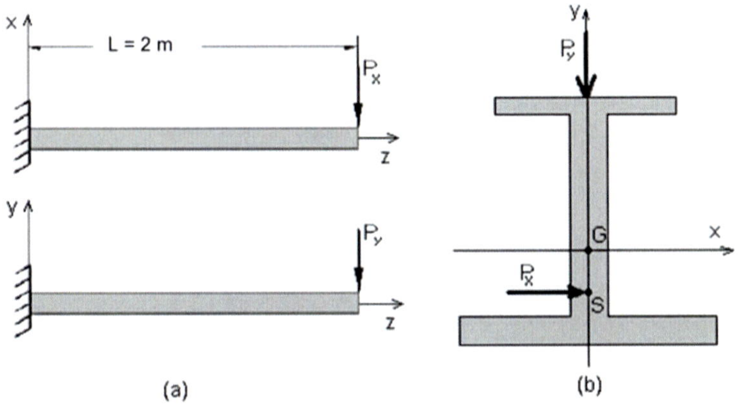

Fig. 9.1 **a** Beam loaded in the two planes x-z and y-z. **b** Cross section of the beam in the plane x-y. The centre of gravity and the shear centre of the cross section are indicated by G and S, respectively. The length of the beam, for all of the simulations performed in this chapter is L = 2 m

Table 9.1 Loads, steel material properties (figures into brackets refer to the aluminium alloy)

Parameter/External load	Mean value	Standard deviation
Horizontal force P_x $[N]$	0	5000
Vertical force P_y $[N]$	10000	5000
Elastic modulus E $[\frac{N}{mm^2}]$	209000 (72000)	7500 (1400)
Yield stress R_p $[\frac{N}{mm^2}]$	300 (200)	5 (3)
Material density ρ $[\frac{kg}{mm^3}]$	$7.5 \cdot 10^{-6}$ $(2.7 \cdot 10^{-6})$	0 (0)

mean values and standard deviations of the materials' properties have been obtained by means of a set of experimental tests. The loads and the material properties are shown in Table 9.1, the figures into brackets refer to the aluminium alloy.

Due to the symmetry of the loading condition, it is obvious to consider a symmetric shape for the cross section with respect to the vertical axis. The lines of action of the horizontal and vertical loads point to the shear centre of the cross section and therefore no twisting moment is present. In actual situations, it is, however, unusual that the beam is loaded exactly on the shear centre, so in general there will be a twisting moment, too. Since however, for commonly used cross sections, the shear centre is close to the centre of gravity, the value of the twisting moment can be neglected, see [170]. Bi-axial bending therefore constitutes the stress situation on the structure. According to [170], the shear actions and the shear stress have been neglected. In fact the shear actions have a vanishing influence on the deflection of the beam. Moreover the shear stress is quite small and it is zero where the normal stress reaches its maximum (or minimum), so, in this case, it doesn't influence the maximum stress at the fixed end of the beam. The maximum stress at the fixed end

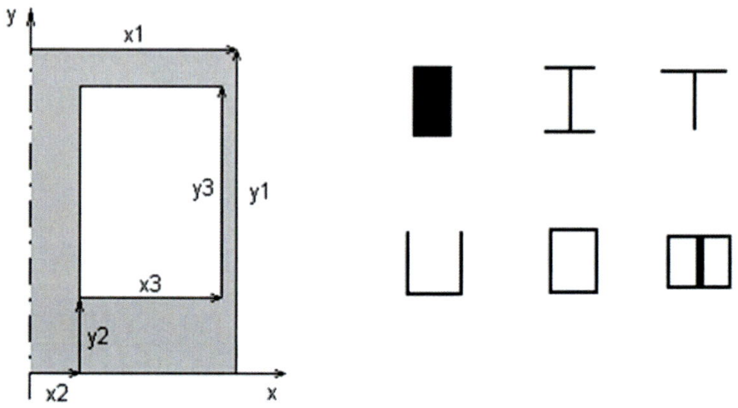

Fig. 9.2 Definition of the cross section of the cantilever beam. The depicted cross section refers to one half of the actual cross section. By properly varying $x_1, y_1, x_2, y_2, x_3, y_3$ the different cross sections on the right can be obtained

must be less than (or equal to) the admissible stress. An ideal fixed constraint is modelled, i.e. no load intensity factors have been considered at the constrained end of the cantilever.

9.2.1 *Design Variables*

The cross section of the beam has been defined by means of a minimum set of design variables, being able to describe the beam cross section types which are more used in usual applications. As already observed, the cross section is symmetric with respect to the vertical axis, it is hence sufficient to describe one half of the cross section by means of the selected design variables ($x_1, y_1, x_2, y_2, x_3, y_3$).The six variables shown in Fig. 9.2 have been considered. The idea is to represent half of the cross section through a rectangle to which another rectangle is subtracted.

The following types of cross sections can be obtained:

- rectangular solid, with $x_2 = y_2 = x_3 = y_3 = 0$.
- T-shape, with $y_2 = 0$ or $y_2 = y_1 - y_3$, and $x_2 = x_1 - x_3$.
- I-shape, with $x_2 = x_1 - x_3$.
- C-shape, with $x_2 = 0$ and $y_2 = y_1 - y_3$ or $y_2 = 0$.
- rectangular hollow, with $x_2 = 0$.
- rectangular hollow with two cells, with $x_1 \neq y_1 \neq x_2 \neq y_2 \neq x_3 \neq y_3$.

The variables x_2, x_3 and y_2, y_3 are normalised with respect to x_1 and y_1, respectively,

$$p_2 = \frac{x_2}{x_1}, \quad p_3 = \frac{x_3}{x_1}, \quad q_2 = \frac{y_2}{y_1}, \quad q_3 = \frac{y_3}{y_1} \tag{9.5}$$

Table 9.2 Design variables ranges

Design variable	Lower bound	Upper bound
x_1 (mm)	0	300
y_1 (mm)	0	600
p_2, p_3, q_2, q_3	0	1

The cross section dimensions should be considered as stochastic variables due to the usual manufacturing tolerances. However the influence of the stochastic variations of the design variables on the objective functions and constraints has proved to be negligible. The dimensional tolerances (ISO IT 7 manufacturing quality) cause, in fact, a worst case variation of the objective functions (both in mean and standard deviation) lower than 0.5%. For this reason all the design variables have been treated as deterministic variables. The design variables of the problem and their ranges of variation are summarised in Table 9.2.

The geometrical constraints on the variables that describe the beam cross section are the following:

$$x_2 + x_3 - x_1 < 0 \qquad y_2 + y_3 - y_1 < 0 \tag{9.6}$$

9.2.2 *Performance Indices*

The system performances (objectives) to be optimised (minimised) are the mass of the beam m and the vertical deflection of the free end δ_y.

The mass of the beam m can be simply determined as follows:

$$m = \rho \cdot L \cdot A \tag{9.7}$$

where A is the area of the cross section, ρ the material density and L the length of the beam. Since the beam mass depends only on deterministic quantities, it will be a deterministic quantity too. Its standard deviation is therefore zero.

The vertical deflection δ_y of the free end of the beam depends only on the bending generated by the vertical force P_y

$$\delta_y = \frac{1}{3} \frac{P_y L^3}{E I_x} \tag{9.8}$$

where I_x is the moment of inertia of the beam cross section with respect to the axis x. The elastic modulus E of the material and the vertical force P_y are the stochastic parameters which have to be taken into account for the computation of the vertical deflection. No gravitational effects have been considered.

The standard deviation of the vertical deflection can hence be estimated as

$$\sigma_{\delta_y} \cong \sqrt{\left(\frac{\partial \delta_y}{\partial P_y} \cdot \sigma_{P_y}\right)^2 + \left(\frac{\partial \delta_y}{\partial E} \cdot \sigma_E\right)^2} = \delta_y \cdot \sqrt{\left(\frac{\sigma_{P_y}}{P_y}\right)^2 + \left(\frac{\sigma_E}{E}\right)^2} \tag{9.9}$$

since the stochastic process defining σ_{P_y} and σ_E are not correlated.

The stochastic objective function, see Eq. 9.2, is therefore

$$\delta_y + \alpha \cdot \sigma_{\delta_y} \tag{9.10}$$

9.2.3 Design Constraints

The optimisation procedure involves the following constraints to be satisfied, namely:

- maximum deflection in the horizontal direction.
- static resistance at the fixed end.
- lateral-torsional buckling.
- local buckling of the beam cross section.

9.2.3.1 Horizontal Maximum Deflection

The horizontal deflection of the free end of the beam caused by the horizontal force P_x is not considered as an objective function but as a design constraint. Horizontal forces are generally smaller than the vertical ones, so they are considered as a perturbation caused by the uncertainty in the load application direction. It is hence sufficient to choose a maximum value $\delta_{x\,max}$ that does not have to be exceeded.

The constraint is defined as

$$\frac{1}{3}\frac{P_x L^3}{E I_y} - \delta_{x\,max} < 0 \tag{9.11}$$

where I_y is the moment of inertia of the beam cross section with respect to the axis y.

The standard deviation of the constraint can be written analytically as (see Eq. 9.3)

$$\sigma_{deflection_x} \cong \sqrt{\left(\frac{1}{3}\frac{L^3}{3 E I_y} \cdot \sigma_{P_x}\right)^2 + \left(\frac{1}{3}\frac{P_x L^3}{E^2 I_y} \cdot \sigma_E\right)^2} \tag{9.12}$$

The stochastic constraint function, Eq. 9.4, is therefore

$$\frac{1}{3}\frac{P_x L^3}{E I_y} - \delta_{x\,max} + \alpha \cdot \sigma_{deflection_x} \tag{9.13}$$

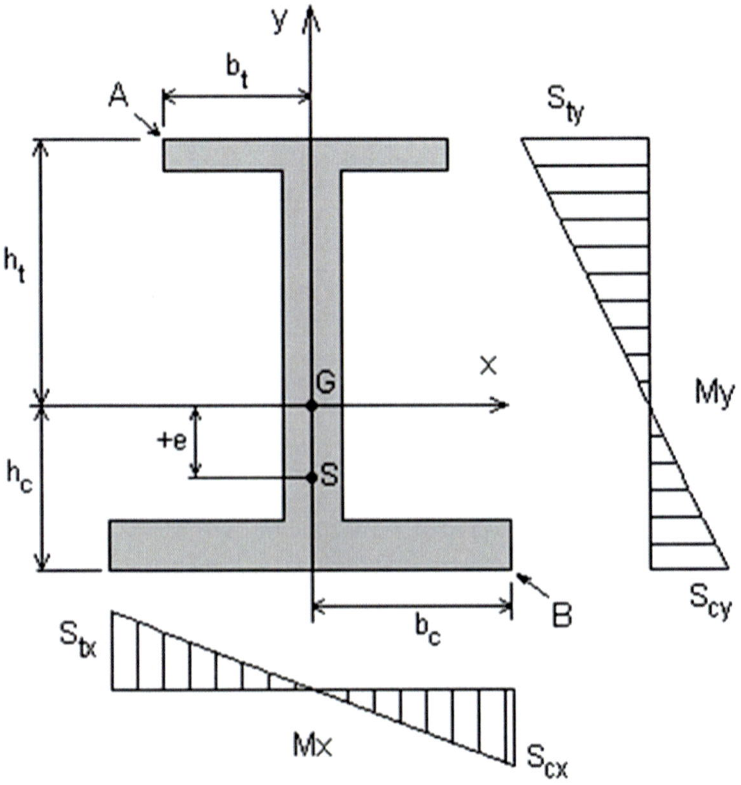

Fig. 9.3 Cross-sectional stress distribution

9.2.3.2 Static Resistance

Static resistance at the fixed beam end, the most stressed one, is guaranteed by limiting the maximum tensile and compressive stress, obtained by the superposition of the effects of the two bending moments. By supposing that the horizontal and vertical forces act as shown in Fig. 9.1, the maximum tensile stress is reached at the upper left corner, named A in Fig. 9.3. The maximum tensile actions are added due to the superposition of two bending moments. In the same way, the maximum compressive stress is reached at the lower right corner, named B in Fig. 9.3.

The tensile stresses S_{ty} and S_{tx} and compressive stresses S_{cy} and S_{cx} caused by forces P_y and P_x are

$$S_{ty} = \frac{P_y L}{I_x} h_t \qquad S_{tx} = \frac{P_x L}{I_y} b_t \tag{9.14}$$

$$S_{cy} = \frac{P_y L}{I_x} h_c \qquad S_{cx} = \frac{P_x L}{I_y} b_c \tag{9.15}$$

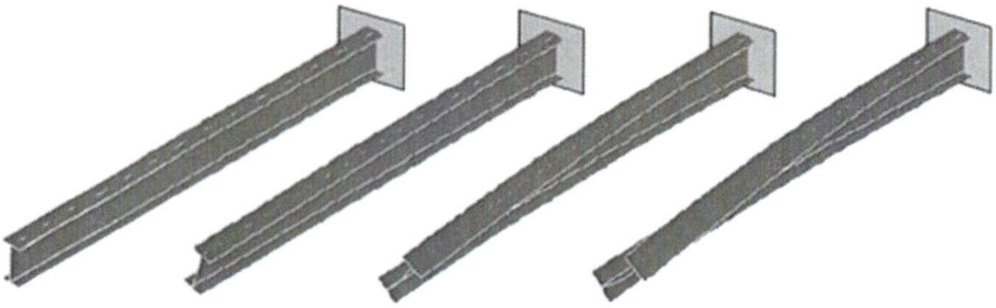

Fig. 9.4 Representation of the lateral-torsional buckling of a cantilever I-beam loaded at the free end. Load increases from left to right. The beam twists and bends in the plane of lower stiffness

The stress at the points A and B must be less than (or equal to) the admissible stress R_p

$$\begin{aligned} A: &\quad \frac{P_y L}{I_x} h_t + \frac{P_x L}{I_y} b_t - R_p < 0 \\ B: &\quad \frac{P_y L}{I_x} h_c + \frac{P_x L}{I_y} b_c - R_p < 0 \end{aligned} \tag{9.16}$$

The standard deviations of the stress constraints are

$$\begin{aligned} \sigma_A &= \sqrt{\left(\frac{\sigma_{P_y} L}{I_x} h_t\right)^2 + \left(\frac{\sigma_{P_x} L}{I_y} b_t\right)^2 + \left(\sigma_{R_p}\right)^2} \\ \sigma_B &= \sqrt{\left(\frac{\sigma_{P_y} L}{I_x} h_c\right)^2 + \left(\frac{\sigma_{P_x} L}{I_y} b_c\right)^2 + \left(\sigma_{R_p}\right)^2} \end{aligned} \tag{9.17}$$

The stochastic constraint functions, see Eq. 9.4, are therefore

$$\begin{aligned} A: &\quad \frac{P_y L}{I_x} h_t + \frac{P_x L}{I_y} b_t - R_p + \alpha \cdot \sigma_A < 0 \\ B: &\quad \frac{P_y L}{I_x} h_c + \frac{P_x L}{I_y} b_c - R_p + \alpha \cdot \sigma_B < 0 \end{aligned} \tag{9.18}$$

9.2.3.3 Lateral-Torsional Buckling

A cantilever beam subject to bending can show a form of instability called Lateral-Torsional Buckling (LTB). The beam twists and bends in the plane with smaller moment of inertia moving therefore laterally. In Fig. 9.4 the deformation of a cantilever I-beam loaded at the free end is shown.

The formula for computing the load which causes the LTB[1] limits the maximum bending moment that can be applied on the beam. This formula, reported in [52, 264], is a generalisation of the equations given in [15, 44] and it is proved to be sufficiently accurate for cross sections with double or single symmetry, under a great variety of end constraints and load conditions:

$$M_{LTB} = C_1 \frac{\pi^2 E I_y}{L_0^2} \left[C_2 g + C_3 j + \sqrt{(C_2 g + C_3 j)^2 + \frac{C_w}{I_y} + \frac{G J L_0^2}{\pi^2 E I_y}} \right] \tag{9.19}$$

where

- C_1, C_2, C_3 are coefficients depending on load conditions and end constraints,
- L_0 is the effective unbraced length,
- g is the distance between the loading point and the shear centre (positive if the load is below the shear centre).

$$j = e + \frac{1}{2I_x} \sum_A y \left(x^2 + y^2\right) dA \tag{9.20}$$

where e is the distance between the shear centre and the c.g. of the cross section, positive if the shear centre lies in the compressed part of the cross section; the co-ordinates x and y are centred in the c.g. of the cross section, see Fig. 9.3; G is the shear modulus. The values for C_1, C_2 and C_3 are reported in [52, 264] and they read

$$C_1 = 1.3 \quad C_2 = 0.55 \quad C_3 = 2.5 \tag{9.21}$$

For a cantilever beam the effective unbraced length is equal to

$$L_0 = 2L \tag{9.22}$$

The shear modulus G is a function of Young's modulus E and of Poisson's coefficient ν (equal to 0.3),

$$G = \frac{E}{2(1+\nu)} \tag{9.23}$$

By substituting the above values in Eq. 9.19

[1] Many researches have been carried out on this particular type of instability [44, 52, 264]. It has been shown that the main factors governing the phenomenon are the moment of inertia with respect to lateral bending I_y, the torsional moment of inertia J and the warping constant C_w, see Eq. 9.24. These quantities describe the stiffness of the section with respect to lateral bending and torsion. Remarkable importance has also the position of the shear centre and the point of application of the vertical load. These factors influence the entity of the twisting action when the buckling begins.

$$M_{LTB} = 3.21\frac{EI_y}{L^2}\left[0.55g + 2.5j + \sqrt{(0.55g + 6.25j)^2 + \frac{C_w}{I_y} + 0.156\frac{JL^2}{I_y}}\right] \tag{9.24}$$

This formula gives the value of the bending moment that causes instability in the beam subject to mono-axial bending (caused by the force P_y). If the force P_x is the only applied load, it causes a bending moment in the direction of the smaller cross-sectional inertia moment. No instability can be found and the beam bends until its failure. Therefore, if $I_x > I_y$ this instability cannot occur and the constraint is not applied.

In presence of a combination of the two bending moments (M_x, M_y) the following formula, reported in [15, 199, 264] can be used

$$\frac{M_x}{M_{xc}} + \frac{M_y}{M_{yc}} < 1 \tag{9.25}$$

where M_{xc} and M_{yc} are the maximum sustainable bending moments in the two planes. We can suppose that the maximum bending moment in the y-direction is limited by the buckling instability (as shown before). The maximum bending moment in the x-direction is instead limited by the maximum stress at the fixed end. Therefore Eq. 9.25 can be written as

$$\frac{P_x L}{M_{break_x}} + \frac{P_y L}{M_{LTB}} - 1 < 0 \tag{9.26}$$

where $M_{break_x} = \frac{R_p I_y}{b_{max}}$ and $b_{max} = max\,(b_c, b_t)$ (symbols are described in Fig. 9.3)

The standard deviation of the constraint has the following expression (Eq. 9.3)

$$\sigma_{LTB} = \sqrt{\left(\frac{\sigma_{P_x} L}{M_{break_x}}\right)^2 + \left(\frac{P_y L}{M_{LTB}}\right)^2 \cdot \left[\left(\frac{\sigma_{P_y}}{P_y}\right)^2 + \left(\frac{\sigma_E}{E}\right)^2\right]} \tag{9.27}$$

The stochastic constraint functions, by considering that P_x has zero mean value, is therefore

$$\frac{P_y L}{M_{LTB}} - 1 + \alpha \cdot \sigma_{LTB} < 0 \tag{9.28}$$

9.2.3.4 Local Buckling of the Beam Cross Section

A second type of instability to which the beam can be subject is a local instability which involves the compressed elements of the cross section. These elements can buckle in presence of a compressive load. We must therefore verify a local buckling constraint for every compressed element in the cross section.

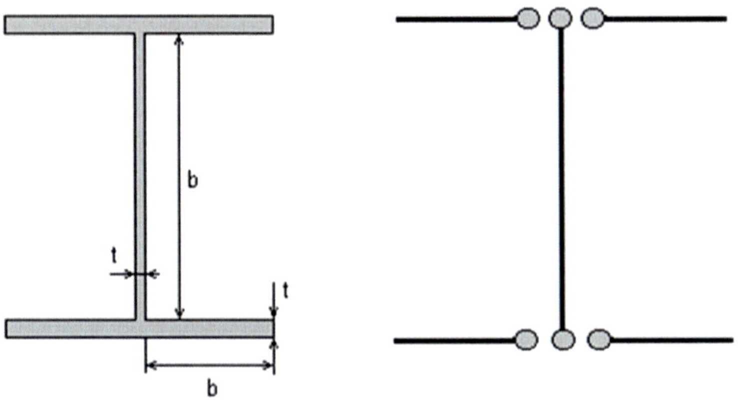

Fig. 9.5 Cross section of an I-beam and its elements (plates) with free or simply supported end

The constraints limit the maximum compressive load in the element under consideration. The element is considered as a compressed plate, isolated from the remaining elements of the cross section. It is assumed [15, 44] that the plate is connected by hinge joints to other plates of the cross section, see Fig. 9.5.

The ultimate compressive load is a function of the thick-length ratio of the plate $\frac{t}{b}$, of the elastic modulus and of a coefficient k (values reported in Table 9.3) that depends on the loading condition and constraints, see [264]

$$S_{C_{max}} = \frac{k\pi^2 E}{12\left(1-\nu^2\right)} \cdot \left(\frac{t}{b}\right)^2 \cong 0.904 \cdot k \cdot E \left(\frac{t}{b}\right)^2 \tag{9.29}$$

The maximum compressive stress acting on the element of the cross section under consideration has the general expression:

$$S_c = \frac{P_x L}{I_y} \cdot d_x + \frac{P_y L}{I_x} \cdot d_y \tag{9.30}$$

The constraint is

$$\frac{P_x L}{I_y} \cdot d_x + \frac{P_y L}{I_x} \cdot d_y - 0.904 \cdot k \cdot E \cdot \left(\frac{t}{b}\right)^2 < 0 \tag{9.31}$$

The standard deviation can be computed as

$$\sigma_{local\ buckling} = \sqrt{\left(\frac{\sigma_{P_x} L}{I_y} \cdot d_x\right)^2 + \left(\frac{\sigma_{P_y} L}{I_x} \cdot d_y\right)^2 + \left(0.904 \cdot k \cdot \sigma_E \cdot \left(\frac{t}{b}\right)^2\right)^2} \tag{9.32}$$

Table 9.3 Optimisation problem statement

	Unloaded edges, simply supported	Top edge free, bottom edge simply supported	Top edge simply supported, bottom edge free
$\sigma_{bottom} = -\sigma_{top}, \sigma_{top} < 0$	23.9	0.85	–
$\sigma_{bottom} = 0, \sigma_{top} < 0$	7.8	0.57	1.70
$\sigma_{bottom} = \sigma_{top}, \sigma_{top} < 0$	4.0	0.42	0.42

Table 9.4 Optimisation problem statement

Name	Category	Reference
$x_1, y_1, p_2, p_3, q_2, q_3$	Deterministic design variable	Table 9.2
P_x, P_y, R_p, E	Stochastic parameters	Table 9.1
Mass	Deterministic objective function	Eq. 9.7
Vertical deflection	Stochastic objective function	Eq. 9.10
Geometrical constraints	Deterministic constraints	Eq. 9.6
Horizontal deflection	Stochastic constraints	Eq. 9.13
Static resistance	Stochastic constraints	Eq. 9.18
Lateral-torsional buckling	Stochastic constraint	Eq. 9.28
Local buckling	Stochastic constraints	Eq. 9.33

The stochastic constraint functions, see Eq. 9.4, is therefore

$$\frac{P_x L}{I_y} \cdot d_x + \frac{P_y L}{I_x} \cdot d_y - 0.904 \cdot k \cdot E \cdot \left(\frac{t}{b}\right)^2 + \alpha \cdot \sigma_{local\ buckling} < 0 \qquad (9.33)$$

The formulated optimisation problem is summarised in Table 9.4. The local buckling constraint must be verified for every compressed element in the cross section.

9.3 Optimal Design of Cantilever Beams

By applying the method described in Sect. 9.2 the cross section of the cantilever beam has been optimised. Six different cross section patterns have been considered, namely, rectangular solid, T-shaped, I-shaped, C-shaped, rectangular hollow (indicated by O) and rectangular hollow with two cells. The maximum horizontal deflection $\delta_{x\ max}$ has been limited to 10 mm. More than 100 Pareto-optimal solutions have been obtained for each type of cross section. The probability risk values β_i for all the constraints and objective functions (see Eqs. 9.2, 9.4) have been chosen equal to 99%. There is

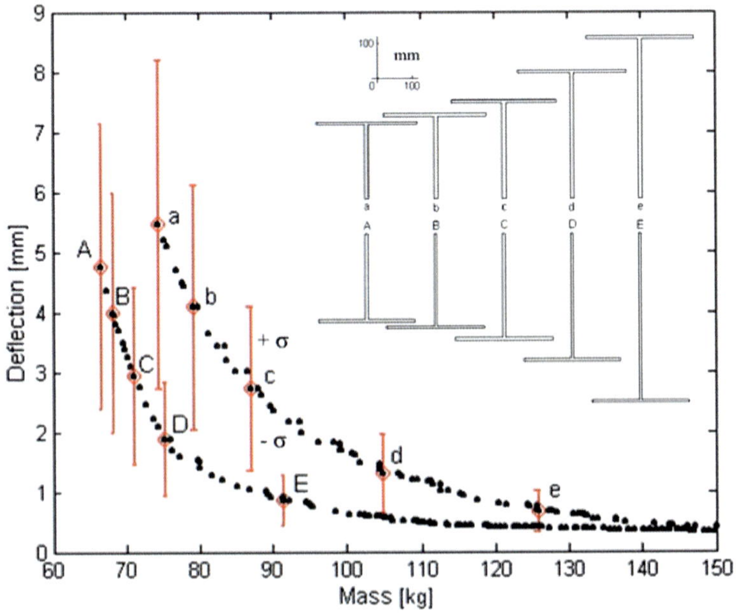

Fig. 9.6 Pareto-optimal sets in the objective functions domain (deflection, mass). T-shaped steel beam cross sections with wing at the top (a, b, c, d, e) and with wing at the bottom (A, B, C, D, E), see Table 4. The variation of each objective function (due to the random variation of P_y, P_x, E, R_p listed in Table 9.1) is reported (range $\pm\sigma$)

therefore a probability of 1% that the optimal solutions will not satisfy a constraint. There is also a probability of 1% that the performance of the optimal solutions will be worse than the expected value.

A detailed analysis of the T-shaped steel cross-sectional beam will be reported. In Fig. 9.6 the expected values (mean) and the variations of the objective functions for the (optimised) T-shaped cross sections are shown. The dimensions of the selected sections are reported in Table 9.5. It is clear that, the higher is the mass of the beam, the more "robust" is the design of the cantilever. It is also evident that there is a remarkable performance variation depending on the upper or lower position of the wing. Figure 9.6 shows that always the best wing position is at the bottom. This can be explained by the following considerations.

When the wing is at the top, the critical part of the cross section is the web which is subject to compression and it must be thicker to avoid buckling. When the wing is at the bottom, the vertical element of the section (web) can be thinner because it is subject mainly to tensile stress and local buckling does not occur. The increase in vertical stiffness is reached mainly by increasing the height and much less effectively by increasing the overall width of the cross section (the other dimensions being kept nearly constant).

Table 9.5 Dimensions in (mm) of the sections a, b, c, d, e, A, B, C, D, E in Fig. 9.6

	A	B	C	D	E	a	b	c	d	e
Overall width	281	285	285	283.5	284.5	293.5	300	309	319	315
Overall height	249.5	268	302	362	479.5	219.5	243.5	281	365	461.5
Web thickness	8.6	8.3	8.0	7.5	7.9	11.2	11.2	11.3	11.5	11.6
Wing thickness	8.0	8.0	7.9	7.8	7.7	8.4	8.3	8.3	8.5	9.3

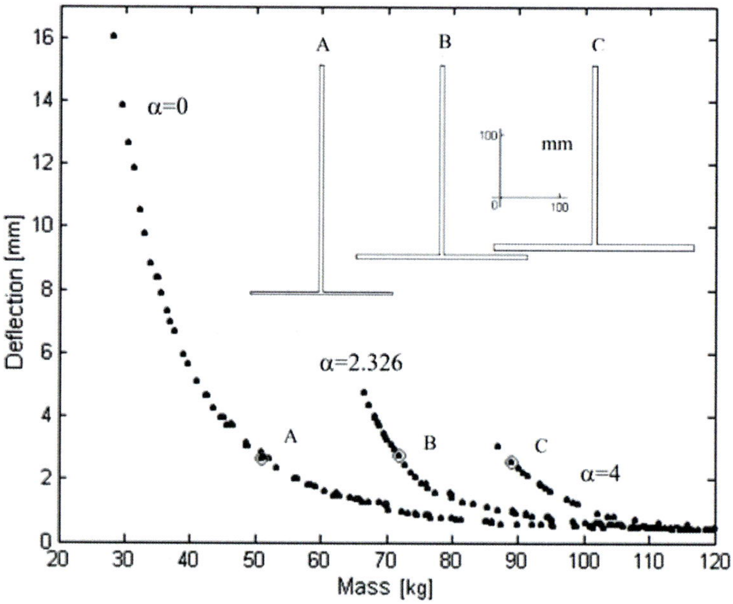

Fig. 9.7 T-shaped cross section of steel beams. Pareto-optimal sets computed at different percentiles ($\alpha = 0, 2.326, 4$). Pareto-optimal solutions A, B, C (see Table 9.6) have approximately the same deflection (3 mm) but the beam designs C and B are far more robust than A

By varying the values of the probabilities β_i, and hence the value of the percentiles α_i, of the objective functions and constraints it is possible to observe that the result of the optimisation changes in a substantial way. In order to show the importance of a robust design approach, Fig. 9.7 shows the optimal curves relative to expected (mean) values of objective functions for T-shaped steel cross section with the wing at the bottom for three different values of the percentiles: 0, 2.326, 4.

Table 9.6 Dimensions in (mm) of the sections a, b, c, d, e, A, B, C, D, E in Fig. 9.7

	A	B	C
Overall width	236	285	334.5
Overall height	368	311	296
Web thickness	7.0	8.0	9.2
Wing thickness	3.1	7.9	9.4

When the percentiles α_i are zero, the stochastic variations of parameters listed in Table 9.1 are neglected and therefore a deterministic optimisation is carried out. A value of $\alpha = 2.326$ corresponds to the probability $\beta_i = 99\%$ that all constraints are satisfied and the expected values of the objective functions are not exceeded in spite of the random variations of parameters (and/or design variables). By increasing the value of percentiles there is a higher safety margin against this stochastic variations. For $\alpha = 4$, there is a probability equal to 0.003% that the constraints and expected values of the objective functions are not satisfied. In Fig. 9.7, three different cross sections (A, B, C) are considered (dimensions reported in Table 9.6). The selected cantilevers (A, B, C) have approximately the same vertical deflection (3mm). The shape of the cross section A turns out to be very slender (less "robust") mainly because the horizontal force is not taken into consideration. By increasing the percentiles α_i, the thickness of all of the elements of the cross section is increased, in order to guarantee the satisfaction of all the design constraints.

In Fig. 9.8, the Pareto-optimal sets related to steel cantilever beams with different types of cross sections (I- C- T- rectangular-solid- and rectangular hollow) are presented and compared. The cross sections that constitute the resulting Pareto-optimal set (I- and rectangular hollow) are the best ones in terms of deflection and mass (see [144, 145] for details on the comparison of Pareto-optimal sets for different design configurations).

In Fig. 9.9 a comparison is shown between the optimal cross sections of cantilever beams made from steel or aluminium alloy. Again the percentile was set equal to $\alpha = 2.326$. Two Pareto-optimal solutions for each material (A, B for aluminium alloy, a, b for steel) have been chosen which show approximately equal deflection. The dimensions of the selected solutions are reported in Table 9.7.

On the contrary, if an additional design constraint is given, e.g. on the overall height of the beam cross section, then steel might be more convenient. Actually the overall height of steel beam cross sections a, b in Table 9.7 are substantially lower that those referring to aluminium alloy.

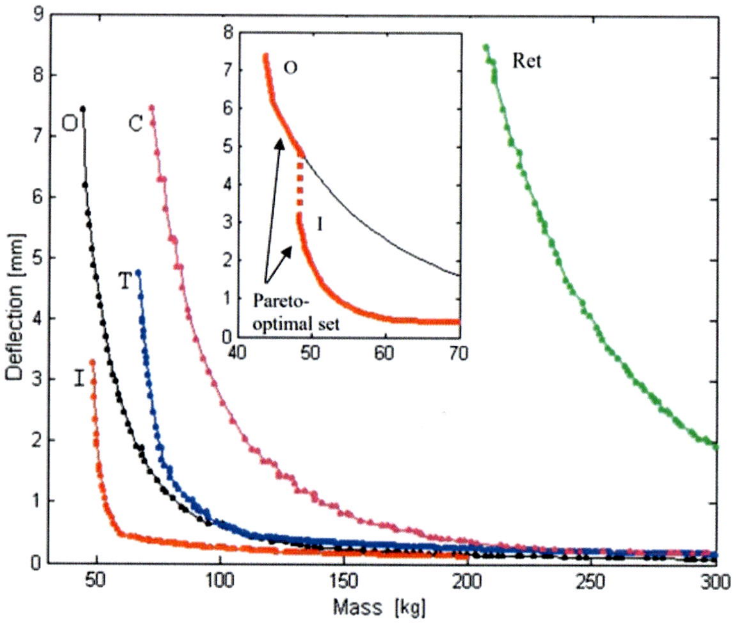

Fig. 9.8 I-, T-, C- ret- (rectangular solid) O- (rectangular hollow) steel cross sections. Pareto-optimal sets in the objective function domain. The Pareto-optimal set is denoted by the thick solid line in the zoomed plot inside the main plot. Notice that the Pareto-optimal set is not continuous (see the dotted line)

Figure 9.9 shows also (with respect to Fig. 9.8) that the considerations on the type of cross section are quite general and almost not dependent on the type of material. The robustness of the solution is higher for I-shaped beams than for O (rectangular hollow) shaped beams (see variations in Fig. 9.9). The relative influence of the uncertainties on the parameters and loads (given in Table 9.1) on the variation of the objective functions and constraints for steel beams is reported in Table 9.8, by considering the solutions belonging to the Pareto-optimal set reported in Fig. 9.8. The strong influence of the applied loads shown in Table 9.8 is mainly due to their large standard deviation reported in Table 9.1.

9.4 Summary

A multi-objective optimisation method suitable to deal with stochastic systems has been introduced and applied in this chapter. The method allows to minimise a vector of objective functions as function of a vector of design variables, and additionally it allows to limit the influence of stochastic variations of the system external disturbances and/or system parameters and/or design variables. Stochastic design

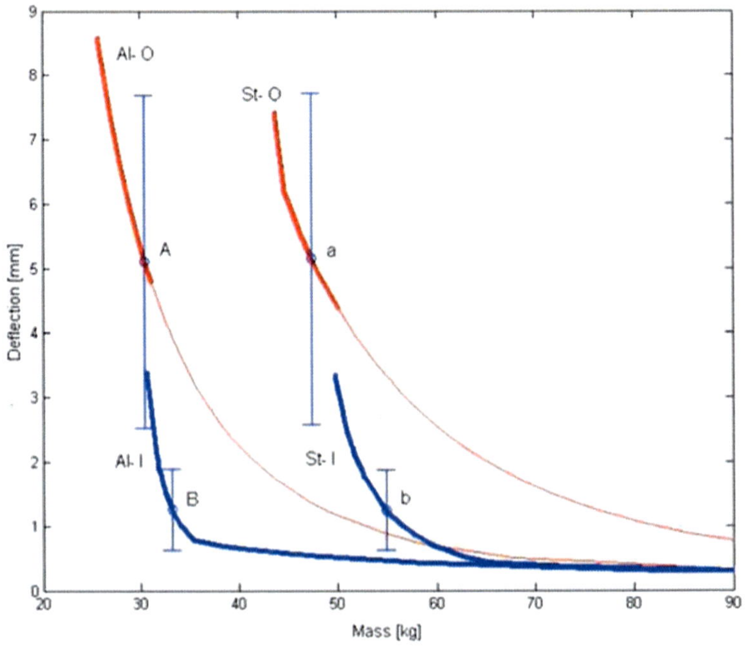

Fig. 9.9 Comparison between aluminium alloy (Al) and steel (St) beams. Pareto-optimal sets for I-, O- (rectangular hollow) cross sections in the objective function domain. Two couples of optimal solutions (A,a and B,b) with approximately the same deflection have been chosen (see Table 9.7). The variation of each objective function is reported (range $\pm\sigma$). The Pareto-optimal sets are denoted by the thick solid lines

Table 9.7 Dimensions in (mm) of the cross sections A, a, B, b in Fig. 9.9

	A	a	B	b
Overall width	263.5	204.5	315	249.5
Overall height	312	247	475	374
Web thickness	6.1	4.3	2.1	1.9
Top wing thickness	2.3	1.7	6.8	4.8
Bottom wing thickness	5	3.0	9.7	6.7

Table 9.8 Relative importance of the parameters and loads variations on the standard deviations of objectives functions and constraints for steel beams. The percentages shown are the mean values on the Pareto-optimal I- and O- cross sections shown in Fig. 9.9. The values of the standard deviations of the parameters are reported in Table 9.1

Objective functions & constraints	I-beam				O-beam			
	σ_{P_y}	σ_{P_x}	σ_E	σ_{R_p}	σ_{P_y}	σ_{P_x}	σ_E	σ_{R_p}
Vertical deflection	50	–	50	–	50	–	50	–
Horizontal deflection	–	100	–	–	–	100	–	–
Tensile stress	20	70	–	10	50	40	–	10
Compressive stress	15	75	–	10	45	45	–	10
LTB	15	85	–	–	2	98	–	–
Web buckling	85	–	15	–	45	45	10	–
Top wing buckling	20	70	10	–	50	40	10	–
Bottom wing buckling	15	75	10	–	45	45	10	–

constraints have been also considered and do represent the truly new contribution to stochastic optimisation theory introduced in this chapter. The remarkable (and novel) feature of the multi-objective optimisation method is the managing of uncertain system parameters and external disturbances introduced by means of a gaussian stochastic process. The optimisation method allows to quantify the "robustness" of the system, i.e. the degradation of system performances due to the uncontrollable variation of stochastic parameters and/or design variables and/or external disturbances. The optimisation method has been applied to find the optimal shape of the cross section of a cantilever beam. The beam was loaded at the free end by a vertical force whose amplitude was subject to stochastic variations of given distribution. It has been assumed that the applied load did not have a perfect vertical direction. The limit strength and the elastic modulus of the two materials under consideration (steel and aluminium alloy) were subject to uncertainty. The mass and the vertical deflection of the free end of the beam had to be minimised, provided that the maximum stress at the fixed end was less/equal to the admissible stress and that the elastic stability (local/global stability) was attained. The beam cross-sectional design according to a purely deterministic optimisation procedure (bare multi-objective programming theory working within a deterministic framework) may make the beam to become totally unreliable, due to the stochastic variation of loads, and system parameters such as material characteristic. The Pareto-optimal beam cross sections have been determined in a stochastic framework. The optimal solutions were derived by giving the probability of attaining a defined performance on each objective. The Pareto-optimal solutions for the cantilever beam problem show that only two cross sections (rectangular hollow (O) and I-shaped) are actually Pareto-optimal both in a deterministic and in a stochastic framework. This is true both for aluminium alloy and for steel cantilever beams. Rectangular hollow sections (O-shaped) should be chosen to obtain the lower mass. Higher stiffness but higher mass is typical of I-shaped cross sections which however are more robust than O-shaped cross section beams. So it has been rigorously proved that higher mass means higher robustness, at least in the considered case. This fact is consistent with common experience and empirical approach which relies on heavy structures to obtain robustness. The novel contribution has been that of defining how much heavy a cantilever beam should be to attain the required robustness, given the requested beam stiffness. Aluminium alloy cantilever beams perform always better than steel ones in terms of both high stiffness and low mass. However the overall dimension of the optimal aluminium alloy beams is higher than that of the optimal steel ones. This conclusion is consistent with what has been stated in the previous chapters. If the room is constrained, and proper stiffness is required, then steel is slightly better than aluminium.

Chapter 10
Multi-objective Optimisation of Truss Structures

The following chapter is derived from [128] and adapted to this book.

In structural design, a well-known problem is the minimisation of the mass of a truss structure under a single loading condition subject to stress and compliance (i.e. stiffness) constraints or, alternatively, the minimisation of the compliance with stress and mass constraints [29, 195].

In the following instead, we will deal with the more "natural" approach; the simultaneous minimisation of mass and compliance by considering a multi-objective formulation. The design variables are the volumes (or the cross sections) of the truss bars and, in general, the member forces (or the node displacements) cannot be statically determined. In Sect. 10.2, we will also deal with the multiple loading conditions.

10.1 Truss Structures Under Single Loading Conditions: Problem Formulation

A generic truss structure with N nodes and s constraint conditions (i.e. fixed nodal displacement) has n degrees of freedom

$$n = N \cdot dim - s \tag{10.1}$$

where dim is the dimension of the design space (2–3 dimensions). Let $\mathbf{x} \in R^m$ be the vector of design variables describing the volumes of the m bars of the truss structure such that $x_i = A_i l_i \geq 0$ with A_i and l_i the cross section and the length of the ith bar, respectively.

Let $\mathbf{f} \in R^n$ be the vector of external nodal loads, the nodal equilibrium can be written as

F. M. Ballo et al., *Optimal Lightweight Construction Principles*,
https://doi.org/10.1007/978-3-030-60835-4_10

$$\mathbf{Dq} = \mathbf{f} \tag{10.2}$$

where $\mathbf{q} \in R^m$ is the vector of the member forces, and $\mathbf{D} \in R^{nxm}$ is the geometric matrix whose column $\mathbf{d_i} \in R^n$ contains the direction cosines of the ith bar. Hence the vector $\mathbf{d_i}$ describes the compatibility condition between the nodal displacement $\mathbf{u} \in R^n$ and the elongation of the ith bar $\Delta l_i \in R$

$$\mathbf{d_i}^T\mathbf{u} = \Delta l_i \rightarrow \mathbf{D}^T\mathbf{u} = \Delta \mathbf{l} \tag{10.3}$$

where $\Delta \mathbf{l} \in R^m$ is the vector containing all the bar elongations. By assuming a linear elastic behaviour, the element stiffness matrix per unit of volume of the ith bar is

$$\mathbf{K_i} = \frac{E_i}{l_i}\mathbf{d_i}\mathbf{d_i}^T \in R^{nxn} \tag{10.4}$$

where E_i is the elastic modulus of the ith bar. Hence the global (symmetric) stiffness matrix is

$$\mathbf{K}(\mathbf{x}) = \sum_{i=1}^{m} x_i \mathbf{K_i} \in R^{nxn} \tag{10.5}$$

and the resulting well-known condition of elastic equilibrium is

$$\mathbf{K}(\mathbf{x})\,\mathbf{u} = \mathbf{f} \tag{10.6}$$

The multi-objective structural optimisation problem can be stated as

$$\begin{aligned} &\min\,(mass,\ compliance) \Longleftrightarrow \min_{\mathbf{x}\in R^m, \mathbf{u}\in R^n} \left(\sum_{i=1}^{m} \rho_i x_i,\ \ \frac{1}{2}\mathbf{f}^T\mathbf{u} \right) \\ &such\ that \quad \mathbf{K}(\mathbf{x})\,\mathbf{u} = \mathbf{f}, \quad \mathbf{x} \geq \mathbf{0} \end{aligned} \tag{10.7}$$

where ρ_i is the density (or the cost per unit of volume) of the ith bar. The first objective is the mass (or the cost) of the structure. The second is the (elastic) energy stored into the structure, an index of the compliance. The minimisation of the compliance implies the maximisation of the stiffness. Since compliance and strain energy are completely equivalent, the problem (10.7) can be rewritten in the following form:

$$\begin{aligned} &\min_{\mathbf{x},\mathbf{q}\in R^m} \left(\sum_{i=1}^{m} \rho_i x_i,\ \ \frac{1}{2}\sum_{i:x_i>0} \frac{l_i^2 q_i^2}{E_i x_i} \right) \\ &such\ that \quad \mathbf{Dq}, \quad q_i = 0 \forall i : x_i = 0, \quad \mathbf{x} \geq 0 \end{aligned} \tag{10.8}$$

Problems (10.7), (10.8) do not include stress constraints because we are not interested in minimal weight design only but in trade-off solutions between mass and compliance.

10.1.1 Optimal Design

To derive the solution of the optimization problem stated in Sect. 10.1, the *L-matrix condition* described in Chap. 3 can be exploited.

By exploiting the equivalence between compliance and potential energy [195] problem (10.7) can be rewritten as

$$\min_{\mathbf{u}\in R^n}\left(\frac{1}{2}\mathbf{f}^T\mathbf{u}\quad|\,\mathbf{K}\left(\mathbf{x}\right)\mathbf{u}=\mathbf{f}\right)=\max_{\mathbf{u}\in R^n}\left(\mathbf{f}^T\mathbf{u}-\frac{1}{2}\mathbf{u}^T\mathbf{K}\left(\mathbf{x}\right)\mathbf{u}\right) \tag{10.9}$$

in which the function to be maximised on the right of Eq. 10.9 represents the potential energy of the truss structure with sign changed. By substituting Eq. 10.9 in Eq. 10.7, we can derive the following unconstrained bi-objective problem equivalent to problems (10.7) and (10.8),

$$\min_{\mathbf{x}\in R^m}\left(\sum_{i=1}^{m}\rho_i x_i,\quad \max_{\mathbf{u}\in R^n}\left(\mathbf{f}^T\mathbf{u}-\frac{1}{2}\mathbf{u}^T\mathbf{K}\left(\mathbf{x}\right)\mathbf{u}\right)\right) \tag{10.10}$$

in which the two objective functions depend only on the vector $\mathbf{x}$

$$\begin{aligned} f_1\left(\mathbf{x}\right) &= \sum_{i=1}^{m}\rho_i x_i \\ f_2\left(\mathbf{x}\right) &= \max_{\mathbf{u}\in R^n}\left(\mathbf{f}^T\mathbf{u}-\frac{1}{2}\mathbf{u}^T\mathbf{K}\left(\mathbf{x}\right)\mathbf{u}\right) \end{aligned} \tag{10.11}$$

The problem has now been converted into a bi-objective unconstrained minimisation problem. Referring to the theory reported in Sect. 3.3, being the number of design variables greater than the number of objective functions ($n > k$), Eq. 3.11 applies.

It is possible to demonstrate (the proof is reported in [128]) that in the particular case of a bi-objective unconstrained minimisation problem Eq. 3.11 reduces to

$$det\left(\mathbf{L}^T\mathbf{L}\right)=0\rightarrow\frac{\partial f_1}{\partial x_i}\frac{\partial f_2}{\partial x_j}=\frac{\partial f_1}{\partial x_j}\frac{\partial f_2}{\partial x_i}\quad\forall i,j \tag{10.12}$$

The derivatives of the objective functions read

$$
\begin{aligned}
\frac{\partial f_1}{\partial x_i} &= \rho_i \\
\frac{\partial f_2}{\partial x_i} &= -\frac{1}{2}\mathbf{u}^T\mathbf{K}_i\mathbf{u} = -\frac{1}{2}E_i\frac{\mathbf{u}^T\mathbf{d}_i}{l_i}\cdot\frac{\mathbf{d}_i^T\mathbf{u}}{l_i} = -\frac{1}{2}E_i\epsilon_i\epsilon_i = -\frac{1}{2}\sigma_i\epsilon_i
\end{aligned}
\tag{10.13}
$$

where ϵ_i in Eq. 10.13 represents the axial strain acting on each bar.

By inspection of Eq. 10.13, we observe that the components of the gradient of first objective function are strictly positive while the ones of the second objective function are null or negative. This means that the second-order sufficiency conditions described in Chap. 3 are satisfied and solution of Eq. 10.11 is Pareto-optimal.

By applying Eq. 10.12, we obtain

$$
\frac{\sigma_i\epsilon_i}{\rho_i} = \frac{\sigma_j\epsilon_j}{\rho_j} \quad \forall i, j \quad \rightarrow \quad \frac{\sigma_i\epsilon_i}{\rho_i} = C = const \quad \forall i : x_i \neq 0 \tag{10.14}
$$

The relations (10.14) show that in the optimal structure, the density of energy per unit of mass is constant throughout the bars that constitute the structure itself: this is the optimality condition for multi-material structures [200]. If the bars of the truss are made of the same material (i.e. $E_i \equiv E,\ \rho_i \equiv \rho,\ \forall i$) the relations (10.14) become

$$
|\sigma_i| = const \quad \forall i : x_i \neq 0 \tag{10.15}
$$

which is the well-known condition of fully stressed design [194].

By means of the optimality conditions (10.14) it is possible to establish a relation between the design variables x_i and the member forces q_i

$$
x_i = \frac{|q_i| l_i}{\sqrt{C\rho_i E_i}} \tag{10.16}
$$

By substituting the expression (10.16) in (10.8) the following problem can be considered

$$
\begin{aligned}
&\min_{\mathbf{q}\in R^m, C\in R^+} \left(\frac{1}{\sqrt{C}\cdot\Psi(\mathbf{q})},\ \frac{1}{2}\sqrt{C}\cdot\Psi(\mathbf{q})\right) \\
&with \quad \Psi(\mathbf{q}) = \sum_{i=1}^{m}\sqrt{\frac{\rho_i}{E_i}}\cdot l_i|q_i|, \quad \mathbf{D}\mathbf{q} = \mathbf{f}
\end{aligned}
\tag{10.17}
$$

which, for constant C, becomes the single-objective problem

$$
\begin{aligned}
&\min_{\mathbf{q}\in R^m} \Psi(\mathbf{q}) \\
&with \quad \mathbf{D}\mathbf{q} = \mathbf{f}
\end{aligned}
\tag{10.18}
$$

Problem (10.18) can be transformed into a linear programming problem [219] by introducing of the variables $\mathbf{q}^+, \mathbf{q}^- > 0$ such that

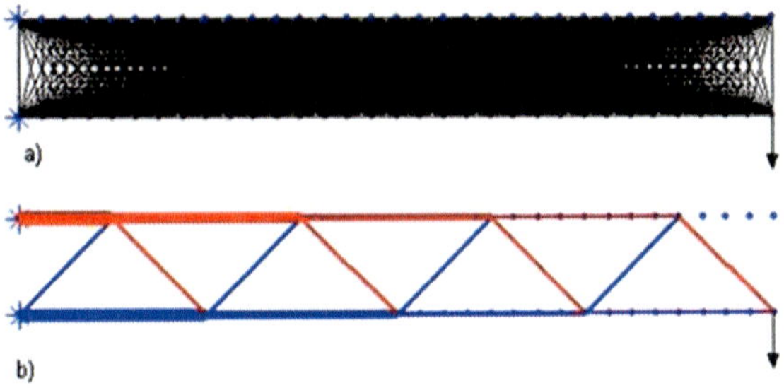

Fig. 10.1 Optimisation of a 2D cantilever truss: in **a** the 66 nodes and the 1153 possible bars, in **b** the solution of the problem (10.18), adapted from [128]

$$\begin{aligned} \mathbf{q}^{+} - \mathbf{q}^{-} = \mathbf{q} \rightarrow \mathbf{q}^{+} = \max(\mathbf{q}, 0) \\ \mathbf{q}^{+} + \mathbf{q}^{-} = |\mathbf{q}| \rightarrow \mathbf{q}^{-} = \min(0, \mathbf{q}) \end{aligned} \tag{10.19}$$

The problem of the simultaneous optimisation of topology and size of a truss structure subject to a single loading condition can therefore be separated in two steps: first we have to solve the topology single-objective problem (10.18) in the variables q_i and then we obtain the Pareto-optimal set by varying the density of energy C (sizing problem). C is related to the cross section of the bars, see Eq. 10.16. This means that, from a topological point of view, mass and stiffness are not conflicting, i.e. the optimal topology for mass is also optimal for stiffness.

As an example, Fig. 10.1 shows the optimisation of a 2D cantilever truss (8×1 [m]) with two rows of 33 nodes each. The nodes on the left are fixed to the ground and a vertical force of 1 kN is applied at the bottom node on the right. The 66 nodes and the 1153 possible bars (except the overlapping ones) are shown in Figure. Figure 10.1a; all the bars are made by steel ($E = 206000$ MPa, $\rho = 7700\,\text{kg/m}^3$). The solution of problem (10.18) is shown in Fig. 10.1b; it consists in 16 bars with the interior ones perpendicular each others.

The Pareto-optimal solutions of problem (10.17) for the cantilever arm are shown on the left of Fig. 10.2: as expected the Pareto-optimal set in the objective domain is hyperbolic. Three different solutions with different values of maximum stress are reported on the right of Fig. 10.2, see Eq. 10.15. Finally, we underline that the latter consideration holds also if it is possible to perform a geometry optimisation, i.e. varying the nodal position by means of the variables $\mathbf{y} \in R^{n \cdot dim}$. The sizing multi-objective problem (10.17) remains actually the same while the topology problem (10.18) becomes the following topology and geometry problem

$$\begin{aligned} \min_{\mathbf{q} \in R^m, \mathbf{y} \in R^{n \cdot dim}} \quad & \Psi(\mathbf{q}) \\ \textit{with} \quad & \mathbf{D}(\mathbf{y})\,\mathbf{q} = \mathbf{f} \end{aligned} \tag{10.20}$$

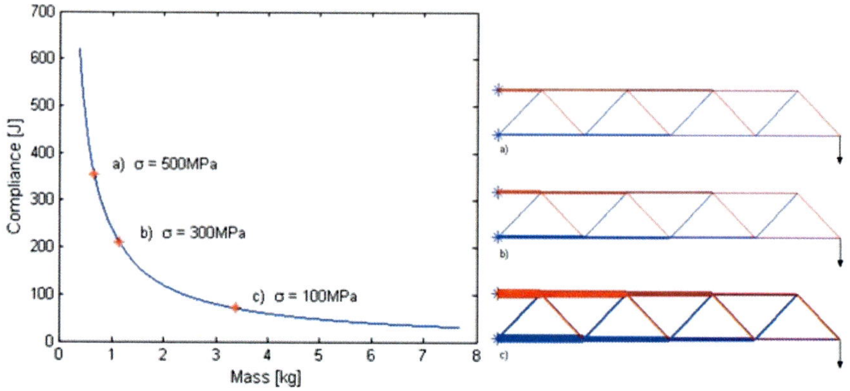

Fig. 10.2 Pareto-optimal set in the objective domain for the cantilever optimisation problem. The maximum stress values (uniform in all the bars) for the three solutions highlighted are **a** 500, **b** 300 and **c** 100 Mpa, respectively. Adapted from [128]

10.2 Truss Structures Under Multiple Loading Conditions: Problem Formulation

The optimisation of truss structures under single load case can be generalised to multiple load cases [1]. The problem is addressed with the worst case approach, where the objective functions are the mass of the truss and the maximum compliance by considering a combination of all the load cases. In this way it is possible to formulate a bi-objective unconstrained problem that can be solved by means of the condition (10.12).

Let $\mathbf{f}_1, \ldots, \mathbf{f}_p \in R^n$ be the vectors of the p external independent load cases that can be applied to the structure in any convex linear combination and let $\delta \in R^p$ be the vector of the weights of this linear combination, the effective loading condition that can be applied on the structure is of the following type:

$$\mathbf{f} = \sum_{k=1}^{p} \delta_k^2 \mathbf{f}_k \quad with \quad \sum_{k=1}^{p} \delta_k^2 = 1 \tag{10.21}$$

The elastic equilibrium for each of the p independent load cases can be expressed as

$$\mathbf{K}(\mathbf{x})\, \mathbf{u}_k = \mathbf{f}_k \quad for\ k = 1, \ldots, p \tag{10.22}$$

where $\mathbf{u}_k \in R^n$ is the vector of nodal displacement for the kth load case whose compliance is therefore $\mathbf{f}_k \mathbf{u}_k$. By considering Eq. 10.21, the compliance J for a generic load case is

$$J(\delta) = \sum_{k=1}^{p} \delta_k^2 \mathbf{f}_k^T \mathbf{u}_k \quad with \quad \sum_{k=1}^{p} \delta_k^2 = 1 \tag{10.23}$$

and the multi-objective optimisation problem (Eq. 10.7) with the worst case approach can be stated as

$$\min_{\mathbf{x}\in R^m,\mathbf{u}\in R^{n\cdot p}} \left(\sum_{i=1}^{m} \rho_i x_i, \quad \max_{\delta\in R^p} \frac{1}{2} \sum_{k=1}^{p} \delta_k^2 \mathbf{f}_k^T \mathbf{u}_k \right)$$

$$\textit{such that} \quad \mathbf{K}(\mathbf{x})\,\mathbf{u}_k = \mathbf{f}_k \quad \forall k, \quad \mathbf{x} \geq 0 \tag{10.24}$$

$$\sum_{k=1}^{p} \delta_k^2 = 1$$

The first objective is the mass (or the cost) of the structure and the second represents half of the worst case (maximum) compliance of the structure by considering all the possible load conditions. The optimal value of the vector δ^2 corresponds to the most critical load condition of Eq. 10.21 since it has the maximum compliance.

By means of the equivalence between compliance and strain energy, problem (10.24) can be transformed into the following:

$$\min_{\mathbf{x}\in R^m,\mathbf{q}\in R^{mxp}} \left(\sum_{i=1}^{m} \rho_i x_i, \quad \max_{\delta\in R^p} \frac{1}{2} \sum_{k=1}^{p} \left(\delta_k^2 \sum_{i:x_>0} \frac{l_i^2 q_{i,k}^2}{E_i x_i} \right) \right)$$

$$\textit{such that} \quad \mathbf{D}\mathbf{q}_k = \mathbf{f}_k \quad \forall k, \quad q_{i,k} = 0 \ \forall i : x_i = 0, \quad \mathbf{x} \geq 0 \tag{10.25}$$

$$\sum_{k=1}^{p} \delta_k^2 = 1$$

where $\mathbf{q}_k \in R^m$ is the vector of member forces in the kth load condition whose ith element is $q_{i,k}$.

Relations (10.9) and (10.10), applied to the problem (10.24), allow to substitute the strain energy with the potential energy in order to remove the elastic equilibrium constraints

$$\min_{\mathbf{x}\in R^m} \left(\sum_{i=1}^{m} \rho_i x_i, \quad \max_{\mathbf{u}\in R^{nxp}\,\delta\in R^p} \sum_{k=1}^{p} \delta_{\mathbf{k}}{}^2 \left(\mathbf{f}_k^T \mathbf{u}_k - \frac{1}{2} \mathbf{u}_k^T \mathbf{K}(\mathbf{x})\, \mathbf{u}_k \right) \right) \tag{10.26}$$

$$\textit{with} \quad \sum_{k=1}^{p} \delta_k^2 = 1$$

10.2.1 Optimal Design

The second objective function of problem (10.26) can be read as the maximisation in the variables $\mathbf{u} \in R^{nxp}$ and $\delta \in R^p$ of a function parameterised in $\mathbf{x} \in R^m$ subject to an equality constraint. Therefore, for any $\mathbf{x} \in R^m$, it is possible to consider the Lagrangian function $\mathcal{L}(\mathbf{x}, \mathbf{u}, \delta, \lambda)$ and to derive the following stationary conditions with respect to the variables $\mathbf{u}$ and δ and to the multiplier λ

$$\begin{aligned}
&\max_{\mathbf{u}\in R^{nxp},\delta\in R^{p}} \left(\sum_{k=1}^{p} \delta_k^2 \left(\mathbf{f}_k\mathbf{u}_k - \frac{1}{2}\mathbf{u}_k^T\mathbf{K}(\mathbf{x})\,\mathbf{u}_k\right)\right)\\
&\quad with \quad \sum_{k=1}^{p} \delta_k^2 = 1\\
&\quad\downarrow\\
&\mathcal{L}(\mathbf{x},\mathbf{u},\delta,\lambda) = \sum_{k=1}^{p} \delta_k^2 \left(\mathbf{f}_k\mathbf{u}_k - \frac{1}{2}\mathbf{u}_k^T\mathbf{K}(\mathbf{x})\,\mathbf{u}_k\right) + \lambda\left(1 - \sum_{k=1}^{p}\delta_k^2\right)\\
&\begin{cases} \nabla_{\mathbf{u}}\mathcal{L} & = 0\\ \nabla_{\delta}\mathcal{L} & = 0\\ \nabla_{\lambda}\mathcal{L} & = 0\end{cases}
\end{aligned} \tag{10.27}$$

The stationary conditions are

$$\left.\begin{aligned}
&\frac{\partial\mathcal{L}}{\partial\mathbf{u}_k} = 0 \rightarrow \mathbf{K}(\mathbf{x})\mathbf{u}_k = \mathbf{f}_k \quad \forall k\\
&\frac{\partial\mathcal{L}}{\partial\delta_k} = 0 \rightarrow 2\delta_k\left(\mathbf{f}_k^T\mathbf{u}_k - \frac{1}{2}\mathbf{u}_k^T\mathbf{K}(\mathbf{x})\mathbf{u}_k - \lambda\right) = 0 \quad \forall k\\
&\frac{\partial\mathcal{L}}{\partial\lambda} = 0 \rightarrow \sum_{k=1}^{p}\delta_k^2 = 1
\end{aligned}\right\} \rightarrow \delta_k\mathbf{f}_k^T\mathbf{u}_k = 2\delta_k\lambda \quad \forall k \tag{10.28}$$

By substituting the relations (10.28) into the second objective function of problem (10.24) representing half of the maximum compliance between all the possible load conditions acting on the structure, see Eq. 10.23, we have

$$\max_{\delta\in R^p}\frac{1}{2}J(\delta) = \max_{\delta\in R^p}\frac{1}{2}\sum_{k=1}^{p}\delta_k^2\mathbf{f}_k^T\mathbf{u}_k = \max_{\delta\in R^p}\sum_{k=1}^{p}\delta_k^2\lambda = \lambda\cdot\max\sum_{k=1}^{p}\delta_k^2 = \lambda \tag{10.29}$$

Relations (10.28) can therefore be rewritten as

$$\delta_k\mathbf{f}_k^T\mathbf{u}_k = \delta_k\cdot\max_{\delta\in R^p}(J) \rightarrow \begin{cases} \delta_k \neq 0, \quad if \quad \mathbf{f}_k^T\mathbf{u}_k = \max\limits_{1\le j\le p}\left(\mathbf{f}_j^T\mathbf{u}_j\right)\\[2ex] \delta_k = 0, \quad if \quad \mathbf{f}_k^T\mathbf{u}_k < \max\limits_{1\le j\le p}\left(\mathbf{f}_j^T\mathbf{u}_j\right)\end{cases} \tag{10.30}$$

The physical meaning of relations (10.30) is clear: in the worst case approach all the p independent load cases must have the same compliance (or energy) J. If there is any loading condition that cannot have the same (maximum) level of compliance, the corresponding weight δ_k^2 must be equal to zero. In this way, these load cases become not active and have therefore no influence at all on the structure.

On the other hand, for any feasible value of the vector δ, the problem (10.26) can be considered as an unconstrained bi-objective optimisation problem parameterised in δ and, therefore, can be solved by means of the relations (10.12). The derivatives of the objective functions with respect to the design variable x_i are

$$
\begin{aligned}
\frac{\partial f_1}{\partial x_i} &= \rho_i \\
\frac{\partial f_2}{\partial x_i} &= -\frac{1}{2}\sum_{k=1}^{p} \delta_k^2 \mathbf{u}_k^T \mathbf{K}_i \mathbf{u}_k = -\frac{1}{2}\sum_{k=1}^{p} \delta_k^2 \sigma_{i,k}\epsilon_{i,k}
\end{aligned}
\tag{10.31}
$$

We can notice that the derivatives of the first objective function are strictly positive while the ones of the second objective function are null or negative. The solutions of Eq. 10.12 will be therefore Pareto-optimal points

$$
\sum_{k=1}^{p} \delta_k^2 \frac{\sigma_{i,k}\epsilon_{i,k}}{\rho_i} = \sum_{k=1}^{p} \delta_k^2 \frac{\sigma_{j,k}\epsilon_{j,k}}{\rho_j} \quad \forall i,j \quad \rightarrow \quad \sum_{k=1}^{p} \delta_k^2 \frac{\sigma_{i,k}\epsilon_{i,k}}{\rho_i} = C = const \;\; \forall i : x_i \neq 0
\tag{10.32}
$$

where $\sigma_{i,k}$, $\epsilon_{i,k}$ are stress and strain of the ith bar in the kth load case, respectively.

Hence optimality conditions (10.30), (10.32) show that, for an optimal solution, the energy (or compliance) of the structure is the same for each independent active load case (i.e. with $\delta_k \neq 0$) and, moreover, the sum of the density of energy per unit of mass along all the independent active load cases weighted with the multipliers δ^2 is the same for each bar present in the optimal structure. The optimal value of the vector δ^2 gives also information on the influence of each load condition on the resulting design: large value of δ_k^2 means that the corresponding load condition is severe and it has greater influence on the resulting design while a value close to zero means that the load condition is not critical. This concept can be considered a generalisation of the optimality condition (10.14) for the single load case. It is interesting to underline that if the bars of the truss are made with the same material (i.e. $E_i \equiv E$, $\rho_i \equiv \rho \; \forall i$), the optimal designs are in general no more "fully stressed" but they have the same Root Weighted Mean Square (RWMS) stress since the relation (10.32) becomes

$$
\begin{aligned}
&\sum_{k=1}^{p} \delta_k^2 \sigma_{i,k}^2 = const \;\; \forall i : x_i \neq 0 \\
&\sigma_{RWMS} = \sqrt{\frac{\sum_{k=1}^{p} \delta_k^2 \sigma_{i,k}^2}{\sum_{k=1}^{p} \delta_k^2}} = \sqrt{\sum_{k=1}^{p} \delta_k^2 \sigma_{i,k}^2} = const \;\; \forall i : x_i \neq 0
\end{aligned}
\tag{10.33}
$$

By means of the optimality conditions (10.32) it is possible to establish a relation between the design variables x_i and the member forces $q_{i,k}$ for fixed δ

$$x_i = \frac{l_i \sqrt{\sum_{k=1}^{p} \delta_k^2 q_{i,k}^2}}{\sqrt{C \rho_i E_i}} \quad \forall i \tag{10.34}$$

The substitution of the expression (10.34) in problem (10.25) leads to the following problem:

$$\begin{aligned} &\min_{\mathbf{q} \in R^{mxp}, C \in R^+} \left(\frac{1}{\sqrt{C}} \cdot \Psi(\mathbf{q}), \quad \frac{1}{2}\sqrt{C} \cdot \Psi(\mathbf{q}) \right) \\ &with \quad \Psi(\mathbf{q}) = \max_{\delta \in R^p} \sum_{i=1}^{m} l_i \cdot \sqrt{\frac{\rho_i}{E_i} \sum_{k=1}^{p} \delta_k^2 q_{i,k}^2} \\ &\mathbf{D}\mathbf{q}_k = \mathbf{f}_k \quad \forall k, \quad and \quad \sum_{k=1}^{p} \delta_k^2 = 1 \end{aligned} \tag{10.35}$$

which, for fixed C, becomes the single-objective problem

$$\begin{aligned} &\min_{\mathbf{q} \in R^{mxp}} \Psi(\mathbf{q}) \\ &with \quad \mathbf{D}\mathbf{q}_k = \mathbf{f}_k \quad \forall k, \quad and \quad \sum_{k=1}^{p} \delta_k^2 = 1 \end{aligned} \tag{10.36}$$

The problem of the simultaneous optimisation of topology and size of a truss structure under multiple loads condition can therefore be separated into two steps as for a single load problem. Once the optimal topology is obtained, see Eq. 10.36, the Pareto-optimal set can be derived by simply varying the constant C (related trough Eq. 10.34 to the cross section of the bars).

As an example, Fig. 10.3 shows the optimisation of a simple 2D structure with two rows of 13 nodes each. The dimensions of the structure are 1.5×1 [m]. The nodes at the bottom left and bottom right are fixed and three load cases (F_1 F_2 and F_3) are considered. Each load case consists in a vertical force of 1 kN as shown in

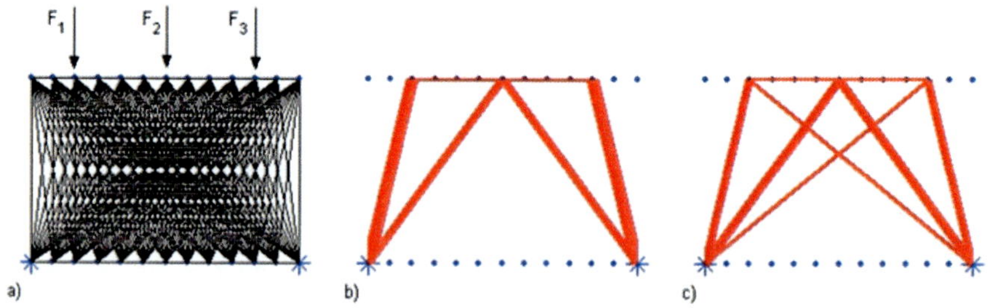

Fig. 10.3 Optimisation of a 2D structure under multiple loads condition: in **a** the 26 nodes, the 193 possible bars and the 3 load cases, in **b** the solution of a single load problem with F_1 F_2 and F_3 acting at the same time, in **c** the solution of the multiple loads problem. Adapted from [128]

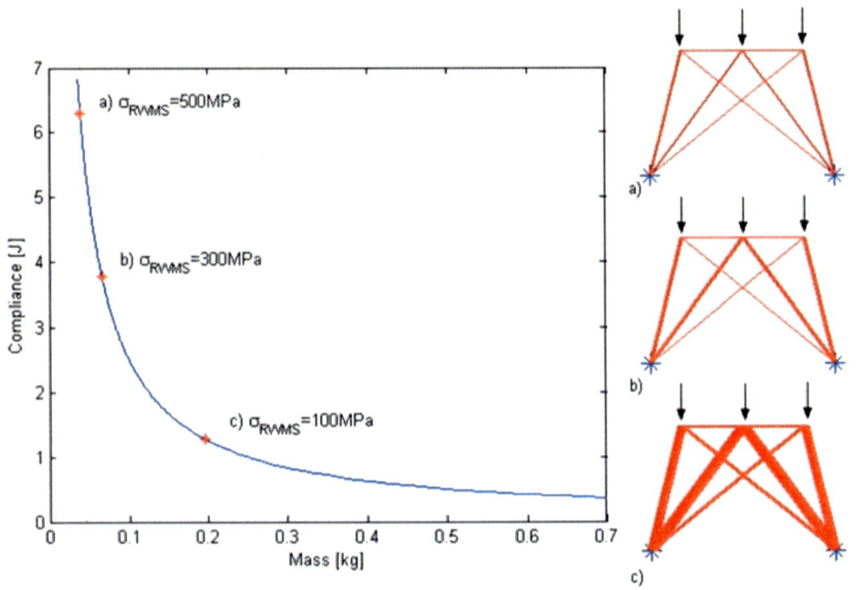

Fig. 10.4 Pareto-optimal set in the objective function domain for a 2D structure under multiple loads condition. The RWMS stress values (uniform in all the bars) for the three solutions highlighted are 500, 300 and 100 MPa. Adapted from [128]

Fig. 10.3a. The 26 nodes and the 193 possible bars (except the overlapping ones) are shown in Fig. 10.3a; all the bars are made by steel ($E = 206000\,\text{MPa}, = 7700\,\frac{\text{kg}}{\text{m}^3}$). The solution of the single load problem (10.18) with the three load cases acting at the same time is shown in Fig. 10.3b; it consists in 6 bars. The solution of the multiple loads problem (10.36) is shown in Fig. 10.3c; it consists in 8 bars with the optimal vector $\delta^2 = [0.1767 \;\; 0.6466 \;\; 0.1767]$. As already observed, the largest value of δ^2 corresponds to the most critical load case. The Pareto-optimal solutions of problem (10.35) is shown on the left of Fig. 10.4: as expected the Pareto-optimal set in the objective function domain is hyperbolic and the compliance for each load case is the same, see Eq. 10.30. Moreover the RWMS stress is constant throughout the bars, see Eq. 10.33. Three different solutions with different RWMS stress values are shown on the right side of Fig. 10.4.

10.3 Summary

In this chapter, the L-matrix condition described by the theory of Chap. 3 has been applied for the analytical derivation of the Pareto-optimal set of the bi-objective mass/compliance minimisation problem applied to truss structures. Some interesting properties of the Pareto-optimal set have been derived and analysed. Both for the case of single external force and for multiple external forces, the paradigms of optimal truss design have been shown.

The density of energy per unit of volume must be constant through the bars. If the bars are of the same material, the well-known condition of fully stressed design is formulated in an optimal design perspective.

The simultaneous optimisation of topology and size of a truss can be made in two steps. At first a topology problem is solved, then Pareto-optimal set is found by varying the density of energy.

Chapter 11
Topology Optimisation of Continuum Structures

The choice of the appropriate structural topology in the conceptual phase is generally the most critical and decisive factor for obtaining an efficient new product. Furthermore for industrial applications, it is always desirable to minimise the total design process duration which is generally proportional to its cost. In this framework *Topology optimisation* approach (often referred to as *layout optimisation* or *generalised shape optimisation [26, 112, 158, 193]* in the literature) provides a very powerful tool for the design.

Generally speaking, three different fields can be distinguished in the framework of structural optimisation, namely, *size optimisation*, *shape optimisation* and *topology optimisation*.

Size optimisation generally deals with variables such as thickness, cross-sectional areas or lengths of members of a predefined structure.

Shape optimisation, on the other hand, is something more general than size optimisation since it also includes shapes and geometric features in the process.

Both of these two methods actually work on a predefined structure [60] which is substantially based on the designer's experience or creativity. In the previous chapters size optimisation has been addressed.

Topology optimisation, instead, is the process of determining the entire structural layout by optimising the shape, location and connectivity of voids inside a given design domain [60]. As one could imagine, topology optimisation is even more general and is able to provide important information in early conceptual and preliminary design phases, where design changes are easy and can be significantly important for the final performances of the product. In fact, given a design domain, information on applied loads and boundary conditions are the only requirements of the topology optimisation approach [28, 60, 74, 157]. Traditionally, the topology design of a product comes from the designer's intuition or experience, or from inspiration from already existing designs [72], this means that at the very early stage of the design process, the designer chooses one or few structural layouts that then are optimised

F. M. Ballo et al., *Optimal Lightweight Construction Principles*,
https://doi.org/10.1007/978-3-030-60835-4_11

and refined up to the final component. As one could understand, with this method the initial choice is not based on a quantitative criterion and therefore is not guaranteed to be an optimal choice. The aim of topology optimisation is precisely that of providing the designer an objective criterion for obtaining an optimal *starting* solution for the entire design process.

Rozvany, in one of his numerous reviews [189], distinguishes two main subfields in topology optimisation of continuum structures: Layout Optimisation (LO) deals with frame structures with a very low volume fraction, while Generalised Shape Optimisation (GSO) is more oriented to higher volume fractions and involves the simultaneous optimisation of the topology and shape of internal boundaries in solid continua.

The first paper on optimisation of frame structures was published by Michell at the beginning of the twentieth century [151]. He introduced the well-known Michell's theory for minimum weight design of truss structures. Michell's optimality criteria had two main limitations [187]:

- The structural domain is *statically determined*, i.e. reactions at the supports can be calculated from equilibrium equations only.
- The admissible stress for members in tension and compression is the same.

Michell's theory has been revised after the 1970s by Rozvany and extended to beam systems (the so-called *grillages*) in several papers [183, 184]. More general optimality conditions have been derived by Hemp in 1973 [101].

Prager and Rozvany developed the *optimal layout theory* in 1977 [173], which became the first general theory on topology optimisation. Basing on that, several analytical solutions for benchmark problems have been obtained by Rozvany in [188] and more recently by Rozvany and other authors in [130–132].

From the late 1980s, numerical methods for solving topology optimisation problems started to arise [25, 221]. As far as finite element-based topology optimisation methods got popular, researchers had to deal with associated numerical problems like the appearance of checkerboards and mesh-dependency of results and local minima [28, 169, 172, 205, 210].

The most important comprehensive surveys on topology optimisation can be found in the works of Rozvany [189, 190] and Eschenauer and Olhoff [74]. The major recent advances in this field can be found in [60].

An interesting paper by Beghini et al. [24] describes the applicability of topology optimisation also in the field of architectural design of buildings, showing a connection between architecture and engineering through topology optimisation.

According to Eschenauer and Olhoff [74], a very rough distinction between two classes of approaches can be made: the so-called material or *Micro-approaches* and *Macro-approaches* as shown in Fig. 11.1.

According to the micro-structure approach, we assume that the amount of material needed for the optimal component is less than the amount that would be needed to fill the full admissible domain. Hence, for the initial design, it is normally chosen to distribute the material evenly in some porous form over the design domain. In this approach, a fixed mesh is used to describe the geometry within the design domain, that

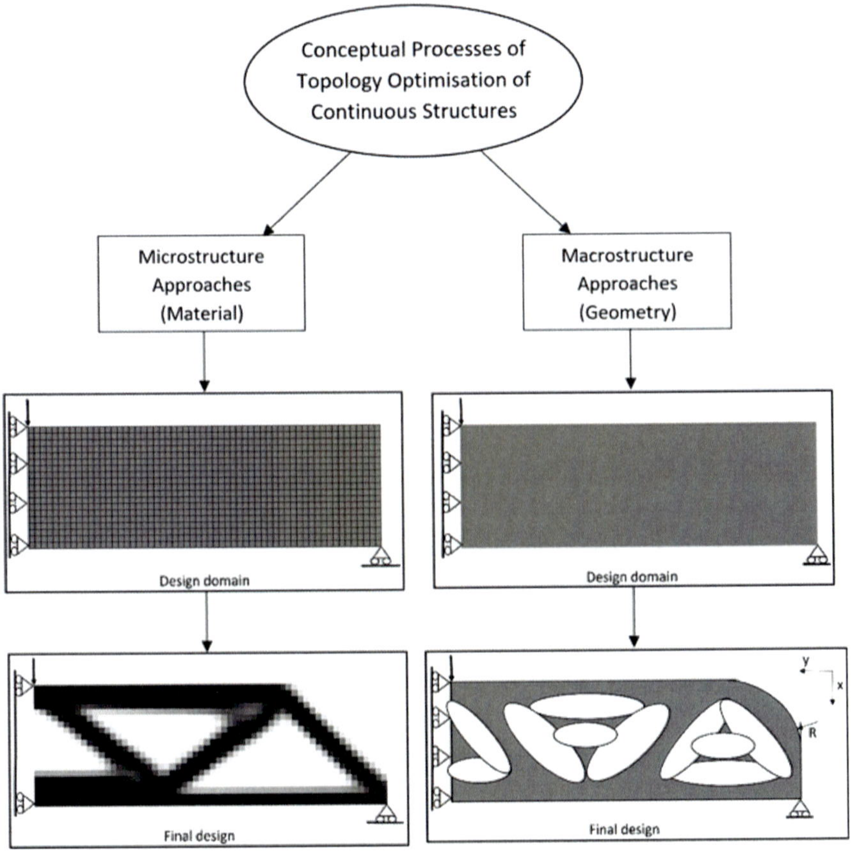

Fig. 11.1 Two main conceptual approaches of topology optimisation—adapted from [74]

is, the finite element mesh does not change during each iteration of the optimisation process. The process of optimisation consists in determining whether each element should contain material or not. The density of material in each element is considered as a single design variable defined between 0 (void) and 1 (solid). For this reason, these class of methods is also known as *density-based methods*, which represent the most widely employed methodologies for structural optimisation [60, 190]. The result is a rough description of outer as well as inner boundaries of the continuous structure, that represents the overall optimum topology design (see Fig. 11.2) [48–50, 114, 115, 159, 192, 193]. Obviously the obtained result needs further and more detailed analysis via shape optimisation [223].

Regarding the macro-structure approach on the other hand, solid isotropic materials are opposed to porous and microstructured ones. In this case topology optimisation is performed in conjunction with a shape optimisation, since shape modifications inside the domain and on its boundaries are handled. Because of that, the finite ele-

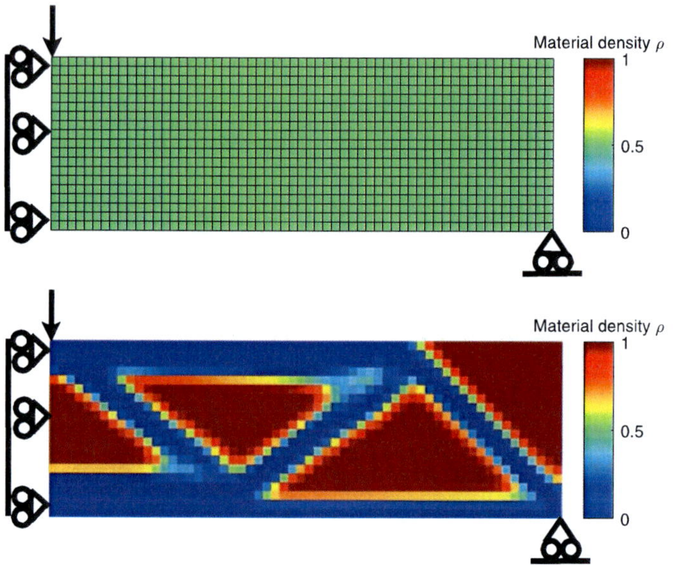

Fig. 11.2 Typical design process in topology optimisation. Top: Starting solution with evenly distributed material. Bottom: Optimal layout

ment mesh cannot be fixed for this kind of approach, leading to an increase in the computational time. Within the macro-structure approach, the topology of a solid body can be changed by growing or degenerating material or by inserting holes [13, 14, 180]. Boundary-based methods can be included in the macro-structure approach. For such kind of problems, the design variables directly control the exterior and interior boundary shapes of the structure [5, 202, 236]. The major disadvantage of these methods is that topology changes are always difficult to handle.

Level set method has proved to be the most promising among boundary-based methods, since it is able to provide important topology changes by using level set functions that are propagated in the design domain [61, 246]. Level set method is also advantageous for solving general multi-material and multi-physics optimisation problems (see [118, 171, 251, 259, 263]).

11.1 Density-Based Methods

Density-based methods represent an important class of topology optimisation approaches. In fact these methods are the most widely used for structural topology optimisation problems. Density-based methods operate on a fixed design domain discretised with finite elements. The main goal is to minimise a given objective function by determining whether each finite element should contain material or not. In structural topology optimisation, this objective is often compliance with a constraint

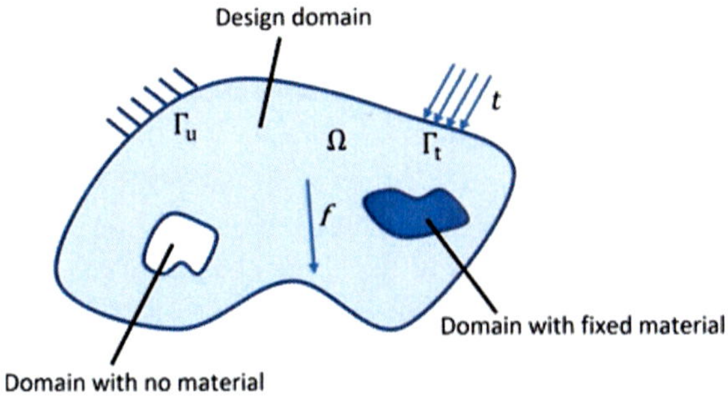

Fig. 11.3 Generalised shape design problem of finding the optimal material distribution

on the maximum amount of material to be used. Density-based methods have also been successfully used in eigenvalue problems [97, 167, 215], stress-based problems [105, 117, 123] and fluid dynamics problems [108, 119]. The result of this approach would be an extremely large-scale integer programming problem with a large computational effort. Therefore, it is desirable to replace discrete variables with continuous ones and find a way to force the solution towards a discrete solid/void solution [60]. The switch to continuous variables is obtained by introducing appropriate interpolation functions, where the continuous design variables are explicitly interpreted as the material density of each finite element. Penalty methods are then employed to force the solution towards an almost discrete *0/1, black/white solid/void* solution.

Referring to Fig. 11.3 the general formulation (continuous form) of a density-based topology optimisation problem can be written as [28]

$$\begin{aligned} &\min \ l(\rho, u) \ \ s.t.: \\ &\int_{\Omega} E_{ijkl}(x)\epsilon_{ij}(u)\epsilon_{kl}(v)d\Omega = \int_{\Omega} fud\Omega + \int_{\Gamma_t} tuds \quad \forall v \in U \\ &g_i(\rho, u) \leq 0 \\ &0 \leq \rho \leq 1 \end{aligned} \tag{11.1}$$

where l is the general cost function, $E_{ijkl}(x)$ is the material stiffness tensor, $\epsilon_{ij} = \frac{1}{2}\left(\frac{\partial u_i}{\partial x_j} + \frac{\partial u_j}{\partial x_i}\right)$ is the linearized strain, u is the displacement and Ω the design domain, f and t are concentrated force and surface traction, respectively. Second line in 11.1 represents the equilibrium equation, while g_i are the remaining design constraints.

Equation 11.1 can be expressed in the discretised form [60] as

$$\begin{aligned} &\min \ l(\boldsymbol{\rho}, \mathbf{u}) \ \ s.t.: \\ &\mathbf{K}(\boldsymbol{\rho})\mathbf{u} = \mathbf{f}(\boldsymbol{\rho}) \\ &g_i(\boldsymbol{\rho}, \mathbf{u}) \leq 0 \\ &0 \leq \boldsymbol{\rho} \leq 1 \end{aligned} \tag{11.2}$$

being $\mathbf{K} = \sum_{e=1}^{N} \mathbf{K}_e$ the global stiffness matrix given by the sum of the N element stiffness matrices $\mathbf{K}_e$.

One of the most important and critical aspects of density-based methods is the selection of the interpolation function and the relative penalization technique, in order to relate the design variables (vector of densities) to physical quantities. In other words, an appropriate expression that relates the density with the structural stiffness of the element has to be chosen. Several material interpolation schemes have been proposed in the literature.

The most famous and widespread method is the so-called SIMP (Solid Isotropic Material with Penalization), originally developed by Bendsøe [27] and Zhou and Rozvany [257]. In this method, density variables are penalised through a power law

$$\eta(\rho) = \frac{E}{E_0} = \rho^p \tag{11.3}$$

where E is Young's modulus at a given value of ρ and E_0 is Young's modulus when $\rho = 1$.

An alternative interpolation scheme was proposed by Stolpe and Svanberg [218] denoted as RAMP (Rational Approximation of Material Properties). The RAMP interpolation is basically a rational expression of the form:

$$\eta(\rho) = \frac{\rho}{1 + q(1 - \rho)} \tag{11.4}$$

A desirable feature of the RAMP scheme is that it has non-zero derivatives at zero density, which reduces numerical problems related to very low values of density in the presence of design dependent loading [60].

Another different interpolation scheme has been presented by Bruns [36]. This scheme differs from the others since it penalizes the volume instead material parameters, meaning that intermediate density material consumes more volume w.r.t. the linear interpolation as shown in 11.5

$$\eta = \frac{V}{V_0} = 1 - \frac{sinh(p(1 - \rho))}{sinh(p)} \tag{11.5}$$

with V the volume at a given value of ρ and V_0 the volume when $\rho = 1$.

A comparison of the three described interpolation schemes is reported in Fig. 11.4.

11.2 Minimum Compliance Problem: Conditions of Optimality

Minimum compliance problem is the simplest and most employed technique for structural optimisation that involves mass minimization and stiffness maximisation.

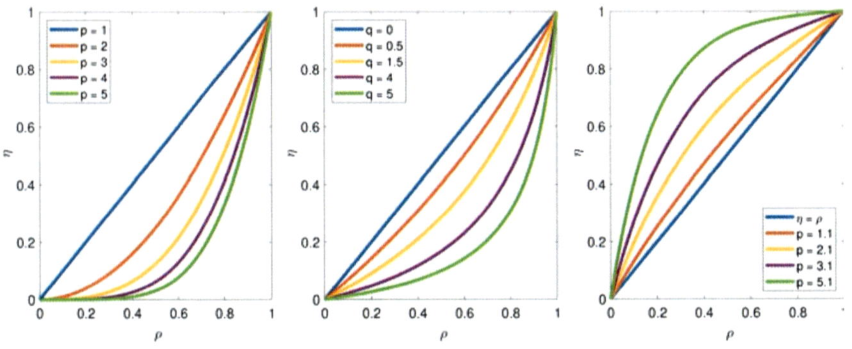

Fig. 11.4 Material interpolation schemes with different penalization factors. Left: SIMP (Eq. 11.3). Middle: RAMP (Eq. 11.4). Right: SINH (Eq. 11.5)

Density methods have proved to be efficient and straightforward in solving such a problem. The use of material interpolation schemes allows to switch from a discrete to a continuous problem. This enables the use of efficient gradient-based minimization algorithm for deriving the solution.

Let us consider a structural optimisation problem where the objective function to be minimised is the total compliance, with a constraint on the maximum amount of material in the design domain. Let us also assume a SIMP interpolation scheme for describing material properties. In the continuum domain this reads

$$\begin{aligned} \min \quad & l(u) = \int_\Omega f u d\Omega + \int_{\Gamma_t} t u ds \;\; s.t.: \\ & a(u, v) = l(v) \quad \forall v \in U, \\ & E_{ijkl}(x) = \rho(x)^p E^0_{ijkl}, \\ & \int_\Omega \rho(x) d\Omega \leq V^*, \quad \rho_{min} \leq \rho \leq 1 \end{aligned} \tag{11.6}$$

where $a(u, v) = \int_\Omega E_{ijkl}(x)\epsilon_{ij}(u)\epsilon_{kl}(v)d\Omega$ is the compliance function. The constrained minimization problem of 11.6 can be solved by using the Lagrange multipliers. The Lagrange functional reads

$$L = l(u) - (a(u, \tilde{u}) - l(\tilde{u})) + \Lambda \left(\int_\Omega \rho(x) d\Omega - V^* \right) + \int_\Omega \lambda_1(x) \left(\rho(x) - 1 \right) d\Omega + \int_\Omega \lambda_2(x) \left(\rho_{min} - \rho(x) \right) d\Omega \tag{11.7}$$

where $\tilde{u}$ is the Lagrange multiplier for the equilibrium equation. Under the assumptions that $\rho \geq \rho_{min} > 0$ the conditions for optimality with respect to variations of the displacement field give that $\tilde{u} = u$ while the condition for ρ is

$$\frac{\partial E_{ijkl}}{\partial \rho}\epsilon_{ij}(u)\epsilon_{kl}(u) = \Lambda + \lambda_1 + \lambda_2 \tag{11.8}$$

For intermediate densities ($\rho_{min} < \rho < 1$) 11.8 gives

$$p\rho^{p-1}E^0_{ijkl}\epsilon_{ij}(u)\epsilon_{kl}(u) = \Lambda \tag{11.9}$$

From 11.8 and 11.9 the following update criteria can be derived [28]

$$\rho^{K+1} = \begin{Bmatrix} max\left\{(1-\zeta)\rho^K, \rho_{min}\right\} & if & \rho^K B_K^\gamma \leq max\left\{(1-\zeta)\rho^K, \rho_{min}\right\} \\ \min\left\{(1+\zeta)\rho^K, 1\right\} & if & \min\left\{(1+\zeta)\rho^K, 1\right\} \leq \rho^K B_K^\gamma \\ \rho^K B_K^\gamma & otherwise & \end{Bmatrix} \tag{11.10}$$

where ρ_K is the density variable at the iteration step K and B_K is given by

$$B_K = \frac{1}{\Lambda_K} p\rho^{p-1}E^0_{ijkl}\epsilon_{ij}(u_K)\epsilon_{kl}(u_K) \tag{11.11}$$

where u_K is the displacement field at step K, γ and ζ are a tuning parameter and a move limit, respectively, and are used to control the rate of convergence of the algorithm (typical values are $\gamma = 0.5$ and $\zeta = 0.2$). The value of the Lagrange multiplier Λ_K has to be adjusted in a nested iteration loop in order to comply the volume constraint.

In the FEM (Finite Element Method) form, the minimum compliance problem reads

$$\begin{aligned} &\min \ \mathbf{f}^T\mathbf{u} \ s.t.: \\ &\mathbf{K}(\rho)\,\mathbf{u} = \mathbf{f} \\ &\sum_{e=1}^{N} v_e\rho_e \leq V^* \\ &\rho_{min} \leq \rho_e \leq 1 \end{aligned} \tag{11.12}$$

where **f** and **u** are the nodal forces and displacements vectors, respectively, **K** is the global stiffness matrix, ρ_e and v_e are the material density and volume of each finite element. A lower bound $\rho_{min} > 0$ is considered in order to avoid singularities in the stiffness matrix for zero densities. By applying the SIMP interpolation scheme, problem 11.12 can be rewritten as

$$\begin{aligned}
&\min \quad \left(c(\rho_e) = \mathbf{f}^T \mathbf{u}\right) \quad s.t.: \\
&\left(\sum_{e=1}^{N} \rho_e^p \mathbf{K}_e\right) \mathbf{u} = \mathbf{f} \\
&\sum_{e=1}^{N} v_e \rho_e \leq V^* \\
&\rho_{min} \leq \rho_e \leq 1
\end{aligned} \tag{11.13}$$

Now for solving the optimization problem *sensitivities* have to be computed, i.e. the derivatives of the compliance objective function with respect to the design variables. Since topology optimization problems typically deal with a large number of design variables with a moderate number of design constraints, the most effective way to compute the sensitivities is the so-called *adjoint method*. The objective function of 11.13 is rewritten by adding the (zero) function $\mathbf{Ku} - \mathbf{f}$

$$c(\rho_e) = \mathbf{f}^T \mathbf{u} - \tilde{\mathbf{u}}^T \left(\mathbf{Ku} - \mathbf{f}\right) \tag{11.14}$$

where $\tilde{\mathbf{u}}$ is an arbitrary fixed real vector. The derivatives of the compliance objective function w.r.t. the density read

$$\begin{aligned}
\frac{\partial c}{\partial \rho_e} &= \mathbf{f}^T \frac{\partial \mathbf{u}}{\partial \rho_e} - \frac{\partial \tilde{\mathbf{u}}}{\rho_e} \mathbf{Ku} + \frac{\partial \tilde{\mathbf{u}}}{\partial \rho_e} \mathbf{f} - \tilde{\mathbf{u}}^T \frac{\partial \mathbf{K}}{\partial \rho_e} \mathbf{u} - \tilde{\mathbf{u}}^T \mathbf{K} \frac{\partial \mathbf{u}}{\partial \rho_e} \\
&= \left(\mathbf{f}^T - \tilde{\mathbf{u}}^T \mathbf{K}\right) \frac{\partial \mathbf{u}}{\partial \rho_e} - \tilde{\mathbf{u}}^T \frac{\partial \mathbf{K}}{\partial \rho_e} \mathbf{u}
\end{aligned} \tag{11.15}$$

which reduces to

$$\frac{\partial c}{\partial \rho_e} = -\tilde{\mathbf{u}} \frac{\mathbf{K}}{\partial \rho_e} \mathbf{u} \tag{11.16}$$

when $\tilde{\mathbf{u}}$ satisfies the adjoint equation $\mathbf{f}^T - \tilde{\mathbf{u}}^T \mathbf{K} = \mathbf{0}$.

From last condition (11.16), we directly obtain $\tilde{\mathbf{u}} = \mathbf{u}$ and the problem is said to be *self-adjoint* (in general the adjoint equation requires additional computations to be solved). For a minimum compliance problem the sensitivities can be explicitly computed and, introducing also the SIMP relation, read

$$\frac{\partial c}{\partial \rho_e} = -p \rho_e^{p-1} \mathbf{u}^T \mathbf{K}_e \mathbf{u} \tag{11.17}$$

Relation 11.17 confirms that, for these kind of problems, sensitivities can be straightforwardly computed without requiring additional computations. Moreover derivatives are negative for all elements and this somehow confirms the physical intuition that additional material decreases compliance, i.e. provides a stiffer structure. Recalling Eq. 11.13, the density update scheme reads [209]

$$\rho^{K+1} = \left\{ \begin{array}{rl} max\left\{(1-\zeta)\rho^K, \rho_{min}\right\} & if \quad \rho^K B_K^\gamma \le max\left\{(1-\zeta)\rho^K, \rho_{min}\right\} \\ \min\left\{(1+\zeta)\rho^K, 1\right\} & if \quad \min\left\{(1+\zeta)\rho^K, 1\right\} \le \rho^K B_K^\gamma \\ \rho^K B_K^\gamma & otherwise \end{array} \right\}$$

$$with \quad B_K = \frac{p\rho_e^{p-1}\mathbf{u}^T\mathbf{K}_e\mathbf{u}}{\Lambda_K} \tag{11.18}$$

11.3 Regularisation Techniques

It is well recognised that the general topology optimization problem formulated in the continuum domain lacks existence of solution [28]. This is principally due to the fact that the introduction of more holes without changing the total volume of the structure generally increases its efficiency. In the limit, this results in obtaining optimal *microstructures* that are typically not isotropic and therefore cannot be represented with the general solid isotropic continuum formulation (the continuum problem is said to be *ill-posed*). Many attempts based on *homogenization* techniques have been done for overcoming this problem [3, 4], but the obtained solutions often lack sufficient penalization.

In computational implementations, this ill-posedness of the continuum problem is seen as a numerical instability, where a larger number of holes appears as a finer element mesh is employed. An example of this mesh-dependence of solutions can be seen in Fig. 11.5. Ideally, mesh refinement should result in a better description of the optimal structural topology and not in a change in the topology itself.

The two primary ways to mitigate this problem are filtering and constraint techniques. Filtering methods are applied via direct modification of density variables or sensitivities, while constraint methods use local or global-level constraints on internal boundaries of the structure [60].

11.3.1 Filtering Techniques

Filtering techniques can be applied either on the density variables or on sensitivities.

Adding a filter on the density variables means modifying the densities of finite elements by adding relations between each element and its neighbours. Design variables are therefore modified according to the following relation:

$$(\rho \otimes S)(x) = \frac{1}{< S >}\int_\Omega \rho(y)S(x-y)dy, \qquad < S >= \int_{R^n} S(y)dy \tag{11.19}$$

where S is a convolution kernel that basically defines the area around each element where the filter applies and reads

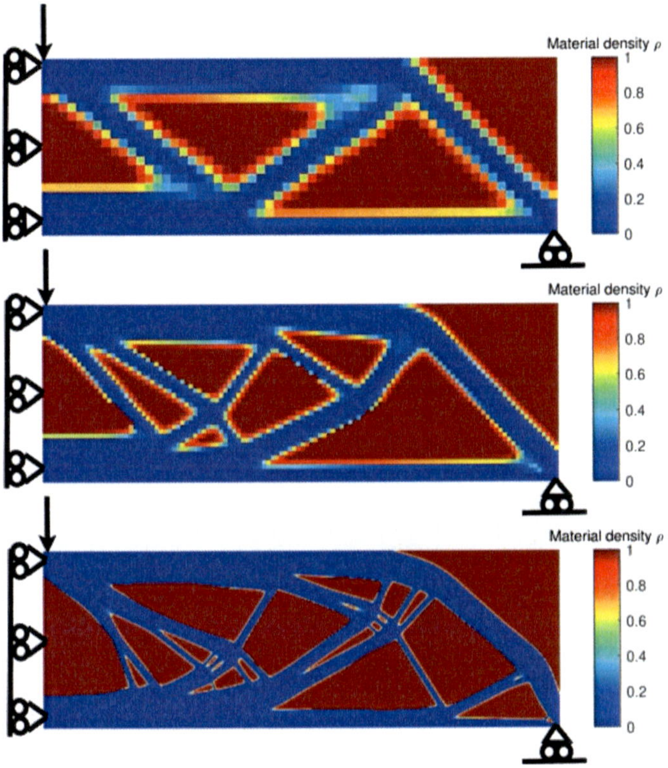

Fig. 11.5 Dependence of the optimal solution on mesh refinement. Top: 60 × 20 elements. Middle: 120 × 40 elements. Bottom: 480 × 160 elements

$$S(x) = \left\{ \begin{array}{cc} 1 - \frac{\|x\|}{r} & if \quad \|x\| \leq r \\ 0 & otherwise \end{array} \right\} \tag{11.20}$$

with r a radius identifying the dimension of the space around the element.

Filtering the sensitivity is probably the most efficient and employed method for assuring mesh-independency [6, 28, 74]. In this case, the filter is applied on the derivatives of the objective function w.r.t. the density variables according to the following scheme:

$$\frac{\hat{\partial f}}{\partial \rho_k} = \frac{1}{\rho_k \sum_{i=1}^{N} \hat{H}_i} \sum_{i=1}^{N} \hat{H}_i \rho_i \frac{\partial f}{\partial \rho_i} \tag{11.21}$$

where N is the total number of finite elements and $\hat{H}_i$ are weighting factors that identifies the neighbourhood of element k

$$\hat{H}_i = r_{min} - dist(k, i), \quad \{i \in N| \; dist(k, i) \leq r_{min}\} \tag{11.22}$$

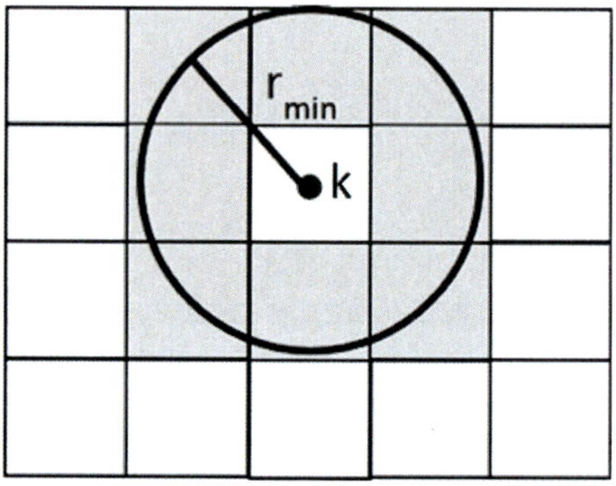

Fig. 11.6 Neighbours of element k for sensitivity filter

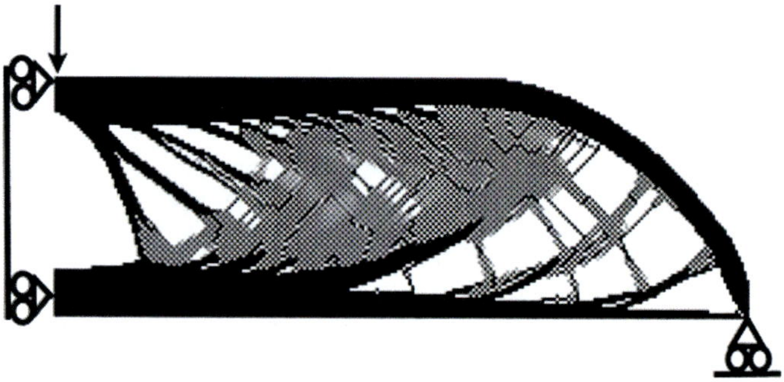

Fig. 11.7 The checkerboard problem on the MBB-beam

where the operator $dist(k, i)$ defines the distance between element k and i (see Fig. 11.6).

Recently, improved filtering techniques based on Helmholtz differential equations have been proposed in [122].

Filtering techniques have become popular for solving numerical instabilities of FE-based topology optimization problems also because they are effective in mitigating checkerboards too. Checkerboard is a numerical instability that appears as a checkerboard-fashion structure, i.e. a series of alternating solids and voids connected on one corner (Fig. 11.7) [28].

11.3.2 Constraining Techniques and Projection Methods

The alternative approach to filtering techniques for reducing mesh-dependence is to restrict the set of admissible solutions by adding topological constraints. One way is the so-called *perimeter control*, that is, roughly speaking, adding constraints on the total perimeter of inner and outer boundaries of the structural layout. Adding

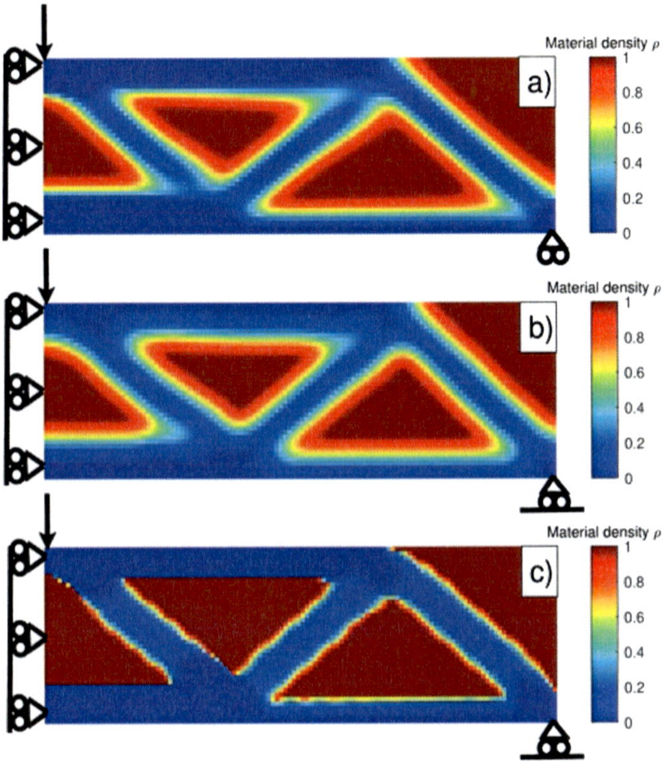

Fig. 11.8 Results of the MBB-cantilever problem with different regularisation techniques. **a** sensitivity filter **b** density filter **c** Heaviside projection filter

constraints on the perimeter obviously limits the number of holes in the design domain. For the SIMP method, a technical way to do that is to impose an upper bound on the total variation $TV(\rho)$ of the density like

$$TV(\rho) = \int_{R^n} \|\nabla \rho\| dx \leq P^* \tag{11.23}$$

An alternative expression of the total variation $TV(\rho)$ and more detailed investigations on perimeter control-based and other regularisation techniques can be found in [169, 258].

Another possibility for controlling the variation of densities is imposing a constraint on the gradient

$$\left| \frac{\partial \rho}{\partial x_i} \right| \leq G \quad i = 1, 2, (3) \tag{11.24}$$

The slope constraint on the density of 11.24 implies that a transition from void to void through full material has to take place on a distance that is longer than $2/G$.

The main drawback of this approach is that a large number of extra constraints are added to the problem, thus increasing the computational effort. For this reason an alternative option could be a global gradient constraint as

$$\|\rho\|_{H^1} = \sqrt{\int_{\Omega} \left(\rho^2 + \|\nabla\rho\|^2\right) d\Omega} \leq M \tag{11.25}$$

that is the norm of function ρ in the Sobolev space $H^1(\Omega)$ has to be upper limited by M.

Finally, more recent methods based on *projection* of the densities through Heaviside functions have been introduced by Guest et al. [95], Kawamoto et al. [107]. These projection overcomes the formation of grey transition material between solid and void regions, that is a direct consequence of filtering techniques as shown in Fig. 11.8.

Chapter 12
Concurrent Topological Optimisation of Two Bodies Sharing Design Space

The topological optimisation of engineering components is a well-discussed topic. In the literature many papers and books can be found defining the basis of topological optimisation and presenting states of the art and perspectives (see, for instance, [28, 74, 96, 190, 255, 260]). Many methods have been developed to face topology optimisation problems, among them, the most used, especially in available software, is the SIMP method [190].

Classical problems of topology optimisation involve the optimal design of single components [28]. For such kind of problems many examples and even free subroutines can be found [209]. More recently, multi-domain problems have been considered. An application of multi-domain optimisation is the design of multi-phase or multi-material structures [224]. In this class of problems, a single component is designed taking into account that the component can be realised by using more than one material. Level set-based approaches have been extensively used to solve such kind of problems [136, 146, 237] and recently became very attractive for the optimal design of innovative mechanical metamaterials [232, 239, 240].

The optimisation of multi-component structures has been considered in some papers. In most cases [174, 245, 252, 253], the optimisation refers to the definition of the optimal structural layout of a part of the structure in which are embedded rigid structures or voids that can change position or orientation during the optimisation process but maintain their original geometries. In few cases, small changes in the embedded components are considered [53]. Actuators can be embedded to obtain smart structures [238].

Referring to the simultaneous optimisation of two bodies, in case sharing part of the design domain, some applications of the level set method can be found. In [121] and [220], the focus is on the optimisation of the interface between two, or more, phases (components) in order to obtain some prescribed interaction force. The solution of such problems involves also the division of the design space between the two components. In general, level set-based approaches for multi-material problems

F. M. Ballo et al., *Optimal Lightweight Construction Principles*,
https://doi.org/10.1007/978-3-030-60835-4_12

have the potential to solve the simultaneous optimisation of two bodies. In [254] a completely different approach, the moving morphable components approach [254], which seems to be adaptable to solve such problems is discussed.

In this chapter a method to solve problems in which two different components share part of the design space is presented. The method is developed in the framework of the Solid Isotropic Microstructure (or Material) with Penalization for intermediate densities (SIMP, [190]) approach. This kind of approach will guarantee an easy implementation of the proposed method to three-dimensional problems. A numerical algorithm based on the classical density-based approach for the optimal material distribution in a given design space and able to address the concurrent topological optimisation is proposed. The algorithm will also consider the optimal allocation of material and voids inside a shared region of the design domains of the two bodies.

12.1 Problem Formulation

In this section, the design problem of the concurrent topological optimisation of two bodies sharing a portion of the design space is stated.

In Fig. 12.1 left, the generalised geometries of two bodies sharing (a portion of) the design space are depicted. The design domain of the two bodies are denominated Ω_1 and Ω_2. Ω_{1-2} represents the shared portion of the design space. The space of this region can be occupied by any of the two bodies. Of course, only one body at a time can occupy a given point in the region. The goal of the optimisation procedure is to allocate in a convenient way the shared space (or leave it void) while, at the same time, optimising the material distribution of the two bodies under the applied loads $\mathbf{f_1}$ and $\mathbf{f_2}$ and the given boundary constraints on Γ_1 and Γ_2.

Fig. 12.1 Generalised geometries of the design domains (Ω_1 and Ω_2) of two bodies sharing a portion of the design space. Each body has its own system of applied forces ($\mathbf{f_1}$ and $\mathbf{f_2}$) and boundary constraints (Γ_1 and Γ_2). Ω_{1-2} represents the shared portion of the design space. Left: initial definition of the domains. Right: division of the domains with assignment of the shared portion of the design space to each body

It must be emphasised that the portion of the shared domain Ω_{1-2} to be assigned to each body is not given a priori, but must be allocated during the optimisation process and can change according to the evolution of the material distributions of the two bodies themselves.

In the following, the problem will be addressed under the hypotheses of

- linear elastic bodies with small deformations;
- the two bodies do not interact in the shared portion of the domain (i.e. no contact is considered between the two bodies in the shared portion of the domain, interactions between the two bodies outside this region are possible);
- each body is made by only one isotropic material;
- both materials have the same reference density, but they can have different elastic moduli;
- loads and boundary conditions can be applied in any point of the domains, even in the shared portion;
- the structural problem is formulated by the finite element theory.

Let us call Ω_1^* and Ω_2^*, see Fig. 12.1 right, the two unshared parts of the domains Ω_1 and Ω_2, respectively (s.t. $\Omega_1 = \Omega_1^* + \Omega_{1-2}$ and $\Omega_1 = \Omega_1^* + \Omega_{1-2}$). The concurrent optimisation problem can be stated as

$$\begin{aligned}
&Find \min_{\mathbf{u}_1,\mathbf{u}_2,\rho} \left(l_1\left(\mathbf{u}_1\right) + l_2\left(\mathbf{u}_2\right)\right) \\
&s.t. : a_1\left(\mathbf{u}_1, \mathbf{v}_1\right) = l_1\left(\mathbf{u}_1\right) \quad and \quad a_2\left(\mathbf{u}_2, \mathbf{v}_1\right) = l_1\left(\mathbf{u}_2\right) \quad \forall \mathbf{v}_1, \mathbf{v}_2 \in U_1, U_2 \\
&\qquad E^1_{ijkl}\left(\mathbf{x}_1\right) = \rho\left(\mathbf{x}_1\right)^p E^{*1}_{ijkl}, \qquad \mathbf{x}_1 \in \Omega_1^* \cup \Omega_{1-2}^{(1)} = \overline{\Omega_1} \\
&\qquad E^2_{ijkl}\left(\mathbf{x}_2\right) = \rho\left(\mathbf{x}_2\right)^p E^{*2}_{ijkl}, \qquad \mathbf{x}_2 \in \Omega_2^* \cup \Omega_{1-2}^{(2)} = \overline{\Omega_2} \\
&\qquad \Omega_{1-2}^{(1)} \cup \Omega_{1-2}^{(2)} = \Omega_{1-2} \quad and \quad \Omega_{1-2}^{(1)} \cap \Omega_{1-2}^{(2)} = \emptyset \\
&\qquad \int_{\overline{\Omega_1}} \rho\left(\mathbf{x}_1\right) d\Omega + \int_{\overline{\Omega_2}} \rho\left(\mathbf{x}_2\right) d\Omega \leq V \qquad 0 \leq \rho \leq 1 \\
&where : l_1\left(\mathbf{u}_1\right) = \int_{\overline{\Omega_1}} \mathbf{f}_1 \mathbf{u}_1 d\overline{\Omega_1} \quad and \quad l_2\left(\mathbf{u}_2\right) = \int_{\overline{\Omega_2}} \mathbf{f}_2 \mathbf{u}_2 d\overline{\Omega_2} \\
&\qquad a_1\left(\mathbf{u}_1, \mathbf{v}_1\right) = \int_{\overline{\Omega_1}} E^1_{ijkl}\left(\mathbf{x}_1\right) \epsilon_{ij}\left(\mathbf{u}_1\right) \epsilon_{kl}\left(\mathbf{v}_1\right) d\Omega_1 \\
&\qquad a_2\left(\mathbf{u}_2, \mathbf{v}_2\right) = \int_{\overline{\Omega_2}} E^2_{ijkl}\left(\mathbf{x}_2\right) \epsilon_{ij}\left(\mathbf{u}_2\right) \epsilon_{kl}\left(\mathbf{v}_2\right) d\Omega_2 \\
&\qquad \overline{\Omega_1} \ and \ \overline{\Omega_2} \ connected
\end{aligned} \tag{12.1}$$

Being l_1 and l_2 the compliance functions of the two bodies, a_1 and a_2 the internal virtual work functions of the two bodies, $\mathbf{u}_1$ and $\mathbf{u}_2$ the displacement fields, $\mathbf{v}_1$ and $\mathbf{v}_2$ the virtual displacement fields, $\mathbf{x}_1$ and $\mathbf{x}_2$ the points of the domains, $\mathbf{f}_1$ and $\mathbf{f}_2$

the applied loads, E^1_{ijkl} and E^2_{ijkl} the material stiffness tensors, E^{*1}_{ijkl} and E^{*2}_{ijkl} the stiffness tensors of the reference materials and ρ the pseudo-density.

In the problem stated in Eq. 12.1 the design variables are represented by the two pseudo-densities fields $\rho\,(\mathbf{x_1})$ and $\rho\,(\mathbf{x_2})$ and by the domain $\Omega^{(1)}_{1-2}$. In fact, in addition to the optimal material distribution inside the two domains, also the division of the common domain has to be found. Thus, domain $\Omega^{(1)}_{1-2}$ is a design variable in the sense that in the solution of the stated problem, each point of Ω_{1-2} has to be assigned either to $\Omega^{(1)}_{1-2}$ or $\Omega^{(2)}_{1-2}$. The domain $\Omega^{(2)}_{1-2}$ is not a design variable, as it is trivially given by $\Omega^{(2)}_{1-2} = \Omega_{1-2} \cap \Omega^{(1)}_{1-2}$.

The two domains $\overline{\Omega_1}$ and $\overline{\Omega_2}$ represent the regions where the two bodies are topologically optimised at any cycle of the optimisation procedure. Each individual domain, i.e. $\overline{\Omega_1}$ and $\overline{\Omega_2}$, should consist of one geometrical region, without disconnected parts. This requirement ensures physically realisable solutions.

Given the formulation of the problem of Eq. 12.1, the reference stiffness E^{*1}_{ijkl} and E^{*2}_{ijkl} are multiplying parameters allowing the inclusion in the problem of different stiffness for the materials of the two bodies. The pseudo-density ρ is a design variable. In the considered formulation, the materials of the two bodies actually have the same reference density. Different densities could be included in the problem by a proper normalisation procedure.

12.2 Numerical Solution

For its numerical solution, the problem of Eq. 12.1 is written in a discretised form as

$$
\begin{aligned}
&Find \min_{\mathbf{u_1},\mathbf{u_2},\rho} \left(\mathbf{f_1^T u_1} + \mathbf{f_2^T u_2}\right) \\
&s.t. : \mathbf{K}\left(\mathbf{E^1_e}\right)\mathbf{u_1} = \mathbf{f_1} \;\; and \;\; \mathbf{K}\left(\mathbf{E^2_e}\right)\mathbf{u_2} = \mathbf{f_2} \\
&\qquad \mathbf{E^1_e} = \rho\,(\mathbf{x_e})^p\,\mathbf{E^{*1}_e}, \qquad \mathbf{x_e} \in \overline{\Omega_1} \\
&\qquad \mathbf{E^2_e} = \rho\,(\mathbf{x_e})^p\,\mathbf{E^{*2}_e}, \qquad \mathbf{x_e} \in \overline{\Omega_2} \\
&\qquad \Omega^{(1)}_{1-2} \cup \Omega^{(2)}_{1-2} = \Omega_{1-2} \;\; and \;\; \Omega^{(1)}_{1-2} \cap \Omega^{(2)}_{1-2} = \emptyset \\
&\qquad \sum_{e=1}^{N_1} \rho\,(\mathbf{x_e}) + \sum_{e=1}^{N_2} \rho\,(\mathbf{x_e}) \le V \qquad 0 < \rho_{min} \le \rho \le 1
\end{aligned}
\tag{12.2}
$$

$$
\begin{aligned}
&where : \mathbf{K}\left(\mathbf{E^1_e}\right) = \sum_{e=1}^{N_1} \mathbf{K_e}\left(\mathbf{E^1_e}\right) \;\; and \;\; \mathbf{K}\left(\mathbf{E^2_e}\right) = \sum_{e=1}^{N_2} \mathbf{K_e}\left(\mathbf{E^2_e}\right) \\
&\qquad \overline{\Omega_1} \;\; and \;\; \overline{\Omega_2} \;\; connected
\end{aligned}
$$

From Eq. 12.2, it can be observed that the only requirements on the two domains $\overline{\Omega_1}$ and $\overline{\Omega_2}$ are that their union is equal to the union of Ω_1 and Ω_2 and that their intersection is a null set. The division of the common part of the domain is not given and the solution algorithm can allocate the common part of the domain during the optimisation process. As a consequence, the geometries of the two domains $\overline{\Omega_1}$ and $\overline{\Omega_2}$ are not fixed but can change during the optimisation. The allocation of the elements of the common domain to one or the other body is controlled by the local sensitivities of each element. Each element is assigned to the domain where it can give the greater contribution to the reduction of the global compliance of the system.

This situation is very different with respect to the usual framework of the density-based topology optimisation, where the design domain (and thus the mesh) is fixed [28]. Although this fundamental difference, the proposed solution approach is based on the classical problem of the optimal material distribution by means of the definition of the pseudo-density field and the SIMP method [28, 190]. In other words, after identifying the proper division of the domains $\overline{\Omega_1}$ and $\overline{\Omega_2}$, the common SIMP method is employed for optimising the material distributions in each domain. In this way, promptly available algorithms for the solution of a wide variety of problems can be adapted to the considered problem.

The solution algorithm is depicted in the diagram of Fig. 12.2. The algorithm is basically divided into two parts, indicated in the figure by the dashed rectangles labelled **A** and **B**.

Sub-algorithm **A** represents a standard topology optimisation algorithm. This sub-algorithm follows the solution of the finite element model of the two bodies (*block 1* in Fig. 12.2). Sub-algorithm **B** is the innovative part of the proposed algorithm and has the function of assigning each element of the shared part of the domain (Ω_{1-2}) to one of the two bodies. Following the diagram of Fig. 12.2, the concurrent optimisation algorithm is now described.

12.2.1 Block 1: *Solve the Two FE Models on Ω_1 and Ω_2*

The finite element models of the two bodies, with the corresponding properties, loads and constraints, are constructed and solved independently over the two domains Ω_1 and Ω_2. It can be noted that Ω_1 and Ω_2 are the whole domain spaces of the two bodies, including the shared part. By building and solving the finite element models always on the two sets Ω_1 and Ω_2, it is possible to use a fixed mesh for all the optimisation process. However, in this way, the elements of the shared set Ω_{1-2} are considered in both models. In the subsequent steps, a procedure to penalise the densities of one of the two bodies on each element of the shared set will be considered. By this penalization, each element actually contributes only to the structural stiffness of one of the two bodies.

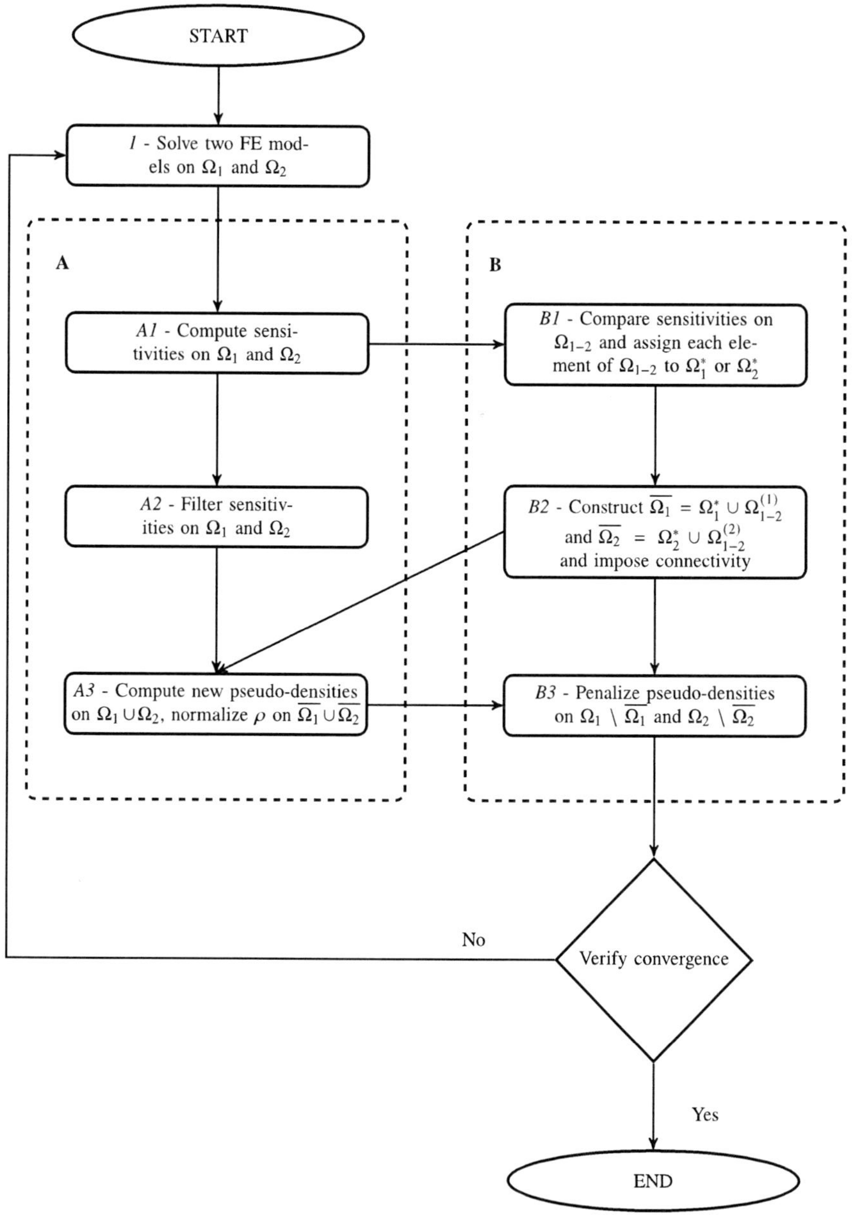

Fig. 12.2 Diagram of the algorithm for the concurrent topological optimization

12.2.2 Sub-algorithm A: Topological Optimisation Over Ω_1 and Ω_2

This sub-algorithm is a standard topology optimisation algorithm employed to compute the pseudo-density field in the two domains. The optimisation algorithm presented in [209] and in Sect. 11.2 of this book is used, with the only minor modification consisting in the substitution of the bisection method used to solve the Lagrange multipliers on the density by a Trust-Region Dogleg method [55]. The steps of the topology optimisation algorithm are the following:

- *Block A1: computation of the sensitivities on Ω_1 and Ω_2.* For each element of Ω_1 and Ω_2, the sensitivities of the total compliance of each element with respect to the pseudo-density are computed as

$$\frac{\partial l}{\partial x_e} = -p\rho\left(x_e\right)^{p-1}\mathbf{u_e^T}\mathbf{K_e^*}\mathbf{u_e} \tag{12.3}$$

  where l is the total compliance, x_e the location of the element, $\mathbf{u_e}$ the element displacements and $\mathbf{K_e^*}$ the reference stiffness matrix of the element. For the penalization exponent p, a value of 3 is used.
  Equation 12.3 is computed on all the elements of Ω_1 and Ω_2. This means that the sensitivities of the elements of the shared domain Ω_{1-2} are computed twice, one time with respect to the solution of the finite element model of the first body on Ω_1 and one time with respect to the solution of the finite element model of the second body on Ω_2.
- *Block A2: filtering of the sensitivities on Ω_1 and Ω_2.* As suggested in [209], a filter on the sensitivities is implemented in order to ensure a regular solution. The filter is formulated as

$$\widehat{\frac{\partial l}{\partial x_e}} = \frac{1}{x_e\sum_{f=1}^{N}\hat{H}_f}\sum_{f=1}^{N}\hat{H}_f x_f\frac{\partial l}{\partial x_f} \tag{12.4}$$

  The weighting factor $\hat{H}_f$ is written as

$$\hat{H}_f = r_{min} - \mathtt{dist}\left(e, f\right), \quad with \quad \left\{f \in N|\mathtt{dist}\left(e, f\right) \leq r_{min}\right\}, \quad e = 1, ..., N \tag{12.5}$$

  where the distance is computed between the centre of the elements e and f. The pseudo-densities are filtered separately on Ω_1 and Ω_2.
- *Computation of the new pseudo-density field on $\Omega_1 \cup \Omega_2$.* The new density field is computed as [28, 209]

$$\rho_e^{new} = \begin{cases} \max\left(\rho_{min}, \rho_e - m\right), & if \quad \rho_e B_e^\eta \leq \max\left(\rho_{min}, \rho_e - m\right) \\ \rho_e B_e^\eta, & if \quad \max\left(\rho_{min}, \rho_e - m\right) < \rho_e B_e^\eta < \min\left(1, \rho_e + m\right) \\ \min\left(1, \rho_e + m\right) & if \quad \min\left(1, \rho_e + m\right) \leq \rho_e B_e^\eta \end{cases} \tag{12.6}$$

where m is a positive move limit, ρ_{min} is the minimum pseudo-density for any element to avoid numerical singularities, η is a numerical damping coefficient and B_e represent the optimality condition

$$B_e = \frac{-\frac{\partial l}{\partial x_e}}{\lambda \frac{\partial V}{\partial x_e}} \tag{12.7}$$

λ is a Lagrange multiplier computed by a Trust-Region Dogleg algorithm (and not by a bisection method as in [209]).
As for the sensitivities, also the new pseudo-density field is computed twice for the elements of Ω_{1-2}.
The computation of the new pseudo-density field is performed on the whole domain $\Omega_1 \cup \Omega_2$. In this way, the total mass is not assigned a priori but is divided among the two bodies accordingly to the computed sensitivities. To guarantee that the total mass on the two physical domains $\overline{\Omega_1}$ and $\overline{\Omega_2}$ matches the target on the design volume fraction, the Lagrange multiplier λ is computed over $\overline{\Omega_1} \cup \overline{\Omega_2}$ and not $\Omega_1 \cup \Omega_2$.

The output of the sub-algorithm **A** is the new pseudo-densities fields on the two domains Ω_1 and Ω_2.

12.2.3 Sub-algorithm B: Computation and Enforcement of the Connectedness of the Two Domains $\overline{\Omega_1}$ and $\overline{\Omega_2}$

Sub-algorithm **B** is devoted to the assignment of each element of the shared part of the design space Ω_{1-2} to either of the two bodies. This assignment is performed according to the relative sensitivity of each element with respect to the total compliance of the two bodies. In other words, each shared element is assigned to the body that provides the higher variation on the compliance. Moreover, this sub-algorithm regularizes the two domains by enforcing the connectedness condition. The sub-algorithm comprises the following steps:

- *Block B1: Comparison of the sensitivities on Ω_{1-2} and assignment of each element of Ω_{1-2} to Ω_1^* or Ω_2^*.* For each element of Ω_{1-2} the two values of the sensitivities computed in *Block A1* are compared. Each element is then assigned to Ω_{1-2}^1 if the sensitivity computed with respect to Ω_1 is greater than the sensitivity computed with respect to Ω_2 or to Ω_{1-2}^2 otherwise.
- *Block B2: Construction of $\overline{\Omega_1} = \Omega_1^* \cup \Omega_{1-2}^{(1)}$ and $\overline{\Omega_2} = \Omega_2^* \cup \Omega_{1-2}^{(2)}$ and connectivity analysis.* The sets Ω_{1-2}^1 and Ω_{1-2}^2 built in the previous step are united with Ω_1^* and Ω_2^*, respectively, to create the sets $\overline{\Omega_1}$ and $\overline{\Omega_2}$. At this point, the two sets $\overline{\Omega_1}$ and $\overline{\Omega_2}$ have been created exclusively on the basis of the relative sensitivities computed on the elements. There is a high probability that the resulting sets are disconnected and thus not physically admissible.

An intuitive definition of connectedness in space can be given as follows. A set is said to be connected if two points of the set can be connected by a curve belonging to the set. In this paper, a more restrictive definition of connectedness is used. In fact, by the previous definition, a surface (or a three-dimensional region of space) is considered connected if two parts of the surface have in common just one point. This is not sufficient to have a physical meaning of the structure. We say that to be considered connected, and have a physical meaning, all the elements of the region must share at least one edge in a two-dimensional space or a face in a three-dimensional space with the rest of the region. By considering this definition, each element of the two sets is analyzed. The elements are grouped according to their connectivity with other elements and the whole region is mapped. Finally, if from this map a part of a region is not connected, it is moved from one set to the other. In this analysis, both domains are treated simultaneously, i.e. the connectivity analysis does not depend on the order in which the two domains are considered.

- *Block B3: Penalization of the pseudo-densities on* $\Omega_1 \setminus \overline{\Omega_1}$ *and* $\Omega_2 \setminus \overline{\Omega_2}$. The new pseudo-density field computed in the block *A3* refers to the whole domain $\Omega_1 \cup \Omega_2$. That is, the elements of Ω_1 but not belonging to $\overline{\Omega_1}$ and of Ω_2 but not belonging to $\overline{\Omega_2}$ have a value of the pseudo-density that depends on their sensitivities and could contribute to the stiffness of the system. To make their contribution vanishing and avoid singularities in the stiffness matrices of the two systems, the pseudo-densities of these elements (belonging to the two sets $\Omega_1 \setminus \overline{\Omega_1}$ and $\Omega_2 \setminus \overline{\Omega_2}$) are multiplied by a penalization factor fixed to 0.05. In any case, the minimum pseudo-density of any element is ρ_{min}.

The outputs of the sub-algorithm **B** are the two pseudo-density fields on Ω_1 and Ω_2, where the elements of $\Omega_1 \setminus \overline{\Omega_1}$ and $\Omega_2 \setminus \overline{\Omega_2}$ have been penalized in order to make their contribution negligible, and the maps of the two sets $\overline{\Omega_1}$ and $\overline{\Omega_2}$.

The optimisation algorithm is then arrested if the maximum difference in the pseudo-density of the elements is below a certain threshold or if a maximum number of iterations is reached.

12.3 Numerical Examples

In this section, three numerical examples are presented to show the potentialities of the presented approach. The first example refers to a situation in which the shared part of the design domains can be divided in order to obtain, for the two bodies, the known solution to two problems of topology optimisation already published in the literature. In the second problem, a quite complex symmetric situation is solved. The problem can test the ability of the method to maintain the symmetry in the solution of the problem. Finally, the third example refers to the structural optimisation of a dovetail guide.

Additionally, the proposed method has been successfully employed on a more complex three-dimensional problem related to the topological optimisation of a road vehicle brake calliper and wheel assembly. The application and relative results are presented in Sect. 13.4 of this book.

For example 1 and 2, a mesh of bilinear quadrilateral elements of dimension 1×1 is used. Since the optimisation problem is a nonconvex problem, the solution depends on the initial conditions. In the following examples, the initial condition is given by a uniform pseudo-density equal to the required volume fraction for all the elements. The shared domain is initially assigned to both bodies, i.e. $\overline{\Omega_1} = \Omega_1$ and $\overline{\Omega_2} = \Omega_2$ at the first step of the iteration. The minimum pseudo-density is 0.001 and the filter distance ρ_{min} is equal to 1.5 times the characteristic dimension of the elements.

12.3.1 Example 1: Literature Cases

In this example two problems, whose optimal solution is known from the literature, are considered. The two domains of the problems are overlapped in such a way that the shared part of the domains can be divided in order to obtain, for both problems, the optimal solution. The example is depicted in Fig. 12.3. The problem corresponding to the domain Ω_1 is the very well known problem of the cantilever with an end load (see, for instance, [28] p. 49). The second problem defined on Ω_2 is a three support structure loaded with a distributed load presented in [37] at p. 342. An overall volume fraction is 0.4.

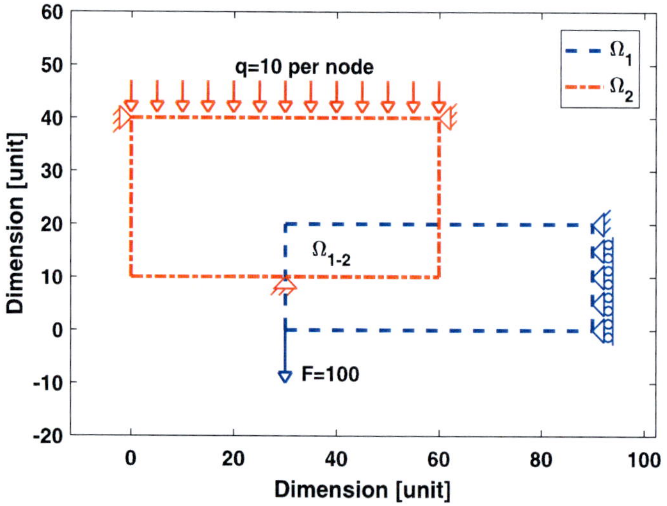

Fig. 12.3 Example 1. Definition of two optimisation problems from the literature. The problem defined over Ω_1 is taken from [28] at p. 49. The problem defined over Ω_2 is taken from [37] at p. 342

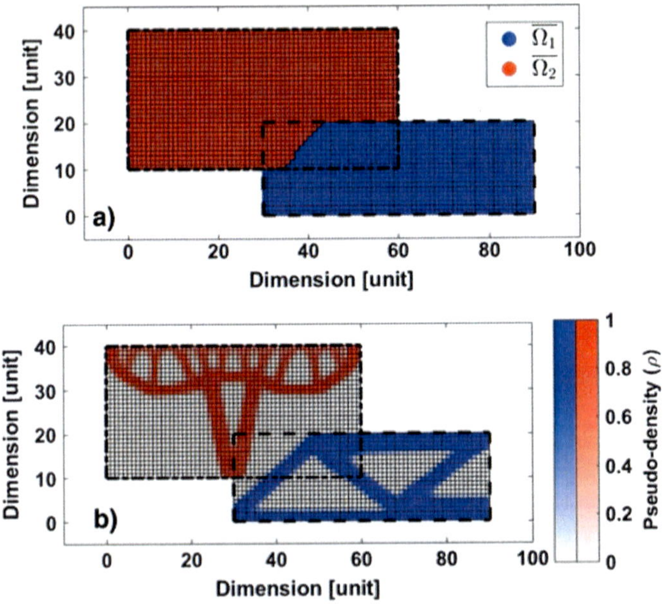

Fig. 12.4 Example 1. Solution of two optimisation problems from the literature. The obtained solutions correspond to the solutions reported in the literature (see [28] at p. 49 and [37] at p. 342). **A** Division of the design space. **B** Optimised structures

In Fig. 12.4 the results of the concurrent topological optimisation are shown. In Fig. 12.4A, the division of the common part of the design space is reported and in Fig. 12.4B the two optimised structures can be seen. The two optimised structures correspond to the optimal solutions reported in the literature. In particular, the cantilever is actually the well-known solution to the problem. The solution of the three supports structure is very close to the solution reported in [37]. The small differences are due to a very different mesh size, volume fraction and filter settings.

According to the division of the domains of Fig. 12.4A, the domain $\overline{\Omega_1}$ has 1112 elements, while the domain $\overline{\Omega_2}$ 1588. The volume fractions are, respectively, 0.53 (corresponding to 591.5 units of mass) and 0.31 (corresponding to 488.5 units of mass). An overall volume fraction of 0.4 is as set as target, corresponding to 1080 units of mass over the 2700 elements of $\overline{\Omega_1} \cup \overline{\Omega_2}$.

Referring to the division of the design space, it can be observed that the design space is divided in a way which allows the optimal material distributions of the two structures (Fig 12.4A). The obtained design space division is not intuitive without any a priori knowledge of the solution. In Fig. 12.5 the results for the optimisation of the problem of Fig. 12.3 are reported for four reasonable a priori divisions of the design space. In absence of any other information, four division of the design space are considered, namely:

- All of the common design Ω_{1-2} space is attributed to Ω_1, so that $\overline{\Omega_1} = \Omega_1$.
- All of the common design Ω_{1-2} space is attributed to Ω_2, so that $\overline{\Omega_2} = \Omega_2$.
- Each element of the common design Ω_{1-2} space is attributed to the closest domain.
- The common design Ω_{1-2} space is divided in half along the diagonal and each half of the common domain is attributed to the closest domain.

The four described divisions are depicted on the left column of Fig. 12.5. On the right side of Fig. 12.5 the four corresponding solutions are depicted. The compliances of the structures obatained on the two domains Ω_1 and Ω_2 in the four conditions are shown in Fig. 12.6. It can be seen that the solution obtained by the concurrent topological optimisation algorithm is always dominating the other solutions, having

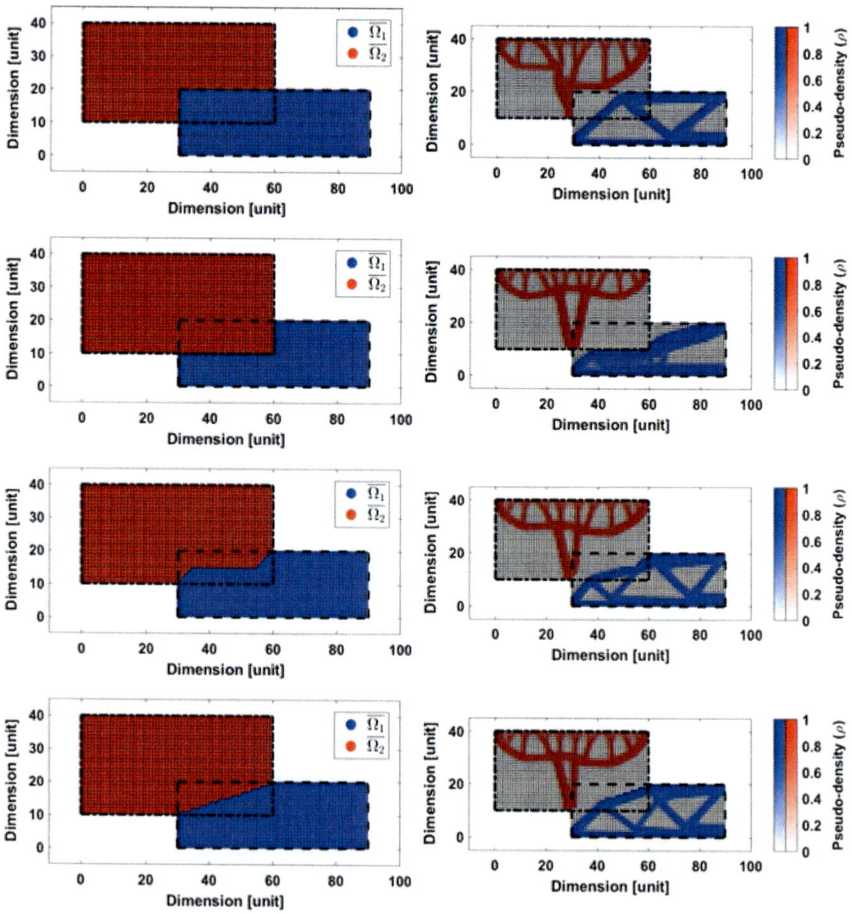

Fig. 12.5 Example 1. Solutions to the problem of Fig. 12.3 for four reasonable a priori divisions of the common domain Ω_{1-2}. From top to bottom: $\overline{\Omega_1} = \Omega_1$, $\overline{\Omega_2} = \Omega_2$, division by the minimum distance, division along the diagonal. Left: divisions of the design space. Right: optimised structures

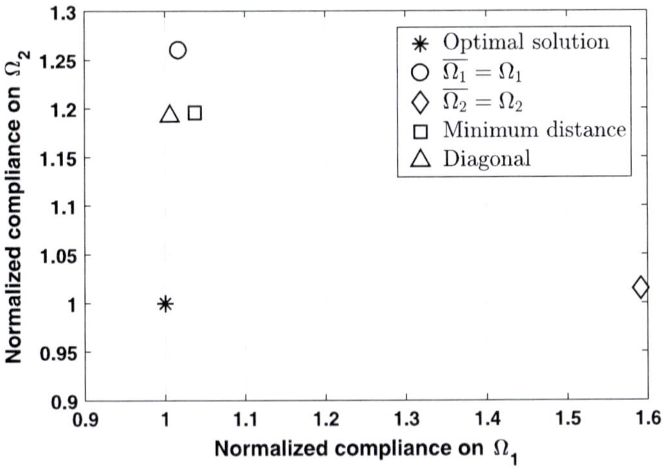

Fig. 12.6 Example 1. Comparison of the normalised compliances of the two structures obtained by the optimal solution of the concurrent topological optimisation algorithm and by considering four reasonable a priori divisions of the common domain Ω_{1-2} ($\overline{\Omega_1} = \Omega_1$, $\overline{\Omega_2} = \Omega_2$, division by the minimum distance, division along the diagonal)

Table 12.1 Results of the optimisation problem of Fig. 12.3

Common domain division	Mass on Ω_1	Mass on Ω_2	Compliance on Ω_1	Compliance on Ω_2	Total compliance
Optimal solution	591.5	488.5	$8.36 \cdot 10^6$	$5.66 \cdot 10^6$	$14.01 \cdot 10^6$
$\overline{\Omega_1} = \Omega_1$	581.0	499.0	$8.50 \cdot 10^6$	$7.13 \cdot 10^6$	$15.62 \cdot 10^6$
$\overline{\Omega_2} = \Omega_2$	595.4	484.6	$13.30 \cdot 10^6$	$5.74 \cdot 10^6$	$19.04 \cdot 10^6$
Minimum distance	592.9	487.1	$8.67 \cdot 10^6$	$6.76 \cdot 10^6$	$15.43 \cdot 10^6$
Diagonal	592.8	487.2	$8.41 \cdot 10^6$	$6.74 \cdot 10^6$	$15.15 \cdot 10^6$

the minimum value of both the compliances. It is interesting to note that the optimal solution has lower compliance of the division $\overline{\Omega_1} = \Omega_1$ on Ω_1 and of the division $\overline{\Omega_2} = \Omega_2$ on Ω_2 even if these divisions can have the optimal material distributions on those domains. This is due to the mass repartition according to the sensitivities (*Block A3* in Sect. 12.2) which redistributes the mass in order to reduce the total compliance. In other words, the algorithm assigns a higher value of mass, with respect to the optimal solution, to the domain in which the optimal material distribution cannot be achieved in order to compensate for the less efficiency of the obtained material distribution (see Table 12.1).

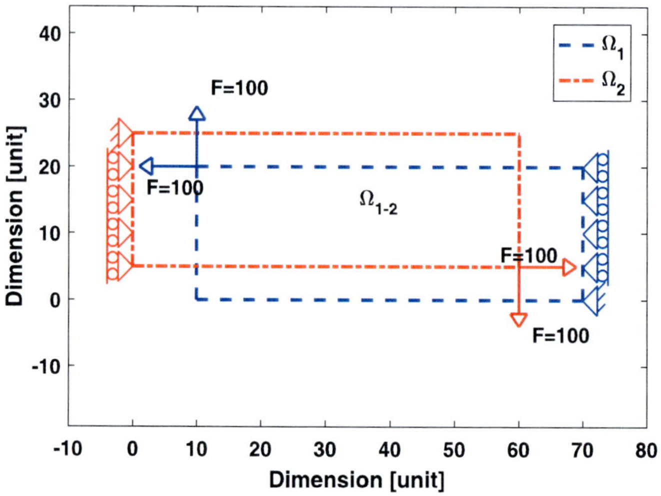

Fig. 12.7 Example 2. Definition of two optimisation problems rotated of 180°. Each problem is a cantilever with a fixed end and two forces applied to one corner of the opposite end. Notice that the forces are applied inside the shared part of the domains

12.3.2 Example 2: Symmetric Structure

The considered example is shown in Fig. 12.7. This example consists of two identical structures rotated of 180°. Each structure is a cantilever with a fixed end and two forces applied to one corner of the opposite end. The corner where the forces are applied has been chosen in order to have the application points inside the shared part of the domain.

In Fig. 12.8, the results of the optimisation process are depicted. The obtained solution, despite its high level of complexity, shows a very good symmetry. In fact, by analysing Fig. 12.8, it can be observed that the two structures are actually the same structure rotated of 180°. The compliances computed for the two bodies differ of about 0.01% and the mass is the same up to round off error.

By inspecting Fig. 12.8 bottom, it can be noted that a small almost void band is present between the two bodies. This almost void is due to the filter on the sensitivities (step *A2* in Fig. 12.2). In fact, in case an element with pseudo-density close to one is close to the border of the assigned domain, the filter smooths the transition of the density to the almost zero of the neighbour elements. In case the two resulting structures are too close and could touch after deformation, an effect which is not modelled, the parameters of the filter could be used to allow for a certain amount of clearance.

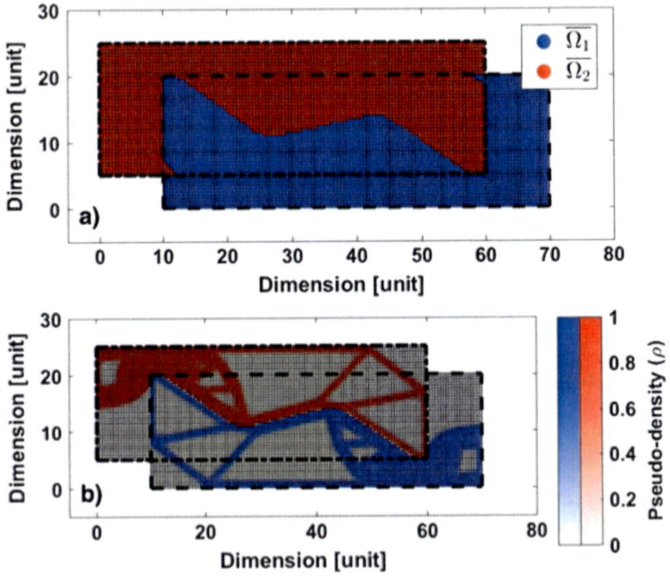

Fig. 12.8 Example 2. Solution of two optimisation problems rotated of 180°. **A** Division of the design space. **B** Optimised structures

12.3.3 *Example 3: Dovetail Guide*

In this example a lightweight dovetail guide has to be designed. The problem is depicted in Fig. 12.9. Both domains are symmetric with respect to the x-axis, where a series of sliding joints are located on both domains to model symmetry. The domain Ω_1 is also symmetric with respect to the line $x = 40$, where again sliding joints are located. The load is applied on the diagonal black line to reproduce the contact between the two parts of the dovetail. Also, to enforce the correct matching of the two parts of the dovetail, the dark grey region of Fig. 12.9 is reserved to the design domain Ω_1, while the light grey region is reserved to Ω_2. The two design regions are then constrained in the x-direction by sliding joints as depicted in the figure. The overall volume fraction is set to 0.4.

The optimised design of the lightweight dovetail is reported in Fig. 12.10. The division of the space domain clearly resembles a dovetail guide. The optimised structures show the typical radial shapes expected in the topological design of non-rectilinear bended structures.

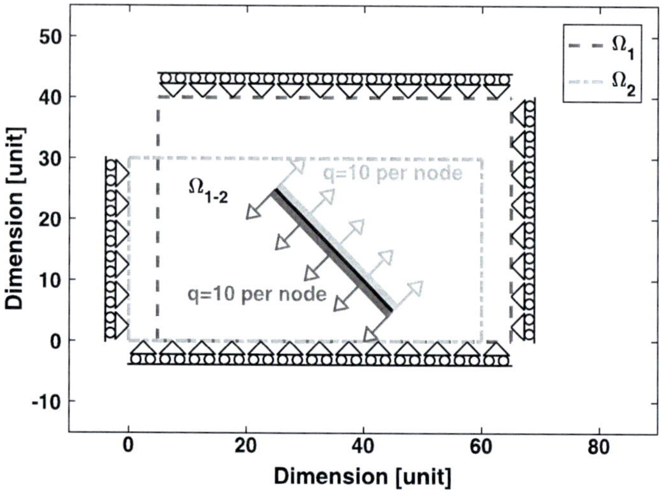

Fig. 12.9 Example 3. Design spaces for the dovetail guide. Dark grey arrows refer to the load applied to Ω_1, while light grey arrows to the load on Ω_2. The load is applied on the black line. The dark grey region is a reserved design space for Ω_1, the light grey region for Ω_2

Fig. 12.10 Example 3. Solution of the design of the lightweight dovetail guide defined in Fig. 12.9. **A** division of the design space. **B** optimised structures

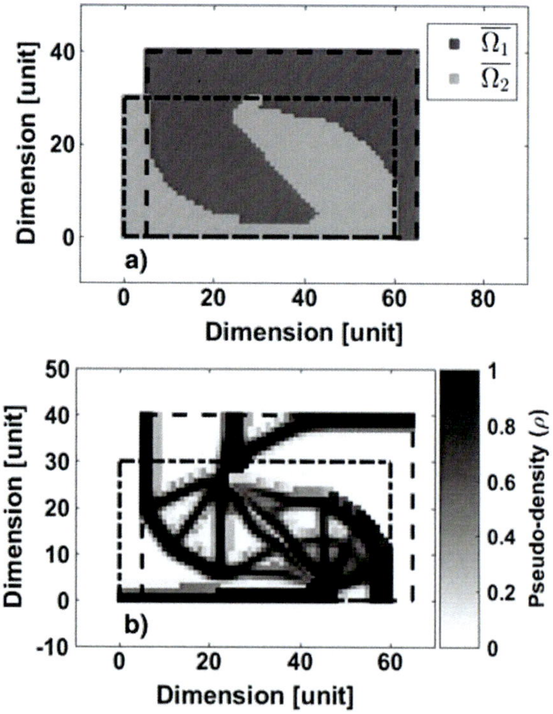

12.4 Summary

A novel topological optimisation problem has been presented. The problem refers to the concurrent topological optimisation of two structures sharing part of the design domain. This kind of problem is of interest in systems design where more than one body has to be considered.

An algorithm for the solution of this kind of problems is proposed. The algorithm is based on the SIMP approach as this approach is the most widely used in currently available software. The main objective has been the derivation of an algorithm for the numerical solution of the problem that could be considered as add-on to current topological optimisation algorithms.

The proposed algorithm assigns the shared space to one or the other body depending on the relative sensitivity to each element to the total compliance of the system. After each element has been assigned to one of the two design domains, the connectedness of the two domains is checked. In case some unconnected regions are present, the assignment of these regions is reversed and the connectedness of the two design domains is enforced. The volume fraction is enforced at system level, i.e. the volume fractions of the two domains can be different, but the total volume fraction complies with the set value. In this way, the available mass can be allocated in the most convenient way between the two bodies.

Three examples have been discussed to show the performances of the proposed algorithm. In the first example, two structures for which the optimal solutions are known from the literature have been considered. The two design domains have been overlapped as to allow the two optimal solutions to be found. The two optimal solutions have been actually got by the concurrent topological optimisation algorithm. Moreover, by imposing some reasonable, but not optimal, division of the design domain, structures with higher compliances have been obtained. In the second example, two identical domains rotated of 180° have been considered. The algorithm has proved to be able to get the same solution on the two domains, respecting the symmetry of the problem. In the third example, the design of a lightweight dovetail guide has been proposed and solved, proving the practical interest of this kind of problems.

Actually, the method is able to handle also more complex three-dimensional problems as it will be shown in Chap. 13, where the concurrent topology optimisation of the brake calliper and wheel assembly of a road vehicle is solved by means of the proposed algorithm.

Chapter 13
Structural Optimisation in Road Vehicle Components Design

Mass minimization is and has always been a central task in vehicle design. Lightweight design is nowadays assuming an even greater importance in road vehicles development. Many automotive companies and vehicle components manufacturers are performing several research activities on new composite materials and high strength aluminium alloys for reducing the overall mass of the vehicles. The weight reduction is a central issue in the design of new road vehicles, not only for improving vehicle performances, but also for the reduction of fuel consumption and emissions, as the new, tighter European standards (Euro VI) on environmental impact require.

Minimising the overall mass of a vehicle means minimising the mass of single components. Together with mass minimization, acceptable structural performances have to be maintained.

Structural optimisation aims to determine the most efficient and lightweight structure. Many parameters, such as cross-sectional dimensions, structural layout, topology, can be involved in this process.

Structural stiffness and integrity are generally common targets in this approach and are also crucial parameters for the safety of vehicles.

Many applications of structural optimisation for improving passive safety of road vehicles can be found in the literature. In fact, as the total mass of the vehicle body is reduced, greater efforts have to be produced in order to maintain acceptable structural performances such as stiffness, integrity and protection of the occupants in case of accidents.

Structural optimisation of a vehicle roof structure has been performed in [104]. A FEM-based optimisation method has been adopted. The thickness of the two pillars and roof rail has been considered as design variables, the total mass of the structure has been minimised within constraints provided by the standards on rollover accidents.

A multilevel decomposition technique has been employed for the optimal design of an aluminium vehicle chassis in [125]. Structural stiffness of the entire vehicle body has been maximised, the cross sections of the structure and the shape of the

F. M. Ballo et al., *Optimal Lightweight Construction Principles*,
https://doi.org/10.1007/978-3-030-60835-4_13

joints connecting the members have been considered as design variables. A significant increment in the bending and torsional stiffness of the vehicle body has been obtained together with a reduction in the overall mass of the vehicle.

Parametric optimisation has been used for optimising vehicle structures in [51]. Thanks to the use of the tailor-welded blank (TWB) manufacturing process, the frame has been divided into several sectors and the thickness of each sector has been optimised for reducing the mass. A multidisciplinary design optimisation approach has been employed; NVH performances and vehicle safety for frontal and lateral impact modes have been maximised.

Optimal design of B-pillar manufactured via the TWB process is discussed in [162]. In this case, a weight reduction of about 30% has been reached through this design approach. Acceptable levels of safety in rollover and lateral crash have been maintained.

Optimal design of TWB-manufactured structural members for improving vehicle passive safety can be found also in [203].

A Successive Heuristic Quadratic Approximation (SHQA) technique has been adopted in [227]. The shock response of the frame structure of a military vehicle subjected to a projectile impact has been minimised. Cross-sectional parameters of the members and joints of the space frame as well the cross sections of the armour have been optimised. The objective functions to minimise were the acceleration at several locations of the structure when subjected to the impact. Constraints on structural integrity of frame members and geometric constraints have been considered in the optimisation process.

Structural optimisation has been employed in assistance to materials selection in [71] for the design of floor sandwich panels of a lightweight vehicle. The mass minimization problem of the entire bottom structure of the vehicle subjected to static loading conditions has been solved. Constraints on structural stiffness, maximum stress and buckling have been considered. Several foam materials for the core of sandwich panels have been analysed and the most efficient has been selected.

Structural optimisation is widely employed also in the design of vehicle parts.

A design process for minimising the mass of vehicle components is presented in [126]. Topology, shape and size optimisation are used in an integrated manner: in a first stage, the optimal structural layout of the component is obtained by means of topology optimisation, in the subsequent stages, shape and size optimisation are employed for the detailed design. Applications to the design of lightweight components such as the lower arm of a McPherson suspension; the lower floor aluminium panel and the front plastic hood can be found in the reference.

A FEM-based size optimisation of a vehicle wheel has been performed in [261]. The total mass of the wheel has been minimised subjected to constraints on structural stiffness and fatigue strength.

Shape optimisation of the upright of a road vehicle has been performed in [110], where the total mass of the component has been minimised subject to constraints on structural stiffness and durability. An optimisation based on global approximation models has been performed. Design variables describing the shape of the component have been identified, finite element simulations for different combinations of the design variables have been performed and objective functions have been computed. A Kriging interpolation method has been used for approximating the response of the system. The obtained model has been then used in the optimisation process.

The same component has been studied in [62, 212], where a reliability-based design approach has been employed, i.e. a probabilistic optimal solution has been derived.

The structural layout of brake discs with internal cooling ribs has been analysed in [213, 233]. The layout of the internal cooling channels has been optimised for minimising brake squeal.

Since optimised solutions are generally not robust, reliability-based design optimisation has rapidly diffused in vehicle engineering. As well explained in [93], the effect of uncertainties in material properties, manufacturing processes, loading conditions, have to be accounted for a robust design. The optimisation problem is therefore reformulated in a probabilistic way in order to obtain a reliable design (see Chap. 9). Examples of reliability-based structural optimisation on vehicle components can be found in [62, 93, 212, 249].

Topology optimisation as well has been widely used for lightweight design of road vehicles [204]. The main advantage of such an approach is that optimal structural layouts can be obtained at an early stage of the design process.

A methodology for the design of high performance vehicles based on topology optimisation is presented in [43]. Firstly an initial structural layout of the entire vehicle frame is derived by means of topology optimisation technique, in the subsequent stages, the structure is refined by means of shape and size optimisation.

An application of topology optimisation for the design of a race motorcycle frame can be found in [17].

Conceptual design of a wheel based on topology optimisation is addressed in [265].

Topology optimisation has been employed for the design of energy-absorbing structures for impact loads in [166]. The structure deformation energy in crash simulation is maximised; constraints on maximum deflection have been introduced. The SIMP (Solid Isotropic Material with Penalization) approach has been followed; material plasticity and geometrical nonlinearities have been taken into account.

Topology and shape optimisation have been used for the design of an air suspension bracket for heavy vehicles in [102]. In the first design step, the classical minimum compliance problem has been solved and the optimal structural layout has been computed by means of the SIMP approach. In a second step, shape optimisation has been performed on the obtained structure in order to satisfy stress constraints and minimise the total mass.

13.1 Multi-objective Structural Optimisation of a Brake Calliper

In this section the structural optimisation of a sport car brake calliper is described.

Fig. 13.1 Quarter-car model for vertical dynamics simulation

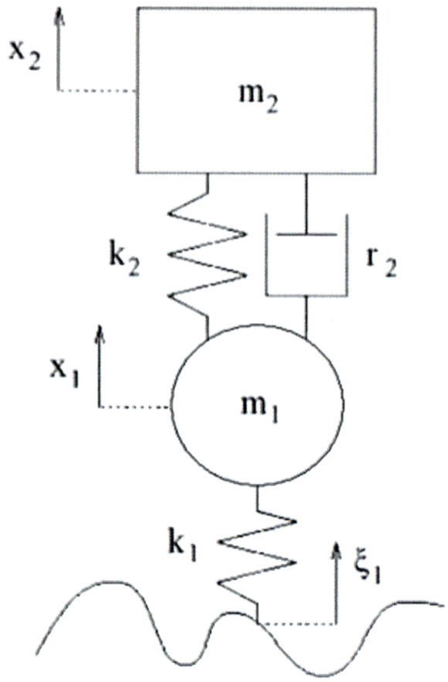

Disc brakes are a combination of a brake calliper and a brake disc. In disc brakes, the clamping forces of the brake calliper are applied to the brake pads in an axial direction by means of hydraulic cylinders. The brake pads then act on the flat surface of the disc. The pistons and the pads are located and retained in the calliper that fits like a saddle on the outer circumference of the disc.

Two kinds of architectures of disc brakes can be distinguished [81, 83]:

- Fixed calliper
- Floating calliper

The calliper constitutes a relevant part of the unsprung mass of the vehicle, therefore it significantly affects its handling performances [85, 153, 196]. In fact if we consider the simple 2 DOF (degrees of freedom) quarter car model of Fig. 13.1 the *road holding*, i.e. the standard deviation of the tyre vertical load, can be expressed as [85]

$$\sigma_{F_z} = \sqrt{\frac{A_b v}{2}\left(\frac{(m_1+m_2)^3}{m_2^2 r_2}k_2^2 - 2\frac{m_1 k_1 (m_1+m_2)}{m_2 r_2}k_2 + \frac{k_1 r_2 (m_1+m_2)^2}{m_2^2} + \frac{k_1^2 m1}{r_2}\right)} \quad (13.1)$$

where m_1 and m_2 are the unsprung and sprung mass, respectively, k_1 is the tyre radial stiffness, k_2 and r_2 are the suspension stiffness and damping factor while v is the vehicle speed and A_b is the road irregularity.

By inspecting 13.1 one may notice that road holding depends mainly on the ratio $\frac{m_1}{m_2}$ which is a common parameter in vehicle design and it is known to be advantageous to keep it small [39, 85, 196].

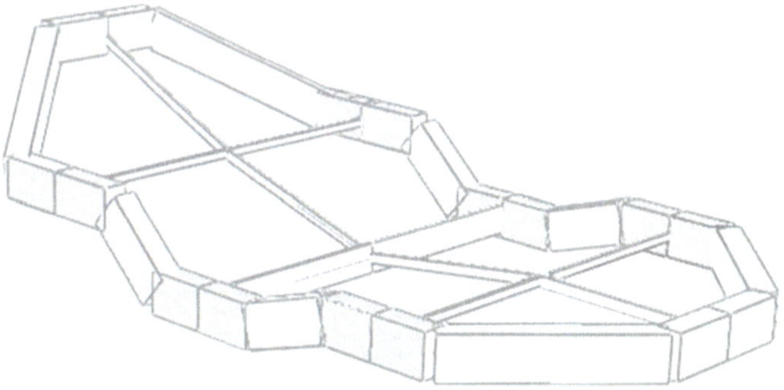

Fig. 13.2 Simplified Finite Element Model of the brake calliper, adapted from [16]—grey elements represent the location of pistons

From the above considerations it is clear the importance of reducing the total mass of the calliper. Besides, an acceptable structural stiffness has to be maintained [35, 176, 201]. A high structural stiffness of the brake calliper allows short pedal stroke, improving comfort, driving feelings and also the safety of the vehicle. Additionally, high structural stiffness could prevent other negative phenomena such as uneven wear of the brake pads and large deformations of the brake calliper [40]. It is therefore clear that minimising the mass and maximising the stiffness should be primary tasks of the designer, resulting in a trade-off between these two objective functions.

A simplified Finite Element (FE) model of the calliper has been developed and validated. The model has been employed for a preliminary structural optimisation of the component. The obtained results show that optimal solutions are characterised by an asymmetric shape. A comparison with symmetric structural layouts has been performed in order to assess if asymmetry were an attribute of optimal solutions.

Finally, the influence of the position of the connection points with the vehicle on the structural efficiency of the component has been investigated by comparing the Pareto-optimal solutions [16].

13.1.1 Simplified FE Model of the Calliper

The calliper is modelled as the frame structure of Fig. 13.2.

The pistons represent the *non-design* domain; thus their positions are derived from the original design and do not change during the optimisation process. Three-dimensional beam elements have been selected for modelling the structure. Obviously this model is a very rough approximation of the actual structure (Fig. 13.3), but it could be useful for a preliminary analysis.

Since the structure is composed mainly of bulky elements (the slenderness ratio is well over $1/15$), the Timoshenko formulation has been adopted, meaning that also the shear contribution has been considered for computing the deformation of the

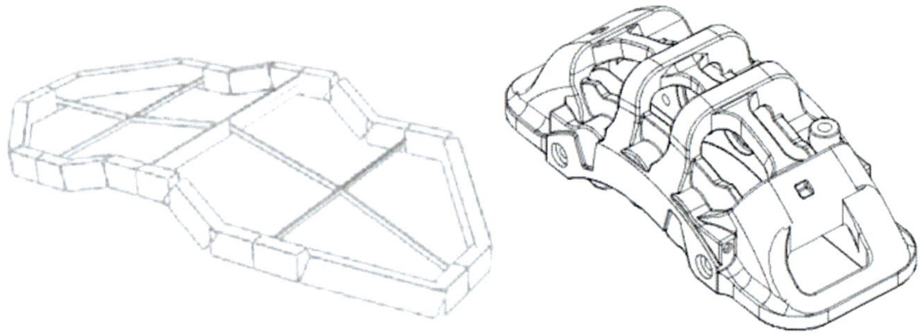

Fig. 13.3 Simplified beam element model (left) and actual sport car brake calliper (adapted from [16])

structure [94]. Each element is composed of 2 nodes with 6 DOF per node [56, 77] and has a rectangular cross section.

The Timoshenko element stiffness matrix for a general rectangular beam element reads [228]

$$
\mathbf{K_e} = \begin{bmatrix}
a_1 & 0 & 0 & 0 & 0 & 0 & -a_1 & 0 & 0 & 0 & 0 & 0 \\
0 & a_2 & 0 & 0 & 0 & a_3 & 0 & -a_2 & 0 & 0 & 0 & a_3 \\
0 & 0 & a_6 & 0 & -a_7 & 0 & 0 & 0 & -a_6 & 0 & -a_7 & 0 \\
0 & 0 & 0 & a_{10} & 0 & 0 & 0 & 0 & 0 & -a_{10} & 0 & 0 \\
0 & 0 & -a_7 & 0 & a_8 & 0 & 0 & 0 & a_7 & 0 & a_9 & 0 \\
0 & a_3 & 0 & 0 & 0 & a_4 & 0 & -a_3 & 0 & 0 & 0 & a_5 \\
-a_1 & 0 & 0 & 0 & 0 & 0 & a_1 & 0 & 0 & 0 & 0 & 0 \\
0 & -a_2 & 0 & 0 & 0 & -a_3 & 0 & a_2 & 0 & 0 & 0 & -a_3 \\
0 & 0 & -a_6 & 0 & a_7 & 0 & 0 & 0 & a_6 & 0 & a_7 & 0 \\
0 & 0 & 0 & -a_{10} & 0 & 0 & 0 & 0 & 0 & a_{10} & 0 & 0 \\
0 & 0 & -a_7 & 0 & a_9 & 0 & 0 & 0 & a_7 & 0 & a_8 & 0 \\
0 & a_3 & 0 & 0 & 0 & a_5 & 0 & -a_3 & 0 & 0 & 0 & a_4
\end{bmatrix} \tag{13.2}
$$

with

$$
\begin{aligned}
a_1 &= \frac{EA}{L} \\
a_2 &= \frac{12EJ_z}{L^3(1+\phi_z)} \\
a_3 &= \frac{6EJ_z}{L^2(1+\phi_z)} \\
a_4 &= \frac{4EJ_z}{L(1+\phi_z)} + \frac{\phi_z EJ_z}{L(1+\phi_z)} \\
a_5 &= \frac{2EJ_z}{L(1+\phi_z)} - \frac{\phi_z EJ_z}{L(1+\phi_z)} \\
a_6 &= \frac{12EJ_y}{L^3\left(1+\phi_y\right)} \\
a_7 &= \frac{6EJ_y}{L^2\left(1+\phi_y\right)} \\
a_8 &= \frac{4EJ_y}{L\left(1+\phi_y\right)} + \frac{\phi_y EJ_y}{L\left(1+\phi_y\right)} \\
a_9 &= \frac{2EJ_y}{L\left(1+\phi_y\right)} - \frac{\phi_y EJ_y}{L\left(1+\phi_y\right)} \\
a_{10} &= \frac{GJ_p}{L} \\
\phi_z &= \frac{12EJ_z\alpha L^2}{GA}, \quad \phi_y = \frac{12EJ_y\alpha L^2}{GA} \\
\alpha &= \frac{6}{5}
\end{aligned}
\tag{13.3}
$$

being E and G Young's modulus and the shear modulus, respectively, A and L the cross-sectional area and the beam length, $J_{y,z}$ and J_p the cross-sectional flexural and torsional moment of inertia, respectively.

The model is depicted in Fig. 13.4. Each piston is described by two elements and the pressure is inserted as an equivalent force on the middle node. The grey arrows represent the equivalent force caused by the pressure in each cylinder while the black arrows are the tangential braking forces caused by the friction between disc and pads.

The calliper is connected to the ground via linear spring elements on nodes 13 and 21, that account for the compliance of the upright at the connection points. The magnitude of the loads and the stiffness of the springs are reported in Tables 13.1 and 13.2.

Before proceeding with the optimisation the developed model has been validated with a commercial FEM software, results are shown in Table 13.3.

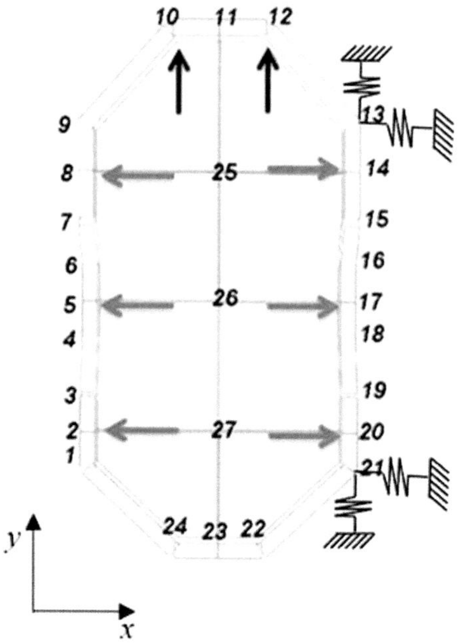

Fig. 13.4 Simplified Finite Element Model, adapted from [16]—node numbers are reported. The grey arrows are the pressure equivalent forces and the black arrows are the tangential forces

Table 13.1 Loads acting on the structure

Node number	F_x(N)	F_y(N)
2	−8000	0
5	−9000	0
8	−13000	0
10	0	15100
12	0	15100
14	13000	0
17	9000	0
20	8000	0

Table 13.2 Springs stiffness

Node number	k_x(N/mm)	k_y(N/mm)	k_z(N/mm)
13	44000	36000	36000
21	26000	53000	20000

Table 13.3 Validation of the developed model—nodal displacements

Node number	Displacement direction	Commercial software (mm)	Model (mm)
2	x	−0.110	−0.110
5	x	−0.200	−0.198
8	x	−0.136	−0.134
10	y	0.151	0.151
11	y	0.071	0.072

13.1.2 Multi-objective Optimisation of the Brake Calliper

The developed model has been employed for the structural optimisation of the calliper. The main goal of the activity is to minimise the total mass of the component while maximising its structural stiffness. A multi-objective optimisation approach has been followed.

13.1.2.1 Design Variables

The design variables that have been considered are the dimensions of the cross sections of the elements and the position of nodes 10, 12, 22 and 24, which are free to move in design space in order to explore different structural layouts. Identical z-coordinate for the mentioned nodes has been considered for the manufacturability of the calliper itself. The dimensions a and b of the cross sections of the perimetral element are included in the set of design variables—the small central elements have fixed cross section (Fig. 13.5). Elements with the same colour in Fig. 13.5 have same cross section. The overall number of design variables is therefore equal to 33.

Bounds on the design variables are reported in Tables 13.4 and 13.5.

13.1.2.2 Objective Functions

Being ρ and V_i the material density and the volume of each finite element, the total mass of the component (to minimise) is given by the sum of the masses of each element

$$m = \sum_{i=1}^{n_{el}} \rho_i \cdot V_i \tag{13.4}$$

The displacements of the loaded nodes have then to be minimised in order to maximise the structural stiffness of the calliper. A 6D (six dimensional) objective functions vector composed by the displacements in x-direction of nodes 2, 5, 8, displacements in y-direction of nodes 10, 12 and the overall mass is defined

Table 13.4 Field of variation of element cross-sectional dimensions

Element numbers	*a* (mm)	*b* (mm)
1,2,3	[30 65]	[20 65]
4,5,6	[32 65]	[20 65]
7,8	[38 70]	[25 65]
9	[25 65]	[17 65]
10,11	[20 65]	[17 65]
12	[25 65]	[17 65]
13,14	[38 70]	[25 65]
15,16,17	[32 65]	[20 65]
18,19,20	[30 65]	[20 65]
21	[20 65]	[17 60]
22,23	[20 65]	[17 60]
24	[20 65]	[17 60]

Table 13.5 Field of variation of the position of movable nodes

Node number	*x* (mm)	*y* (mm)	*z* (mm)
10	[10 36]	[150 220]	[15 42]
12	[69 94]	[150 220]	[15 42]
22	[69 94]	[−85 − 15]	[15 42]
24	[10 36]	[−85 15]	[15 42]

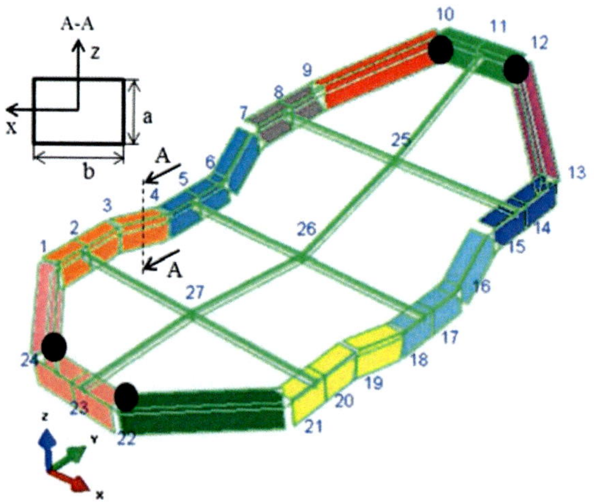

Fig. 13.5 Design variables—elements with the same colours have same cross section, adapted from [16]

$$\mathbf{f} = \begin{bmatrix} f_1 \\ f_2 \\ f_3 \\ f_4 \\ f_5 \\ f_6 \end{bmatrix} = \begin{bmatrix} u_{2x} \\ u_{5x} \\ u_{8x} \\ u_{10y} \\ u_{12y} \\ m \end{bmatrix} \tag{13.5}$$

displacements of nodes 14, 17 and 20 have not been considered for the moment since they are lower than the ones of the left side of the calliper.

In order to reduce the dimensionality of the problem a *reduced* objective functions vector $\mathbf{f}^*$ has been defined also

$$\mathbf{f}^* = \begin{bmatrix} \sum_{i=1}^{5} |f_i| \\ m \end{bmatrix} \tag{13.6}$$

Results have been computed for both cases (the 6D and 2D objective functions vector).

The multi-objective optimisation problem thus reads

$$\begin{aligned} &\min\ \mathbf{f}(\mathbf{x}) \quad s.t. \\ &x_i \ \in\ [LB_i \quad UB_i] \quad i = 1, ..., n_{dv} \end{aligned} \tag{13.7}$$

where LB_i and UB_i are the lower and upper bound of each of the n_{dv} design variables.

The PSI (Parameter Space Investigation) optimisation method [216, 217] has been applied for solving the problem. According to this method, a subset of all the possible combinations of the design variables is investigated: for each combination, the objective functions and the Pareto-optimal solutions are then selected by implementing definition 1.5 of Pareto optimality.

A low discrepancy Sobol sequence (see Sect. 2.2.2) has been used for selecting a “small” number of samples from all possible combinations of design variables well distributed in design variables domain. This method is desirable since the selected samples are uniformly distributed in the design domain. Additionally, the uniformity of the low discrepancy sequence is independent on the number of samples N, which means that even if N is small, the samples are distributed evenly in the space [216, 217].

13.1.3 *Optimal Solutions*

A simulation with 10^7 samples has been performed and the Pareto-optimal solutions have been extracted by implementing the sorting method [143].

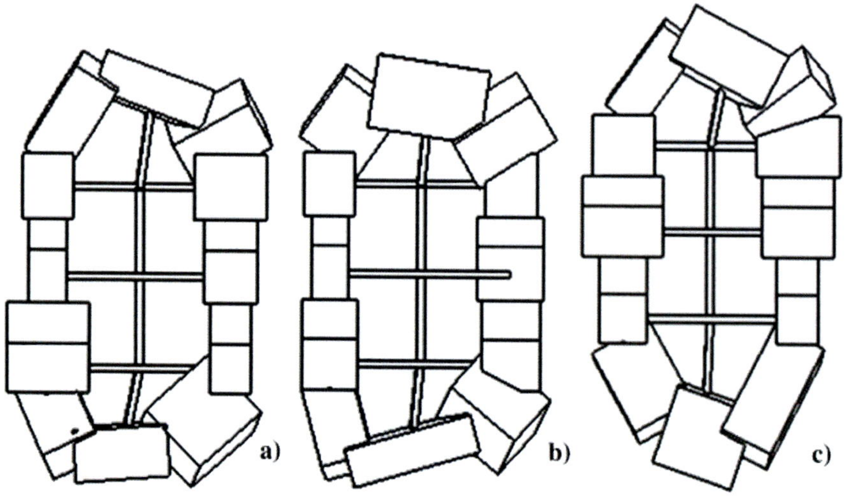

Fig. 13.6 Few solutions extracted from the Pareto-optimal set for the 6D objective functions vector, adapted from [16]

Table 13.6 Numerical values for the objective functions of solutions of Fig. 13.6

	Sol. A	Sol. B	Sol. C
u_{2x} (mm)	−0.136	−0.118	−0.128
u_{5x} (mm)	−0.151	−0.142	−0.128
u_{8x} (mm)	−0.093	−0.100	−0.086
u_{10y} (mm)	0.617	0.519	0.491
u_{12y} (mm)	0.396	0.397	0.397
m (kg)	2.068	2.280	2.333

Figure 13.6 depicts a few Pareto optimal solutions that have been obtained considering the 6D objective functions vector **f**. Numerical values of the objective functions for those solutions are reported in Table 13.6.

The obtained results show that optimal solutions are characterised by an asymmetric structural layout, suggesting that asymmetry could be a prerogative of optimality for this kind of problem. In order to assess that, a rigorous comparison between symmetric and asymmetric *optimal* solutions is required, i.e. a new optimisation problem with constraints on symmetry of the structure is solved and results are compared (see 13.1.4).

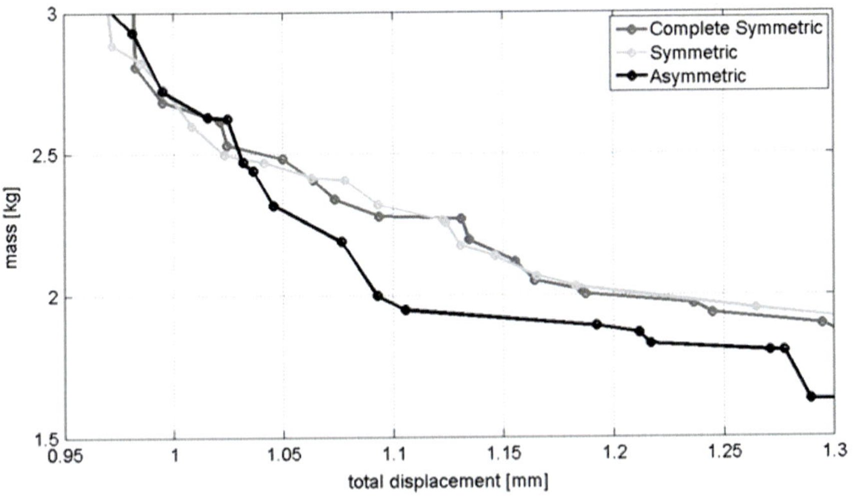

Fig. 13.7 Pareto-optimal solutions in the objective functions domain for the three analysed cases: "Asymmetric" (black line), "Simmetric" (light grey line) and "Complete Symmetric" (dark grey line). Adapted from [16]

13.1.4 *Comparison Between Symmetric and Asymmetric Layouts*

In this section, symmetric and asymmetric structural layouts are compared. Problem 13.7 now has been solved by imposing constraints on symmetry of the structure. Two other cases have been analysed, namely:

- *Symmetric* case. In this case, referring to Fig. 13.5, nodes 12 and 22 are the mirror image of nodes 10 and 24 with respect to the middle plane of the calliper, while cross sections are free.
- *Complete Symmetric* case. In this case, other than the nodes, the right side cross sections are also the mirror image of the left side.

For this analysis the reduced objective functions vector $\mathbf{f}^*$ has been considered in order to better visualise and compare the Pareto-optimal solutions.

Figure 13.7 shows the computed Pareto optimal sets for the three analysed cases. From the comparison of Fig. 13.7 it clearly emerges that asymmetric structural layouts are better than the others. For low values of mass (less than 2.5 kg) the black curve of Fig. 13.7 dominates the others, meaning that the asymmetric solution provides always a lower mass for a prescribed displacement. The *Symmetric* and *Complete Symmetric* design provide almost the same structural efficiency.

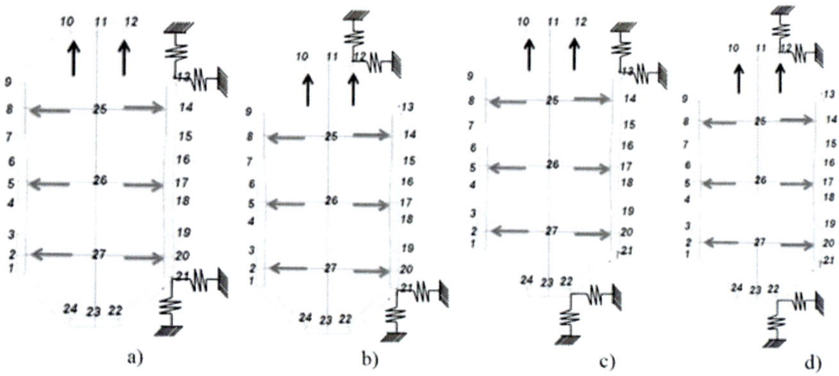

Fig. 13.8 Positions of the connection points considered

13.1.5 Position of the Connection Points

In this section, the effect of the position of the connection points between calliper and upright has been investigated.

The same approach of 13.1.4 has been followed, i.e. Pareto-optimal solutions have been computed for different problems where the position of the constrained nodes has been modified and results have been compared. Four different positions (see Fig. 13.8) of the constrained nodes have been evaluated.

The linear springs in the simplified model have been moved from their original position in order to evaluate the effect on the structural stiffness of the component. Also in this case the reduced objective functions vector has been considered in the optimisation. Additionally, also the displacements of nodes on the right side of the calliper have been included in the objective functions. The problem now reads

$$\begin{aligned} &\min \left(\begin{bmatrix} s = \sum_{i=1}^{8} d_i \\ m \end{bmatrix} \right) \quad s.t. \\ &x_i \in [LB_i \quad UB_i] \quad i = 1, \ldots, n_{dv} \\ &\mathbf{d} = \left[u_{2x}\; u_{5x}\; u_{8x}\; u_{10y}\; u_{12y}\; u_{14x}\; u_{17x}\; u_{20x} \right]^T \end{aligned} \tag{13.8}$$

The comparison of Pareto-optimal solutions is reported in Fig. 13.9.

Results (Fig. 13.9) show that solution (d) of Fig. 13.8 is the best one since the respective Pareto-optimal solutions are dominating the others. Four different solutions extracted from the Pareto-optimal set are reported in Figs. 13.10 and 13.11. Even in this case the obtained structural layouts exhibit an asymmetric shape which comply with results obtained in (13.1.4).

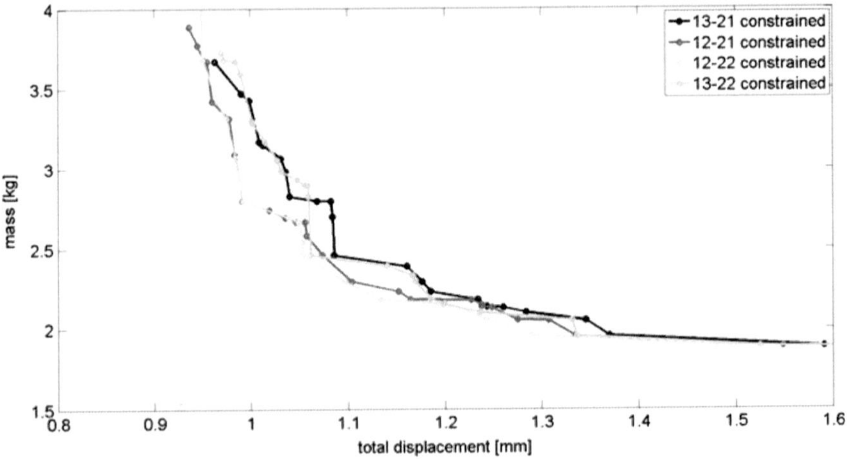

Fig. 13.9 Pareto-optimal solutions in the objective functions domain for different positions of the connections to the upright

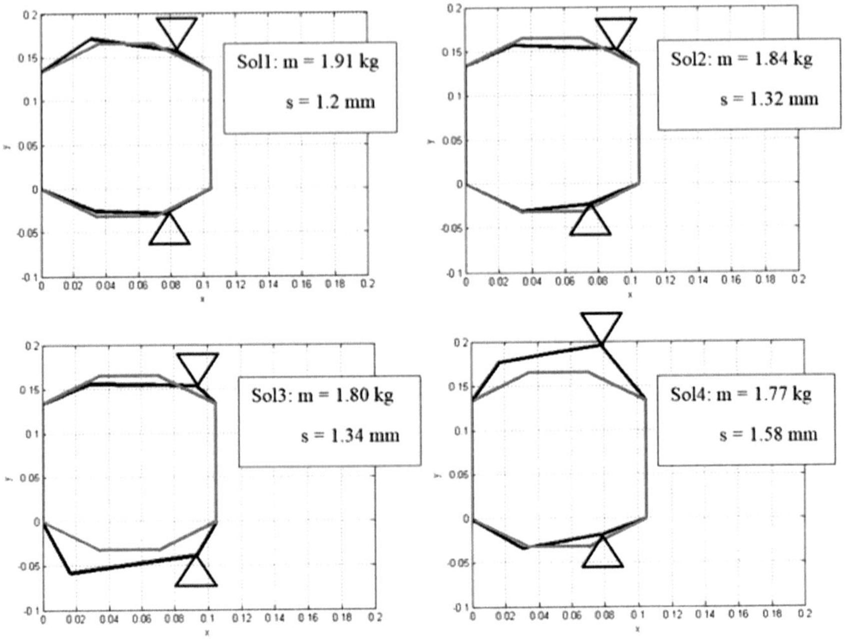

Fig. 13.10 Optimal layout (black) and reference layout (grey)—triangles are the constraints

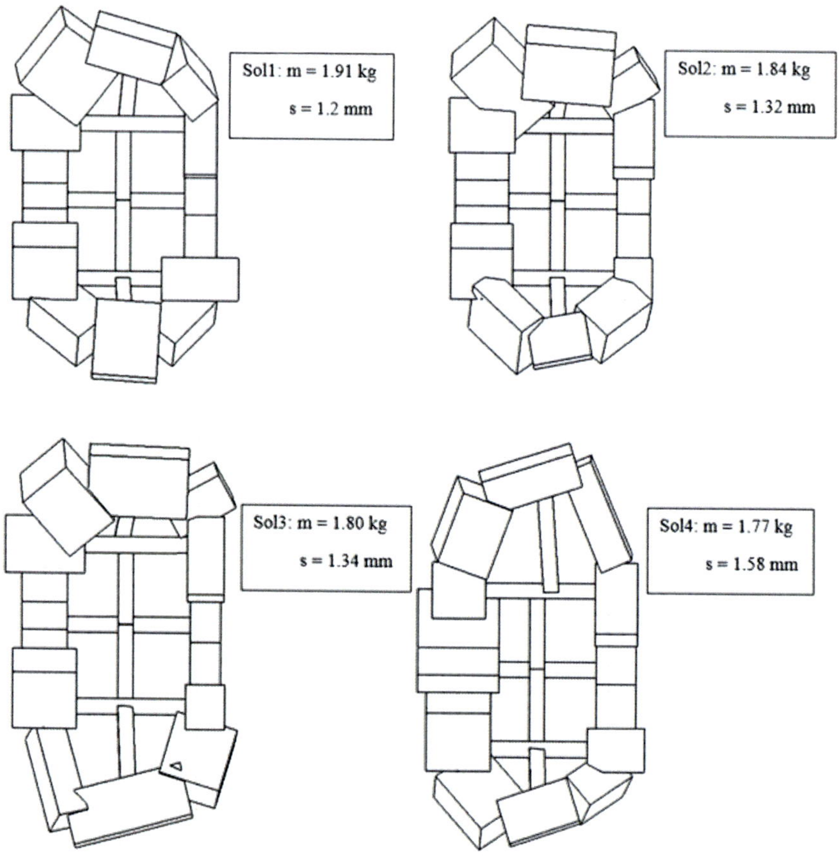

Fig. 13.11 Three-dimensional representation of some of the obtained solutions

13.2 Topology Optimisation of a Brake Calliper and Upright of a Race Car

In this section, topology optimisation approach has been applied for the preliminary design of the front calliper and upright of a race car. The calliper is the same one that has been described in Sect. 13.1. Firstly, topology optimisation has been applied for the structural design of the calliper only and results have been analysed. Secondly, also the front upright has been included in the optimisation and the structural layouts of both the calliper and the upright have been derived. The obtained results could be a good starting point for a more detailed design stage and the development of the new components.

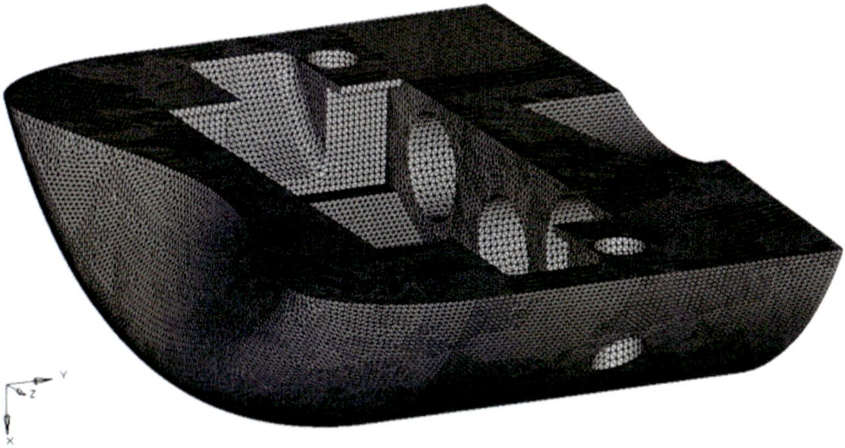

Fig. 13.12 Calliper design domain, given by the total admissible space that the component has to fit

13.2.1 Topology Optimisation of the Brake Calliper

The SIMP approach has been employed for the structural optimisation of the brake calliper. Two different domains have been considered for the component, namely, the *design domain* and the *frozen domain*. The design domain is given by the total admissible volume that the component has to fit. The design domain is discretised with finite elements whose densities are the design variables of the optimisation problem. The frozen domain instead is fixed and is not involved in the optimisation process.

Both the frozen and design domain have been discretised with linear tetrahedrons with an average dimension of 2 mm. The design domain is depicted in Fig. 13.12 and represents the admissible space that the final component has to fit.

The frozen domain is given by the two connections to the upright, the six cylinders where the hydraulic pressure is applied and by the supports of the braking pads as shown in Fig. 13.13.

The calliper is fixed to the ground via the elastic springs of Table 13.2 that account for the structural stiffness of the upright. A tangential braking force of 33400 N has been uniformly applied on the pads supports. This force is given by the friction generated when the pads are pressed against the disc surfaces. An hydraulic pressure of 142 bar is exerted on the six cylinders of Fig. 13.13. The finite element model employed for the optimisation is shown in Fig. 13.14. Aluminium alloy has been employed as a reference material for the calliper.

The SIMP approach has been applied for the optimisation problem. A classical minimum compliance problem has been set up. The total compliance of the structure has been minimised subjected to a constraint on the volume fraction of the final component. An additional constraint denoted as *volumetric displacement* has been

Fig. 13.13 Calliper frozen domain, given by the six cylinders where hydraulic pressure is applied, by the supports of the braking pads and by the connection points to the upright

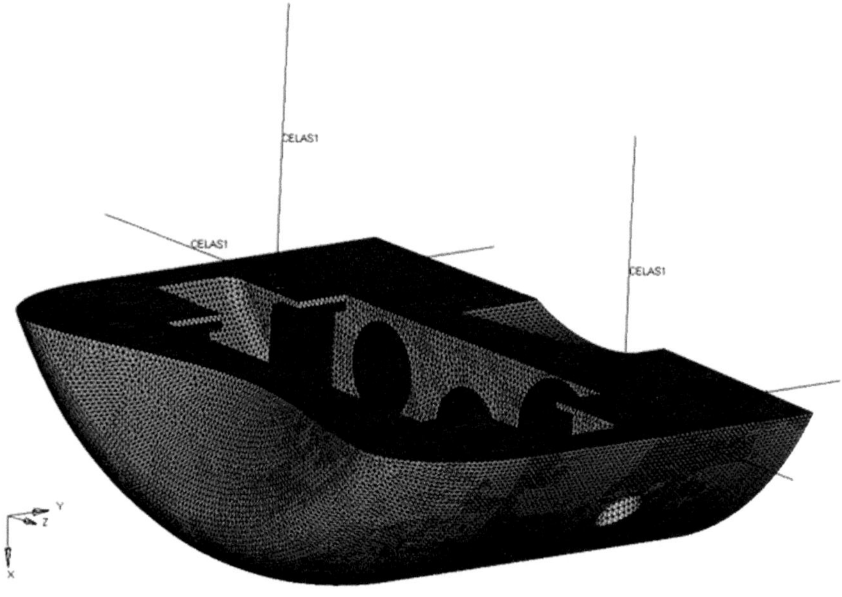

Fig. 13.14 Finite Element model for topology optimisation— light grey elements are the design domain, dark grey elements are frozen domain

considered in the optimisation problem. This constraint is given by the total volume variation in the hydraulic cylinders due to the deformation of the calliper when subjected to braking loads [16]. The volumetric displacement is somehow related to the pedal displacement that the driver feels when activating brakes since it is directly proportional to the amount of fluid that is pressed in the hydraulic cylinders.

The topology optimisation problem can be expressed as

$$\begin{aligned} &\min_{\rho_e} \left(\frac{1}{2} \mathbf{u}^T \mathbf{K}_g \mathbf{u} \right) \quad s.t.: \\ &\mathbf{K}_g = \sum_{e=1}^{N} \rho_e^p \mathbf{K}_e \\ &\sum_{e=1}^{N} \rho_e v_e \leq V^* \\ &D \leq D^* \\ &\rho_{min} \leq \rho_e \leq 1 \end{aligned} \tag{13.9}$$

where **u** and **f** are the displacement and load vectors, respectively, $\mathbf{K}_g$ and $\mathbf{K}_e$ are the global stiffness matrix and the element stiffness matrix, while D is the volumetric displacement.

A penalization factor p equal to 4 has been employed in the SIMP algorithm. Density design variables are iteratively updated until convergence is reached.

The solution converges after 20 iterations and the final structural layout is shown in Figs. 13.15 and 13.16.

The obtained solution has an overall mass of 1.128 kg, with a reduction of 17% with respect to the actual component. The optimal layout shows an asymmetric shape; this result confirms what was obtained in Sect. 13.1 with the multi-objective optimisation. Moreover, the asymmetric layout was also quite expected because of the asymmetric boundary conditions [191].

13.2.2 *Topology Optimisation of the Calliper and the Upright*

In this section the complete system made by the assembly of the brake calliper and the front upright of the vehicle have been considered and optimised by applying the SIMP approach. Two different configurations have been considered and compared, namely, the *horizontal configuration* and the *vertical configuration*. In the horizontal configuration the calliper is mounted below the upright in horizontal position, in the vertical configuration instead the calliper stands in vertical position. In both cases the calliper is connected to the upright via two bolted joints.

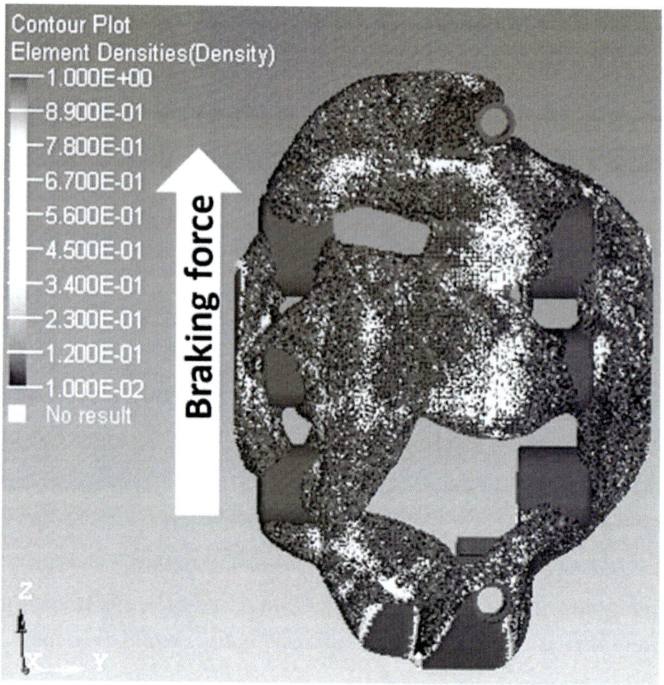

Fig. 13.15 Optimal structural layout of the brake calliper— bottom view

13.2.2.1 Horizontal Configuration

As in Sect. 13.2.1, the design domain and the frozen domain have been defined. The main difference with respect to Sect. 13.2.1 is that now two separated design domains have to be considered, i.e. the one related to the calliper and the one related to the upright. The frozen domain is depicted in Fig. 13.17 and is composed by the frozen domain of the calliper (Fig. 13.13) the hub bearing houses and the connections to the wishbones and the steering rod.

The two design domains are shown in Fig. 13.18.

Design and frozen domains have been discretised with linear tetraedric elements, the average mesh size was 2 mm. Aluminium alloy has been considered as a reference material for both components. The upright is fixed to the ground through three spherical hinges that represent the connections with the two wishbones and the steering rod. Each hinge is rigidly connected to the respective attachments on the upright as shown in Fig. 13.19.

The calliper is loaded as explained in Sect. 13.2.1. Loads acting on the upright come from the tyre/road contact forces. The vertical and lateral force are applied at the contact patch that is rigidly connected to the upright centre as shown in Fig. 13.20. The upright centre is rigidly connected to the bearing houses.

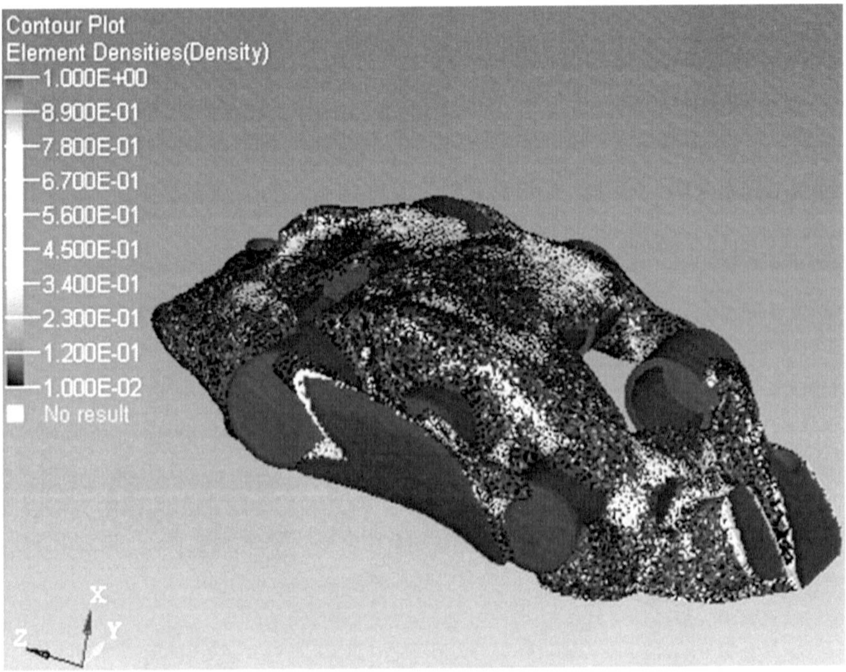

Fig. 13.16 Optimal structural layout of the brake calliper— lateral view

The resultant longitudinal force F_x applied at the upright centre is obtained through the following relation:

$$F_x = F_{long} - F_b \tag{13.10}$$

where F_{long} is the longitudinal tyre force acting at the contact patch, while F_b is the braking force that acts at the pads support. The moment M_z is given by the product of the longitudinal force F_{long} and the steering arm, i.e. the distance between the centre of the contact patch and the centre of the upright in the y-direction.

Several loading conditions have been considered in the optimisation problem—these load cases represent the most critical working situations for the components and are reported in Table 13.7.

The topology optimisation problem reads

Fig. 13.17 Calliper and upright frozen domains

$$
\begin{aligned}
&\min_{\rho_e}\left(\sum_{k=1}^{6}\mathbf{u}_k^T\mathbf{K}_g\mathbf{u}_k\right) \quad s.t.: \\
&\mathbf{K}_g=\sum_{e=1}^{N}\rho_e^p\mathbf{K}_e \\
&V_{calliper}\le V_1^* \\
&V_{upright}\le V_2^* \\
&D_j\le D_j^* \qquad j=1,\dots,3 \\
&\rho_{min}\le\rho_e\le 1
\end{aligned} \tag{13.11}
$$

Fig. 13.18 Calliper and upright design domains

Table 13.7 Load cases considered for structural optimisation

Load case	ID number	F_{long}(N)	F_y(N)	F_z(N)	F_b(N)	Brake pressure (bar)
Straight line braking	1	11400	0	11100	31000	130
Corner-brake1	2	9900	11500	8800	27000	115
Corner-brake2	3	8600	10000	12000	23500	100
Corner	4	0	18900	10100	0	0
Bump	5	0	0	13000	0	0
Kerb	6	0	−20300	4800	0	0

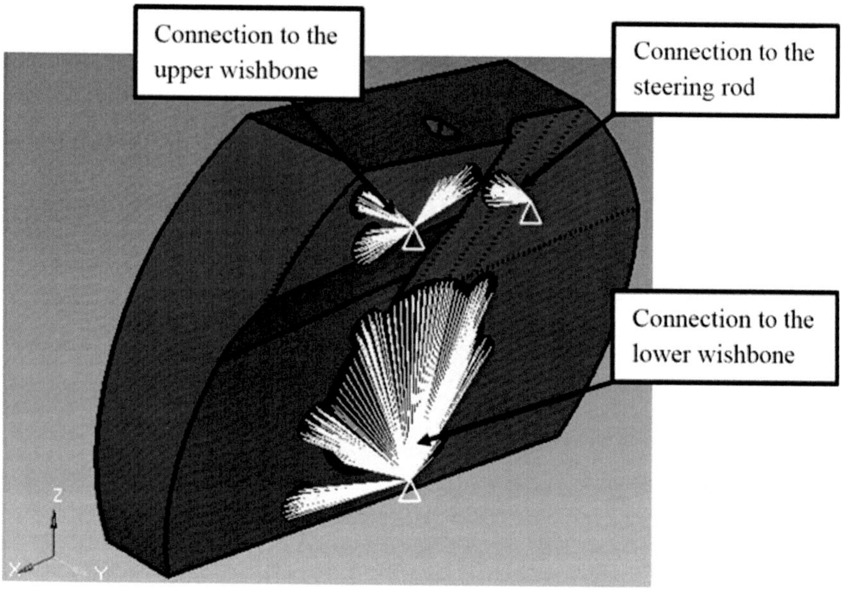

Fig. 13.19 Upright boundary conditions

Fig. 13.20 Tyre contact forces applied to the upright

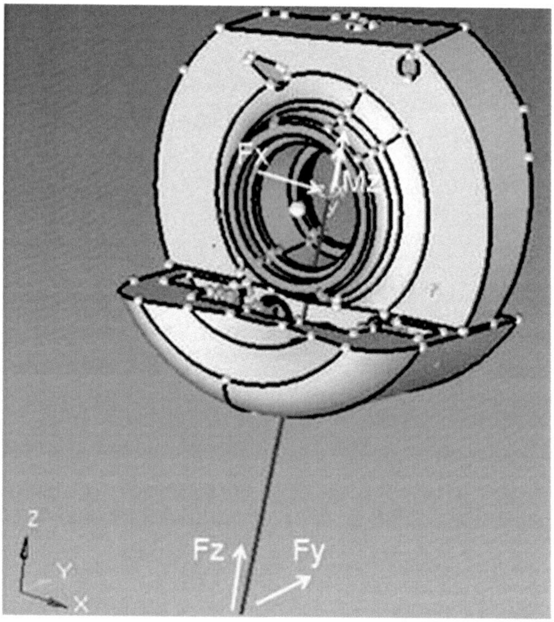

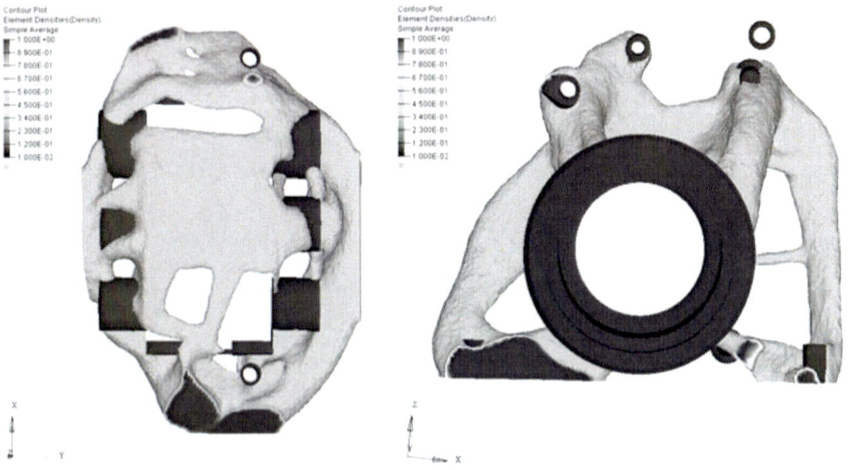

Fig. 13.21 Solution of topology optimisation problem for the horizontal configuration—calliper (left) and upright (right)

Table 13.8 Structural performances of the obtained solution and comparison with the actual component—horizontal configuration

Performance index	Value	Variation w.r.t. actual component
Calliper mass (kg)	1.063	−21.7%
Upright mass (kg)	1.796	−9.3%
D_1 (mm^3)	459	+4.1%
D_2 (mm^3)	272	−28.4%
D_3 (mm^3)	244	−26%

where the objective function is the sum of the compliances of each load case. Third and fourth line of (13.11) are the volume constraints on the calliper and the upright, respectively, while the fifth line is the upper bound on the volume displacement of the calliper for each load case where brake pressure is applied (the first three load cases).

A penalization factor equal to 4 has been considered. Convergence was obtained after 36 iterations, the final solution is shown in Fig. 13.21. The structural performances of the obtained solution are reported in Table 13.8.

The obtained solution is very promising since it provides a significant reduction on the overall mass of the unsprung mass of the vehicle even though a (small) increase in the volume displacement in the straight line braking. A preliminary analysis of the stress field on the obtained structure has been performed and results are shown in Fig. 13.22. The stress levels are quite acceptable in almost all the components, obviously the geometry has to be refined and more detailed evaluation of the stress

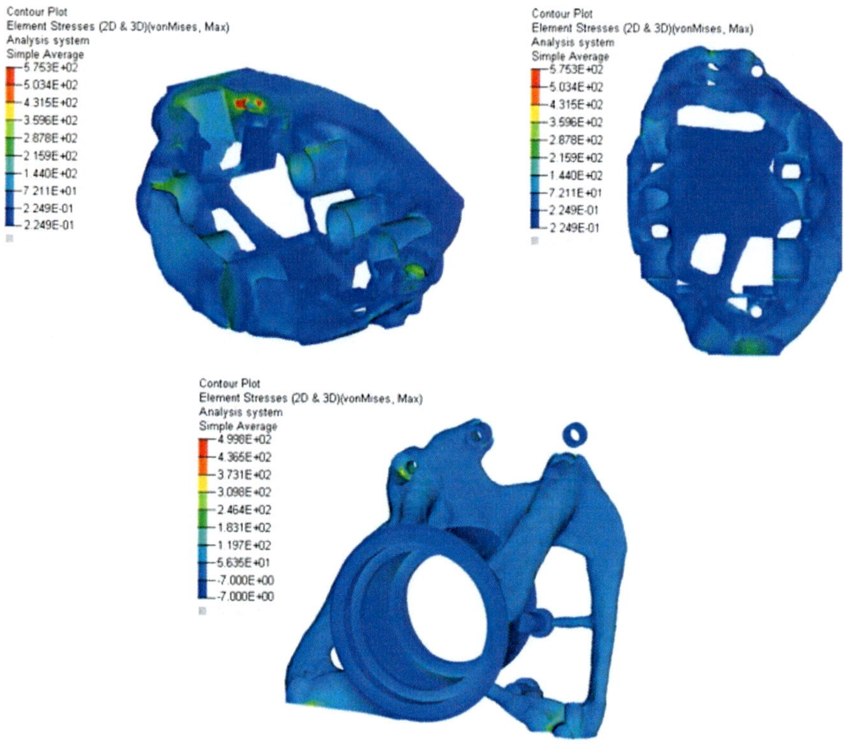

Fig. 13.22 Preliminary stress analysis on the obtained structural layout

field is required, but the advantage is that an optimised starting solution has been obtained.

13.2.2.2 Vertical Configuration

The same procedure has been followed for the vertical configuration. In this case the finite element model looks like the one of Fig. 13.23.

The same loading conditions and parameters used for the horizontal configuration have been considered for this alternative configuration. Convergence is reached after 35 iterations in this case and the solution can be seen in Fig. 13.24.

The stress fields on the two components are reported in Figs. 13.25 and 13.26. The structural layout of the calliper is almost identical for both the vertical and the horizontal configuration. The vertical configuration, however, exhibits better structural performances as it can be observed from Table 13.9.

The vertical configuration has shown a better structural efficiency than the horizontal one. Results show also that the structural layouts of the two callipers are almost identical, meaning that the way that the calliper is connected to the upright

Fig. 13.23 Design domain (light grey) and frozen domain (dark grey) for the topology optimisation of the upright and the calliper in vertical configuration, adapted from [8]

Table 13.9 Structural performances of the obtained solution and comparison with the actual component—vertical configuration

Performance index	Value	Variation w.r.t. actual component
Calliper mass (kg)	1.063	−21.7%
Upright mass (kg)	1.529	−22.8%
D_1 (mm^3)	375	−14.7%
D_2 (mm^3)	201	−46.9%
D_3 (mm^3)	192	−41.8%

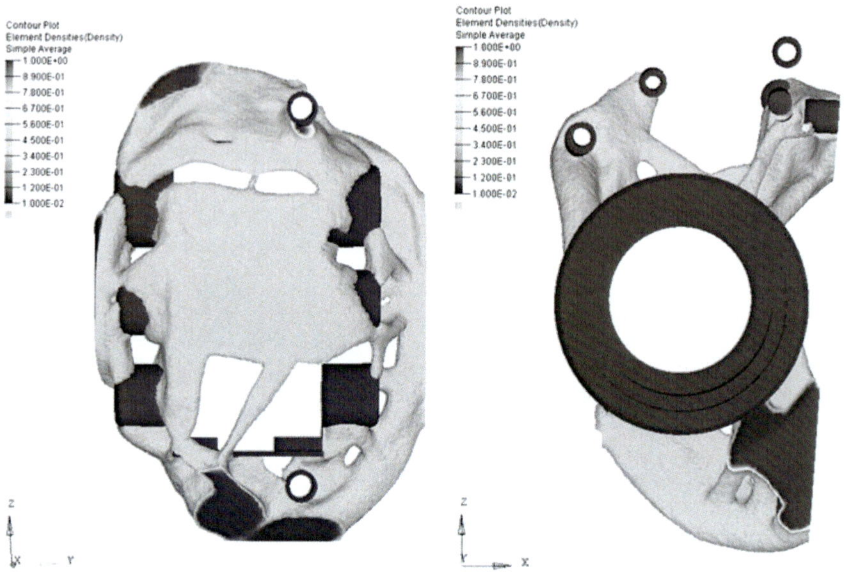

Fig. 13.24 Solution of topology optimisation problem for the vertical configuration—calliper (left) and upright (right), adapted from [8]

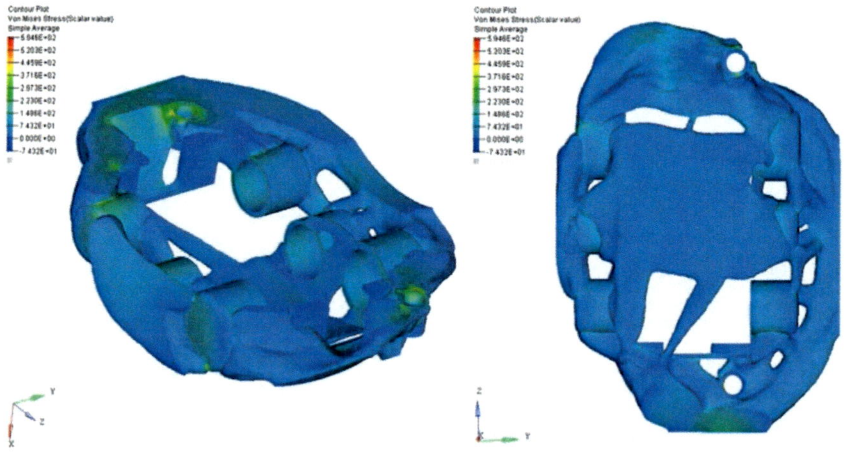

Fig. 13.25 Von Mises stress on the optimised calliper—vertical configuration, adapted from [8]

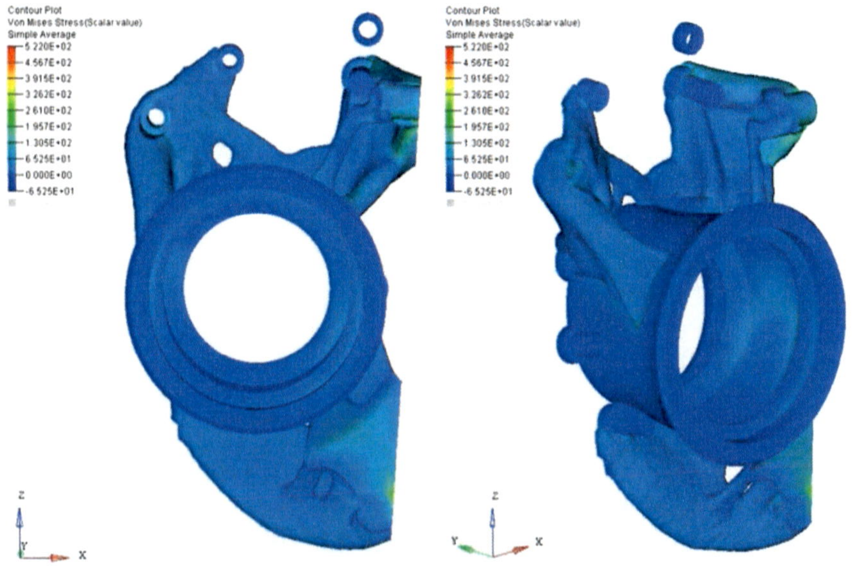

Fig. 13.26 Von Mises stress on the optimised upright—vertical configuration, adapted from [8]

deeply affects the structural behaviour of the components and the vertical configuration is preferable. On the other hand, the horizontal configuration has the important advantage of minimising the height of the masses, thus moving the centre of gravity of the entire vehicle downwards.

13.3 Topology Optimisation of a Motorcycle Wheel

This section deals with structural optimisation of a front wheel of a race motorcycle. The minimum compliance problem has been solved for the considered component. Constraints on vertical, lateral and longitudinal stiffness of the wheel have been considered in the optimisation process.

Loads acting on the component have been obtained from experimental data acquired by means of a specifically developed forces/torques sensor denoted as *Smart Wheel* (the reader is addressed to reference [91] for details on the sensor).

Fig. 13.27 Wheel FEM model for topology optimisation—the dark volume is design domain while the light grey volumes are frozen domain

13.3.1 Finite Element Model of the Wheel

The FEM model employed for the optimisation is shown in Fig. 13.27, the dark volume represents the design domain. The outer rim and the internal hub (light grey volumes) have been considered as frozen domain.

Linear brick elements have been employed for the analysis. Several load cases representing typical riding conditions have been considered. Straight line braking and pure cornering have been analysed.

A uniform pressure that accounts for tyre inflation has been applied to the rim. Radial load acting on the wheel has been applied through the distributed forces in radial and axial (i.e. y-direction in Fig. 13.27) directions depicted in Fig. 13.28. The force distribution of Fig. 13.28 is applied on the wheel rim and has been obtained from a specifically developed analytical model of the tyre structure. The developed model is based on a flexible ring laying on a suspended foundation and allows to calculate the load distribution at the tyre/rim interface in case a vertical load is applied. The reader is addressed to reference [18] for further details on the model.

Longitudinal and lateral (i.e. x- and y-direction in Fig. 13.27) forces have been applied at the contact point, rigidly connected to the rim as shown in Fig. 13.29.

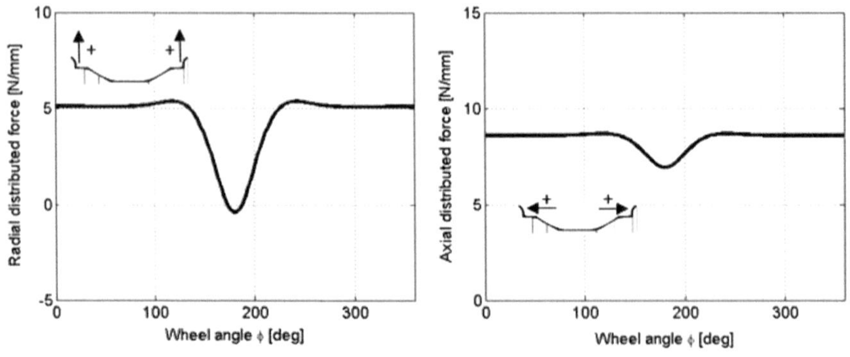

Fig. 13.28 Radial and axial distributed forces acting on the rim when subjected to a vertical load of 2500 N with a tyre inflation pressure of 2.1 bar

Fig. 13.29 Wheel model for TO—longitudinal and lateral loads are applied at the contact point rigidly connected to the rim

Table 13.10 Load cases considered for topology optimisation of the wheel

Load case	ID number	F_x(N)	F_y(N)	F_z(N)	Tyre pressure (bar)
Straight line braking	1	−1500	0	2500	2.1
Corner	2	0	200	1700	2.1

The wheel is fixed at the two bearings centres, rigidly connected to the bearing houses.

The brake discs fixtures have been rigidly connected to a central point and the rotation of this point around the y-axis has been constrained. Such a constraint scheme attempts to replicate the actual coupling between the wheel and the two brake discs (one per side).

The topology optimisation problem reads

$$\begin{aligned} &\min_{\rho_e} \left(\sum_{j=1}^{2} \mathbf{u}_j^T \mathbf{K}_g \mathbf{u}_j \right) \quad s.t.: \\ &\mathbf{K}_g = \sum_{e=1}^{N} \rho_e^p \mathbf{K}_e \\ &V \leq V^* \\ &k_x \geq k_x^* \\ &k_y \geq k_y^* \\ &k_z \geq k_z^* \\ &\rho_{min} \leq \rho_e \leq 1 \end{aligned} \tag{13.12}$$

where the sum of the compliances of the straight line braking and cornering load case is minimised with respect to ρ. The third line is the volume constraint, while the fourth, fifth and sixth lines are the constraints on the wheel longitudinal, lateral and vertical stiffness, respectively. A penalization factor p equal to 4 has been used.

The considered load cases are reported in Table 13.10.

Since the wheel rotates, a periodic structure (w.r.t. y-axis) is needed. Periodicity constraint of the final solution has been imposed in the optimisation process, meaning that the optimal solution will be characterised by a cyclic repetition of a determined structural layout around the wheel rotation axis. Additionally, a draw constraint is included to account for the wheel manufacturing process (the wheel is realised from a forged semifinished that undergoes grinding and milling process).

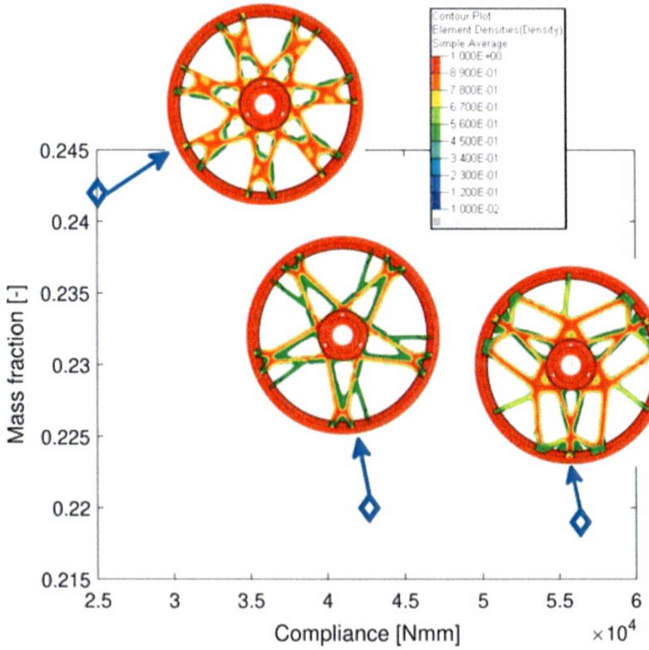

Fig. 13.30 Optimal structural layouts with different periodicity constraints, adapted from [19]

13.3.2 Results

Several optimisation processes have been performed by varying the number of cyclic repetitions. The three-, five- and seven-spoked structural layouts were evaluated. In all the analysed cases, well defined structures have been obtained. Results are shown in the plot of Fig. 13.30, where the respective values of compliance and mass fraction are highlighted.

A trade-off between stiffness (proportional to the number of spokes) and the mass is clear. A stiffer structure is obtained by increasing the number of spokes, a structure with a larger number of spokes has a larger mass. All the obtained solutions are (Pareto)-optimal, i.e. the designer can choose the preferred compromise between stiffness and mass. It will be a task of the designer choosing one of the presented solutions. Passing from the five-spoked to the three-spoked structure, the reduction in mass is very limited compared to the increase in compliance. The three-spoked structure is not the best choice for racing motorcycle wheels.

Table 13.11 Loads applied to the wheel in the four considered load cases

Load case	Vertical force (N)	Longitudinal force (N)	Lateral force (N)
Bump	15000	7500	0
Corner	7000	0	7000
Straight brake	7000	7000	0
Corner brake	8000	5500	5000

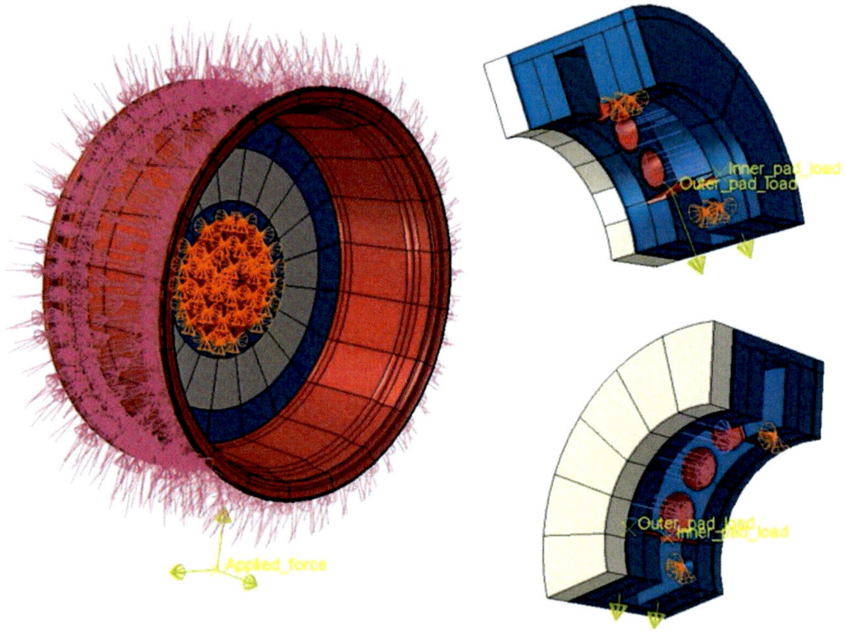

Fig. 13.31 Wheel and brake calliper models. Left wheel. Right: calliper views

13.4 Concurrent Topology Optimisation of a Wheel and Brake Calliper Assembly

In this section, the structural optimisation of a wheel and brake calliper assembly is presented.

In the design of these two components, one of the most critical aspects to be considered is the limited room available. In fact, to improve braking performances of the vehicle, the diameter of the brake disc tends to be maximised and therefore the brake calliper is naturally placed very close to the wheel. In the structural optimisation of these two components, other than finding the optimal structural layout, the problem of the optimal allocation of the (limited) available space among the calliper and the wheel has to be solved.

The problem has been solved by means of the concurrent topology optimisation approach described in Chap. 12.

The wheel and brake calliper are depicted in Fig. 13.31. The calliper is a fixed calliper with three pistons per each side. The calliper is constrained to the ground at the location of the two radial holes in the inner side of the calliper. To the calliper, a pressure load of 6.3 MPa is applied at each piston location and a force of 10 kN is applied on each of the two stops of the pads in the circumferential direction.

Referring to the wheel, four different load cases have been considered. For each load case, the inner side of the wheel hub is constrained to the ground and a pressure of 0.1 MPa is applied to the rim. The load is applied to a reference point corresponding to the centre of the contact patch and connected to the two sides of the flange of the wheel rim for a angle of 72°. The four load cases are described in Table 13.11.

The two models are meshed by trilinear brick elements with reduced integration and solved by a linear solver. For the calliper, 23157 elements have been used, while for the wheel 206640 elements.

In Fig. 13.32, the wheel and brake calliper assembly is shown and the design space and the non-design space are reported. In the picture, only half of the wheel is reported to show the brake calliper model. For both components, the design space is divided into the non-shared part (region that is assigned exclusively to one component) and into the shared part (region that can be assigned to either component during the optimisation process). The non-design domains are reported in red, while the design domains are reported in blue (non-shared parts) and in white (shared parts). It must be observed that since the wheel rotates with respect to the calliper, a complete annular section of the wheel has to be considered as shared domain and not only the part corresponding to the actual position of the calliper. Moreover, when dividing the shared part of the domains, a cylindrical symmetry condition must be considered. The cylindrical symmetry condition is applied as follows. The meshes have been realised along cylindrical paths. In this way, the elements of the shared domain pertaining to each of the two bodies can be grouped according to the circumferential path on which have been generated. The sensitivity of all the elements pertaining to the same circumferential group is summed up and the comparison is realised per groups between the two bodies. Once a circumferential group is assigned to one of the two bodies, the pseudodensity of the elements of the group is computed in the standard way, i.e. the elements of the same group can have different pseudodensities.

Finally, in the optimisation of the wheel, in order to obtain a reasonable material distribution, a cyclic symmetry with the repetition of five sectors is applied. The resulting wheel will be a periodic structure with the repetition of five sectors of 72° each. No manufacturability constraint is considered.

The optimisation problem is solved by using Abaqus™2016 for the solution of the finite element model and a handwritten code to read the results, implement the optimisation algorithm and write the new input file. The handwritten code is realised by using Python and Matlab™.

The results of the optimisation process, with a overall target volume fraction of 0.5, are shown in Figs. 13.33 and 13.34. Figure 13.33 shows the assignment of the shared part of the domain, while Fig. 13.34 shows the optimised material distribution

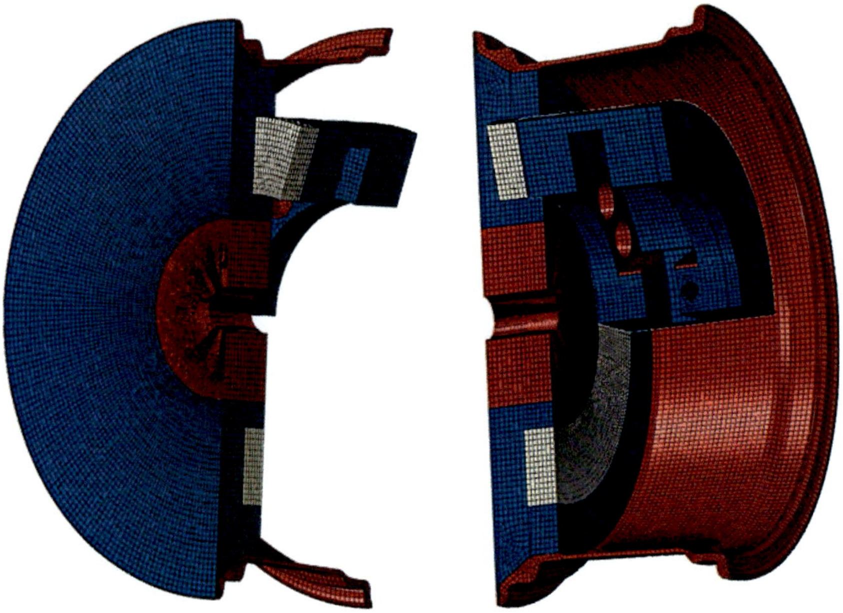

Fig. 13.32 Wheel and brake calliper assembly. Red: non-design domains. Blue: non-shared parts of the design domain. White: shared part of the design domain. Only half of the wheel is depicted. Notice that given the rotation of the wheel, the shared part of the design domain has two different geometries for the two bodies

Table 13.12 Results of the optimisation of the wheel and brake calliper assembly with different divisions of the shared domain

Shared domain division	% of mass on brake calliper	% of mass on wheel	Brake calliper compliance*	Wheel compliance*	Total compliance*
Optimal solution	16.8	83.2	0.256	0.744	1.000
Shared domain assigned wheel	18.0	82.0	0.455	0.682	1.137
Shared domain assigned brake calliper	16.2	83.8	0.257	1.202	1.459

* Normalised with respect to the total compliance of the "Optimal solution"

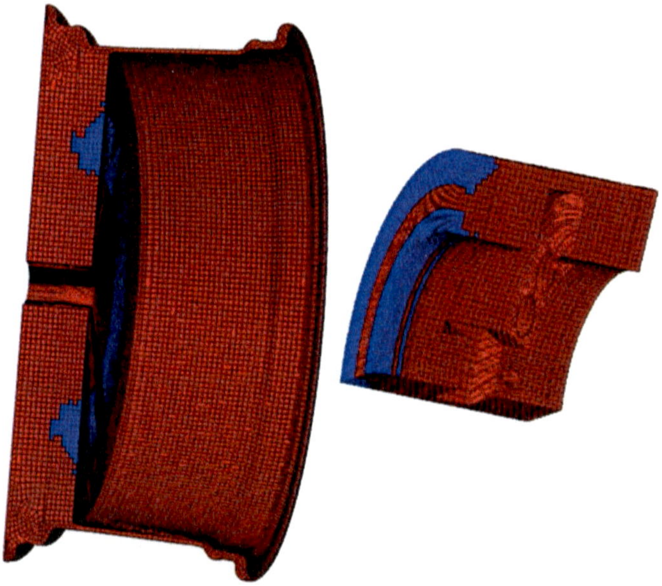

Fig. 13.33 Optimisation results—division of the design space. Red: elements of Ω_i and belonging to $\overline{\Omega_i}$ (with $i = 1, 2$). Blue: elements of Ω_i but not belonging to $\overline{\Omega_i}$ (with $i = 1, 2$). Left: wheel. Right: brake calliper

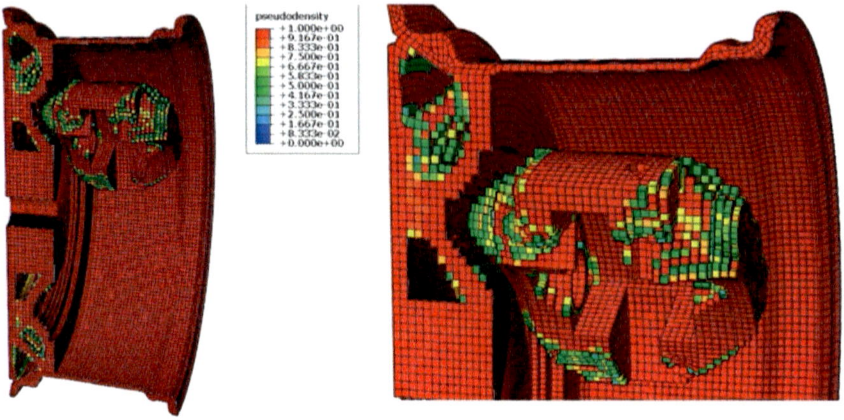

Fig. 13.34 Results of the optimisation of the wheel and brake calliper assembly. The elements with pseudodensity lower than 0.5 have been removed

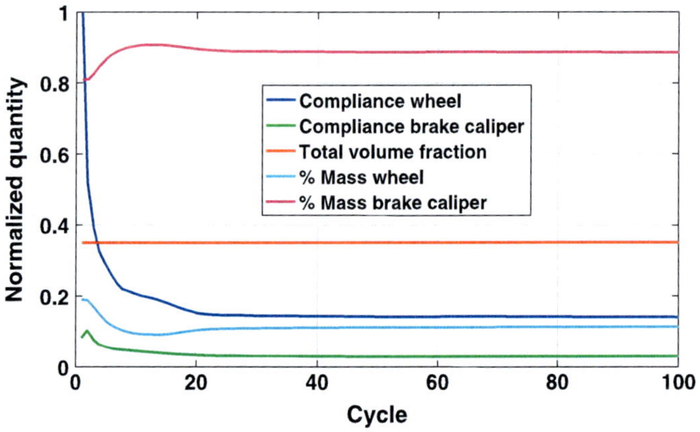

Fig. 13.35 Evolution of relevant variables during the optimisation process

removing all the elements with pseudodensity lower than 0.5. The obtained volume fractions for the two bodies are 0.431 and 0.516 for the brake calliper and wheel, respectively.

In Table 13.12 the results of the optimisation of the wheel and brake calliper assembly for different divisions of the shared part of the domain are reported. The considered divisions are the optimal one, the shared domain assigned completely to the wheel and the shared domain assigned completely to the brake calliper. From Table 13.12, it can be observed that the optimal division shows a lower level of compliance of the other two divisions. Moreover, as found for Example 1 in Sect. 12.3.1, not only the optimal solution has the lowest total compliance, but also the lowest values for the compliances of each of the two bodies. This shows that also in this complex three-dimensional problem the proposed algorithm is able to identify an effective division of the domain.

Figure 13.35 shows the evolution of some relevant variables during the optimisation problem. As expected, the greater variation of the variables is in the first few cycles and then a convergence configuration is reached. From the figure, it can be observed that for each body the mass varies during the optimisation process, while the overall volume fraction is constant.

References

1. Achtziger W (1998) Multiple-load truss topology and sizing optimization: some properties of minimax compliance. J Optim Theory Appl 98(2):255–280. https://doi.org/10.1023/A:1022637216104
2. Aghezzaf B, Hachimi M (1999) Second-order optimality conditions in multiobjective optimization problems. J Optim Theory Appl 102(1):37–50. https://doi.org/10.1023/A:1021834210437
3. Allaire G (2001) Shape optimization by the homogenization method. Springer, Berlin, Heidelberg, New York
4. Allaire G, Jouve F, Maillot H (2004a) Topology optimization for minimum stress design with the homogenization method. Struct Multidiscip Optim 28:87–98. https://doi.org/10.1007/s00158-004-0442-8
5. Allaire G, Jouve F, Toader A (2004b) Structural optimization using sensitivity analysis and a level-set method. J Comput Phys 194:363–393
6. Andreassen E, Clausen A, Schevenels M, Lazarov BS, Sigmund O (2011) Efficient topology optimization in MATLAB using 88 lines of code. Struct Multidiscip Optim 43(1):1–16. https://doi.org/10.1007/s00158-010-0594-7
7. Anton H, Rorres C (2005) Elementary linear algebra. Wiley, Hoboken
8. Arienti R, Cantoni C, Passoni R, Ballo F, Gobbi M, Mastinu G (2017) BRAKE DEVELOPMENT METHODS - Topology optimization of a brake caliper and upright of a race car. In: Pfeffer PDPE (ed) 7th International Munich Chassis Symposium 2016. Springer Fachmedien Wiesbaden, Wiesbaden, pp 655–668
9. Arora J (2004) Introduction to optimal design. Elsevier Academic Press, Cambridge
10. Ashby M (2011) Materials selection in mechanical design, 4th edn. Butterworth-Heinemann, Oxford, UK
11. Askar S, Tiwari A (2011) Finding innovative design principles for multiobjective optimization problems. IEEE Trans Syst Man Cybern-Part C: Appl Rev 41(4):554–559. https://doi.org/10.1109/TSMCC.2010.2081666
12. Askar SS, Tiwari A (2009) Finding exact solutions for multi-objective optimisation problems using a symbolic algorithm. In: IEEE congress on evolutionary computation
13. Atrek E, Agarwal B (1992) Shape optimization of structural design. In: Scientific excellence in supercomputing-The IBM 1990 contest prize papers. University of Georgia, Athens, GA
14. Atrek E, Kodali R (1989) Shape optimization of structural design. In: Prasad B (ed) CAD/CAM robotics and factories of the future, vol 2, pp 11–15. Springer, Berlin

F. M. Ballo et al., *Optimal Lightweight Construction Principles*,
https://doi.org/10.1007/978-3-030-60835-4

15. Ballio G, Mazzolani FM (1988) Steel structures (in Italian). Hoepli, Milano
16. Ballo F, Gobbi M, Mastinu G, Pishdad A (2013) Lightweight design of a brake caliper. In: Proceedings of the 2013 international design engineering techincal conference (IDETC), Portland, Oregon, USA, paper no. DETC2013-12431
17. Ballo F, Gobbi M, Massera M, Mastinu G (2014) A race motorcycle frame: advanced design. In: Proceedings of the 2014 international design engineering technical conferences & computers and information in engineering conference (IDETC/CIE 2014), Buffalo, New York, USA, paper no. DETC2014-34339
18. Ballo F, Gobbi M, Mastinu G, Previati G (2016a) Motorcycle tire modeling for the study of tire-rim interaction. J Mech Design 138(5):051404. https://doi.org/10.1115/1.4032470, http://mechanicaldesign.asmedigitalcollection.asme.org/article.aspx?doi=10.1115/1.4032470
19. Ballo F, Mastinu G, Gobbi M, Milano P (2016b) Lightweight design of a racing motorcycle wheel. In: SAE Technical paper, 10.4271/2016-01-1576.Copyright
20. Ballo F, Gobbi M, Previati G (2017) Optimal design of a beam subject to bending: a basic application. Meccanica 52(15), 3563–3576. https://doi.org/10.1007/s11012-017-0682-5
21. Banichuk NV (1990) Introduction to optimization of structures. Springer, New York
22. Banichuk NV, Ragnedda F, Serra M (2002) Optimum shapes of bar cross-sections. Struct Multidiscip Optim 23(3):222–232. https://doi.org/10.1007/s00158-002-0180-8
23. Bazant ZC, Cedolin L (2010) Stability of structures: elastic, inelastic fracture and damage theories. World Scientific Publishing Company, Singapore
24. Beghini L, Beghini A, Katz N, Baker W, Paulino G (2014) Connecting architecture and engineering through structural topology optimization. Eng Struct 59:716–726. https://doi.org/10.1016/j.engstruct.2013.10.032
25. Bendsøe M, Kikuchi N, (1988) Generating optimal topologies in structural design using a homogenization method. Comput Meth Appl Mech Eng 71:197–224
26. Bendsøe M, Diaz A, Kikuchi N (1993) Topology and generalized layout optimization of elastic structures. In: Bendsøe MP, Mota Soares CA (eds) Topology design of structures, pp 159–205
27. Bendsøe M, (1989) Optimal shape design as a material distribution problem. Struct Optim 1:193–202
28. Bendsøe M, Sigmund O (2003) Topology optimization: theory methods and applications. Springer, Berlin
29. Bendsøe MP (1995) Optimization of structural topology, shape, and material. Springer, Berlin, Heidelberg
30. Benfratello S, Giambanco F, Palizzolo L, Tabbuso P (2013) Optimal design of steel frames accounting for buckling. Meccanica 48(9):2281–2298. https://doi.org/10.1007/s11012-013-9745-4
31. Benoist J (1998) Connectedness of the efficient solutions for vector maximization problems. J Optim Theory Appl 96(3)
32. Beretta S, Citti P, Capitani R, Paolini G (1999) Quality reliability and safety of mechanical structures and systems (in Italian). CUSL, Firenze
33. Björck Å (1996) Numerical methods for least squares problems. Society for industrial and applied mathematics, Philadelphia USA
34. Björkman M, Holmström K (1999) Global optimization using DIRECT algorithm in matlab. Adv Model Optim 1(2):17–37. http://citeseerx.ist.psu.edu/viewdoc/summary?doi=10.1.1.71.6135
35. Breuer B, Bill K et al (2009) Brake technology handbook. SAE international
36. Bruns T (2005) A reevaluation of the simp method with filtering and an alternative formulation for solid-void topology optimization. Struct Multidiscip Optim 30:428–436
37. Buhl T (2002) Simultaneous topology optimization of structure and supports. Struct Multidiscip Optim 23(5):336–346. https://doi.org/10.1007/s00158-002-0194-2
38. Caballero R, Cerdá E, Muñoz MM, Rey L, Stancu-Minasian IM (2001) Efficient solution concepts and their relations in stochastic multiobjective programming. J Optim Theory Appl 110(1):53–74. https://doi.org/10.1023/A:1017591412366

39. Campbell C (1981) Automobile suspensions. Chapman and Hall, London
40. Cantoni C, Cesarini R, Mastinu G, Rocca G, Sicigliano R (2009) Brake comfrot—a review. Int J Veh Mech Mobil 47:901–947
41. Carlsson C, Fullér R (1995) Multiple criteria decision making: The case for interdependence. Computers & Operations Research 22(3):251–260. https://doi.org/10.1016/0305-0548(94)E0023-Z, https://www.sciencedirect.com/science/article/pii/0305054894E0023Z
42. Cavallaro PV, Sadegh AM, Island R (2006) Air-Inflated Fabric Structures Naval Undersea Warfare Center Division. Tech. Rep, November, Naval Undersea Warfare Center Division, Newport, Rhode Island
43. Cavazzuti M, Baldini A, Bertocchi E, Costi D, Torricelli E, Moruzzi P (2011) High performance automotive chassis design: a topology optimization based approach. Structural and Multidisciplinary Optimization 44:45–56
44. Cedolin L, Bazant Z (1991) Stability of Structures. Oxford University Press, New York
45. Chawla N, Deng X (2005) Microstructure and mechanical behavior of porous sintered steels. Materials Science and Engineering: A 390(1–2):98–112. https://doi.org/10.1016/J.MSEA.2004.08.046 https://www.sciencedirect.com/science/article/pii/S0921509304010111?via%3Dihub
46. Chen W, Wiecek MM, Zhang J (1999) Quality utility-a compromise programming approach to robust design. Journal of Mechanical Design, Transactions of the ASME 121(2):179–187. https://doi.org/10.1115/1.2829440
47. Chen W, Sahai A, Messac A, Sundararaj GJ (2000) Exploration of the effectiveness of physical programming in robust design. Journal of Mechanical Design, Transactions of the ASME 122(2):155–163. https://doi.org/10.1115/1.533565
48. Cheng G (1981) On non-smoothness in optimal design of solid, elastic plates. International Journal of Solids and Structures 17:795–810
49. Cheng G, Olhoff N (1981) Regularized formulation for optimal design of axisymmetric plates. International Journal of Solids and Structures 18:153–169
50. Cherkaev A (2000) Variational Methods for Structural Optimization. Springer-Verlag, New York
51. Chuang C, Yang R, Li G, Mallela K (2008) Multidisciplinary design optimization on vehicle tailor rolled blank design. Structural and Multidisciplinary Optimization 35:551–560
52. Clark JW, Hill HN (1962) Lateral buckling of beams and girders. Transactions ASCE 127:Part II
53. Clausen A, Aage N, Sigmund O (2014) Topology optimization with flexible void area. Structural and Multidisciplinary Optimization 50(6):927–943. https://doi.org/10.1007/s00158-014-1109-8
54. Comer RL, Levy S (1963) Deflections of an inflated circular cylindrical cantilever beam. AIAA Journal 1(7):1652–1655. https://doi.org/10.2514/3.1873
55. Conn AR, Gould NIM, Toint PL (2000) Trust Region Methods. Society for Industrial and Applied Mathematics, MPS/SIAM Series on Optimization
56. Cook R (1995) Finite Element Modeling for Stress Analysis. Wiley
57. Cowper GR (1966) The Shear Coefficient in Timoshenko's Beam Theory. ASME J Appl Mech 33(2):335–340. https://doi.org/10.1115/1.3625046
58. Crate H, Batdorf S, Baab G (1944) The effect of internal pressure on the buckling stress of thin-walled circular cylinders under torsion. Tech. Rep. L4E27, National Advisory Committee for Aeronautics, Langley Field, VA, USA
59. Dagli CH, Kusiak A (eds) (1994) Intelligent Systems in Design and Manufacturing. A S M E Press
60. Deaton J, Grandhi R (2014) A survey of structural and multidisciplinary continuum topology optimization: post 2000. Structural and Multidisciplinary Optimization 49:1–38
61. van Dijk N, Maute K, Langelaar M, van Keulen F (2013) Level-set methods for structural topology optimization: a review. Structural and Multidisciplinary Optimization 48:437–472. https://doi.org/10.1007/s00158-013-0912-y

62. d'Ippolito R, Hack M, Donders S, Hermans L, Tzannetakis N, Vandepitte C (2009) Improving the fatigue life of a vheicle knuckle with a reliability-based design optimization approach. Journal of Statistical Planning and Inference 139:1619–1632
63. Donnell L (1935) Stability of thin-walled tubes under torsion. Tech. Rep. 479, National Advisory Committee for Aeronautics, Langley Field, VA, USA
64. Edgeworth FY (ed) (1881) Mathematical Psychics: An Essay on the Application of Mathematics to the Moral Sciences. Paul & co, C.K
65. Eggert RJ, Mayne RW (1993) Probabilistic optimal design using successive surrogate probability density functions. Journal of Mechanical Design, Transactions of the ASME 115(3):385–391. https://doi.org/10.1115/1.2919203
66. Ehrgott M, Klamroth K (1997) Connectedness of efficient solutions in multiple criteria combinatorial optimization. European Journal of Operational Research 97(1):159–166. https://doi.org/10.1016/S0377-2217(96)00116-6 https://www.scopus.com/inward/record.uri?eid=2-s2.0-0031075124&doi=10.1016%2FS0377-2217%2896%2900116-6&partnerID=40&md5=42c359b03201ee7cc8ce468595b0cc57
67. EN-10034 (1995) EN 10034: Structural steel I and H sections. Tolerances on shapes and dimensions
68. EN-10219-2 (2006) EN 10219-2: Cold formed welded structural hollow sections of non-alloy and fine grain steels - Tolerances, dimensions and sectional properties
69. EN-1993-1-1 (2005) EN 1993-1-1 Eurocode3: Design of steel structures-Part 1-1: General rules and rules for buildings
70. Erfani T, Utyuzhnikov SV, Kolo B (2013) A modified directed search domain algorithm for multiobjective engineering and design optimization. Structural and Multidisciplinary Optimization 48(6):1129–1141. https://doi.org/10.1007/s00158-013-0946-1
71. Ermolaeva N, Castro M, Kandachar P (2004) Materials selection for an automotive structure by integrating structural optimization with environmental impact assessment. Materials and Design 25:689–698
72. Eschenauer H (2001) Multidisciplinary optimization procedure in design processes. In: Blachut J, Eschenauer H (eds) Emerging Methods for Multidisciplinary Optimization, CISM International Centre for Mechanical Sciences, vol 425, pp 1–51
73. Eschenauer H, Koski J, Osyczka A (eds) (1990) Multicriteria Design Optimization. Springer-Verlag, Berlin Heidelberg
74. Eschenauer HA, Olhoff N (2001) Topology optimization of continuum structures: A review. Applied Mechanics Reviews 54(4):331. https://doi.org/10.1115/1.1388075 http://appliedmechanicsreviews.asmedigitalcollection.asme.org/article.aspx?articleid=1396619
75. Fadel G, Li Y (2002) Approximating the Pareto curve to help solve biobjective design problems. Structural and Multidisciplinary Optimization 23(4):280–296. https://doi.org/10.1007/s00158-002-0186-2
76. Fang J, Gao Y, Sun G, Zhang Y, Li Q (2014) Parametric analysis and multiobjective optimization for functionally graded foam-filled thin-wall tube under lateral impact. Computational Materials Science 90(Complete):265–275, 10.1016/j.commatsci.2014.03.044
77. Ferreira A (2009) MATLAB Codes for Finite Element analysis. Springer
78. Fichter WB (1966) A theory for inflated thin-wall cylindrical beams. Tech. rep, NASA
79. Floudas CA, Pardalos P (eds) (2009) Encyclopedia of Optimization, 2nd edn. Springer, US
80. Fonseca CM, Fleming PJ (1998) Multiobjective optimization and multiple constraint handling with evolutionary algorithms - Part I: A unified formulation. IEEE Transactions on Systems, Man, and Cybernetics Part A: Systems and Humans 28(1):26–37. https://doi.org/10.1109/3468.650319
81. Genta G, Morello L (2009) The automotive chassis, vol 1: Components design. Springer
82. Gero JS (1990) Design prototypes. A knowledge representation schema for design. AI Magazine 11(4):26–36, https://www.scopus.com/inward/record.uri?eid=2-s2.0-0025544968&partnerID=40&md5=2706bfcd4bbb634aef4fdf3af11408d8
83. GmbH RB (2011) Automotive Handbook, 8th edn. Wiley

84. Gobbi M (2013) A k, k-ϵ optimality selection based multi objective genetic algorithm with applications to vehicle engineering. Optimization and Engineering 14(2):345–360. https://doi.org/10.1007/s11081-011-9185-8
85. Gobbi M, MASTINU G, (2001) ANALYTICAL DESCRIPTION AND OPTIMIZATION OF THE DYNAMIC BEHAVIOUR OF PASSIVELY SUSPENDED ROAD VEHICLES. Journal of Sound and Vibration 245(3):457–481. https://doi.org/10.1006/JSVI.2001.3591 https://www.sciencedirect.com/science/article/pii/S0022460X01935916
86. Gobbi M, Mastinu G (2001) On the optimal design of composite material tubular helical springs. Meccanica 36(5):525–553. https://doi.org/10.1023/A:1015640909013
87. Gobbi M, Mastinu G (2003) Stochastic multi-objective optimisation for the design of vehicle systems. In: Proceedings of the WCSMO5, Italy
88. Gobbi M, Mastinu G, Caudano M (2003) Stochastic multi-objective optimisation of a gearbox synchroniser and selector mechanism. In: American Society of Mechanical Engineers, Design Engineering Division (Publication) DE, vol 116, pp 113–124, 10.1115/IMECE2003-43005, https://www.scopus.com/inward/record.uri?eid=2-s2.0-1942521010&doi=10.1115%2fIMECE2003-43005&partnerID=40&md5=322bbe88e5886138a6a4182e2a4af38f, cited By 7
89. Gobbi M, Levi F, Mastinu G (2006) Multi-objective stochastic optimisation of the suspension system of road vehicles. Journal of Sound and Vibration 298:1055–1072
90. Gobbi M, Levi F, Mastinu G, Previati G (2014) On the analytical derivation of the Pareto-optimal set with an application to structural design. Structural and Multidisciplinary Optimization -:–
91. Gobbi M, Mastinu G, Ballo F, Previati G (2015) Race Motorcycle Smart Wheel. SAE International Journal of Passenger Cars - Mechanical Systems 8(1):119–127. https://doi.org/10.4271/2015-01-1520
92. Gobbi M, Previati G, Ballo F, Mastinu G (2016) Bending of beams of arbitrary cross sections - optimal design by analytical formulae. Structural and Multidisciplinary Optimization. https://doi.org/10.1007/s00158-016-1539-6 http://link.springer.com/10.1007/s00158-016-1539-6
93. Grujicic M, Arakere G, Bell W, Marvi H, Yalavarthy H, Pandurangan B, Haque I, Fadel G (2010) Reliability-based design optimization for durability of ground vehicle suspension system components. Journal of Materials Engineering and Performance 19:301–313
94. Gruttman F, Wagner W (2001) Shear correction factors in timoshenko's beam theory for arbitrary saped cross-sections. Computational Mechanics 27:199–207
95. Guest J, Prévost J, Belytschko T (2004) Achieving minimum length scale in topology optimization using nodal design variables and projection functions. International Journal for Numerical Methods in Engineering 61:238–254. https://doi.org/10.1002/nme.1064
96. Guo X, Cheng GD (2010) Recent development in structural design and optimization. Acta Mechanica Sinica 26(6):807–823. https://doi.org/10.1007/s10409-010-0395-7 http://link.springer.com/10.1007/s10409-010-0395-7, 1010.1724
97. Ha Y, Cho S (2006) Design sensitivity analysis and topology optimization of eigenvalue problems for piezoelectric resonators. Smart Materials and Structures 15:1513–1524
98. Haedir J (2012) An application of analytical multi-objective optimization to truss structures. In: Structural Stability Research Council Annual Stability Conference
99. Haftka R, Gurdal Z (2002) Elements of Structural Optimization. Kluver, Dordrecht
100. Haimes YY, Lasdon LS, Wismer D (1971) On a Bicriterion Formulation of the Problems of Integrated System Identification and System Optimization. IEE Transactions on system, man, and cybernetics 1(3):296–297
101. Hemp W (1973) Optimum structures. Clarendon Press
102. Hongwei Z, Xiaokai C, Yi L, Mingjun Z (2008) Topology optimization for air suspension bracket integrated shape optimization. In: IEEE Vehicle Power and Propulsion Conference (VPPC), Harbin, China
103. Hubka V (1976) Theorie der Konstruktions-Processe. Springer, Berlin
104. Jeong S, Yi S, Kan C, Nagabhushana V, Park G (2008) Structural optimization of an automobile roof structure using equivalent static loads. Proceedings of the Institution of Mechanical Engineers, Part D: Journal of Automobile Engineering 222:1985–1995

105. Jeong S, Park S, Choi D, Yoon G (2012) Topology optimization considering static failures theories for ductile and brittle materials. Computers and Structures 110–111:116–132. https://doi.org/10.1016/j.compstruc.2012.07.007
106. Kasperska R, Magnucki K, Ostwald M (2007) Bicriteria optimization of cold-formed thin-walled beams with monosymmetrical open cross sections under pure bending. Thin-Walled Structures 45(6):563–572. https://doi.org/10.1016/J.TWS.2007.05.002 https://www.sciencedirect.com/science/article/pii/S0263823107001012?via%3Dihub
107. Kawamoto A, Matsumori T, Yamasaki S, Nomura T, Kondoh T, Nishiwaki S (2011) Heaviside projection based topology optimization by a pde-filtered scalar function. Structural and Multidisciplinary Optimization 44:19–24. https://doi.org/10.1007/s00158-010-0562-2
108. Kim C, Sun H (2012) Topology optimization of gas flow channel routes in an automotive fuel cell. international Journal of Automotive Technology 13:783–789, 10.1007/s12239-012-0078-4
109. Kim DS, Lee GM, Lee BS, Cho SJ (2001) Counterexample and Optimality Conditions in Differentiable Multiobjective Programming. Journal of Optimization Theory and Applications 109:187–192
110. Kim G, Han S, Lee K (2014) Structural optimization of a knuckle with consideration of stiffness and durability requirements. The Scientific World Journal 2014:Article ID 763692
111. Kim T, Kim Y (2002) Multiobjective topology optimization of a beam under torsion and distortion. AIAA Journal 40:376–381
112. Kirsch U (1990) On the relationship between optimum structural topologies and geometries. Structural Optimization 2:39–45
113. Klein B (2009) Leichtbau- Konstruktion. Viewg+Teubner, Wiesbaden
114. Kohn R, Strang G (1986a) Optimal design and relaxation of variational problems. Communications on Pure and Applied Mechanics 39:113–137
115. Kohn R, Strang G (1986b) Optimal design in elasticity and plasticity. International Journal for Numerical Methods in Engineering 22:183–188
116. Kolesnikov AM, Zubov LM (2009) Large bending deformations of a cylindrical membrane with internal pressure. ZAMM Zeitschrift fur Angewandte Mathematik und Mechanik 89(4):288–305. https://doi.org/10.1002/zamm.200800182
117. Kočvara M, Stingl M (2012) Solving stress constrained problems in topology and material optimization. Structural and Multidisciplinary Optimization 46:1–15. https://doi.org/10.1007/s00158-012-0762-z
118. Kreissl S, Pingen G, Maute K (2011a) An explicit level set approach for generalized shape optimization of fluids with the lattice boltzmann method. International Journal for Numerical Methods in Engineering 65:496–519. https://doi.org/10.1002/fld.2193
119. Kreissl S, Pingen G, Maute K (2011b) Topology optimization for unsteady flow. International Journal for Numerical Methods in Engineering 87:1229–1253. https://doi.org/10.1002/nme.3151
120. Guo L, SY, Jiao H, (2013) Behavior of thin-walled circular hollow section tubes subjected to bending. Thin-Walled Structures 73:281–289
121. Lawry M, Maute K (2018) Level set shape and topology optimization of finite strain bilateral contact problems. International Journal for Numerical Methods in Engineering 113(8):1340–1369. https://doi.org/10.1002/nme.5582, 1701.06092
122. Lazarov B, Sigmund O (2011) Filters in topology optimization based on helmholtz-type differential equations. International Journal for Numerical Methods in Engineering 86:765–781. https://doi.org/10.1002/nme.3072
123. Le C, Norato J, Bruns T, Ha C, Tortorelli D (2010) Stress-based topology optimization for continua. Structural and Multidisciplinary Optimization 41:605–620. https://doi.org/10.1007/s00158-009-0440-y
124. Le Van A, Wielgosz C (2005) Bending and buckling of inflatable beams: Some new theoretical results. Thin-Walled Structures 43(8):1166–1187. https://doi.org/10.1016/j.tws.2005.03.005
125. Lee D, Lee J (2003) A structural optimization design for an aluminum-intensive vehicle. Proceedings of the Institution of Mechanical Engineers, Part D: Journal of Automobile Engineering 217:771–779

126. Lee S, Lee D, Lee J, Han C, Hedrick K (2007) Integrated process for structural-topological configuration design of weight-reduced vehicle components. Finite Elements in Analysis and Design 43:620–629
127. Levi F, Gobbi M (2006a) An Application of Analytical Multi-Objective Optimization to Truss Structures. In: 11th AIAA/ISSMO Multidisciplinary Analysis and Optimization Conference
128. Levi F, Gobbi M (2006b) An application of analytical multi-objective optimization to truss structures. In: Collection of Technical Papers - 11th AIAA/ISSMO Multidisciplinary Analysis and Optimization Conference, vol 2, pp 823–836, https://www.scopus.com/inward/record.uri?eid=2-s2.0-33846512333&partnerID=40&md5=e44e0af0aa2055bac3bde745b95dc10f
129. Levi F, Gobbi M, Mastinu G (2005) An application of multi-objective stochastic optimisation to structural design. Structural and Multidisciplinary Optimization 29(4):272–284. https://doi.org/10.1007/s00158-004-0456-2
130. Lewinski T, Rozvany G (2007) Exact analytical solutions for some popular benchmark problems in topology optimization ii: three-sided poligonal supports. Structural and Multidisciplinary Optimization 33:337–349
131. Lewinski T, Rozvany G (2008) Exact analytical solutions for some popular benchmark problems in topology optimization iii: L-shaped domains. Structural and Multidisciplinary Optimization 35:165–174
132. Lewinski T, Rozvany G, Sokól T, Bolbotowski K (2013) Exact analytical solutions for some popular benchmark problems in topology optimization iii: L-shaped domains revisited. Structural and Multidisciplinary Optimization 47:937–942
133. Ligarò SS, Barsotti R (2012) Shear deformations in inflated cylindrical beams: An old model revisited. Thin-Walled Structures 60:137–144. https://doi.org/10.1016/j.tws.2012.07.004
134. Lou M, Fang H, Hsia LM (2000) A combined analytical and experimental study on space inflatable booms. IEEE Aerospace Conference Proceedings 2:503–512. https://doi.org/10.1109/AERO.2000.878259
135. Lundquist E (1934) Strength tests on thin-walled duralumin cylinders. Tech. Rep. 473, National Advisory Committee for Aeronautics, Langley Field, VA, USA
136. Luo Z, Tong L, Luo J, Wei P, Wang MY (2009) Design of piezoelectric actuators using a multiphase level set method of piecewise constants. Journal of Computational Physics 228(7):2643–2659. https://doi.org/10.1016/J.JCP.2008.12.019 https://www.sciencedirect.com/science/article/pii/S0021999108006633
137. Lütkepohl H (1996) Handbook of matrices. John Wiley & Sons
138. Mai S, Fleck N, Lu T (2007) Optimal design of box-section sandwich beams in three-point bending. International Journal of Solids and Structures 44:4742–4769
139. Main LA, Peterson SW, Strauss AM (1994) Load-Deflection Behavior of Space-Based Inflatable Fabric Beams. Journal of Aerospace Engineering 7(2):225–238. https://doi.org/10.1061/(ASCE)0893-1321(1994)7:2(225)
140. Malm CG, Davids WG, Peterson ML, Turner AW (2009) Experimental characterization and finite element analysis of inflated fabric beams. Construction and Building Materials 23(5):2027–2034. https://doi.org/10.1016/j.conbuildmat.2008.08.029 http://dx.doi.org/10.1016/j.conbuildmat.2008.08.029
141. Marler RT, Arora JS (2004) Survey of multi-objective optimization methods for engineering. Structural and Multidisciplinary Optimization 26(6):369–395. https://doi.org/10.1007/s00158-003-0368-6
142. Marusciac I (1982) On the fritz john type optimality criterion in multi-objective goal programming. Revue d'Analyse Numerique et de Theorie de l'Approximation 11(1–2):
143. Mastinu G, Gobbi M, Miano C (2006) Optimal Design of Complex Mechanical Systems. Springer-Verlag
144. Mattson C, Messac A (2002) A non-deterministic approach to concept selection using s-pareto frontiers. In: Proceedings of the ASME Design Engineering Technical Conference, vol 2, pp 859–870, https://www.scopus.com/inward/record.uri?eid=2-s2.0-0036976941&partnerID=40&md5=3fd96fb6de0e0125986d3a57f74701a1, cited By 17

145. Mattson CA, Messac A (2003) Concept selection using s-pareto frontiers. AIAA Journal 41(6):1190–1198. https://doi.org/10.2514/2.2063
146. Mei Y, Wang X (2004) A level set method for structural topology optimization and its applications. Advances in Engineering Software 35(7):415–441. https://doi.org/10.1016/J.ADVENGSOFT.2004.06.004 https://www.sciencedirect.com/science/article/pii/S0965997804000705
147. Messac A, Ismail-Yahaya A (2002) Multiobjective robust design using physical programming. Structural and Multidisciplinary Optimization 23(5):357–371. https://doi.org/10.1007/s00158-002-0196-0
148. Meyer CD (2001) Matrix Analysis and Applied Linear Algebra. Society for Industrial and Applied Mathematics (SIAM)
149. Michelelena N, Jiang T, Papalambros P (1995) Design Laboratory Department of Mechanical Engineering and Applied Mechanics The University of Michigan Ann Arbor, MI 48109-2125, USA. Structural and Multidisciplinary Optimization
150. Michelena NF, Papalambros PY (1997) A Hypergraph Framework for Optimal Model-Based Decomposition of Design Problems. Computational Optimization and Applications 8(2):173–196. https://doi.org/10.1023/A:1008673321406 https://www.scopus.com/inward/record.uri?eid=2-s2.0-0031224168&doi=10.1023%2FA%3A1008673321406&partnerID=40&md5=9f4ee25adc62c43ae6ced8c89b55011c
151. Michell A (1904) The limits of economy of material in frame structures. Philosophical Magazine 8:89–597. https://doi.org/10.1080/14786440409463229
152. Miettinen K (1999) Nonlinear Multi-objective Optimization. Kluwer Academic Publishers, Boston
153. Mitschke M, Wallentowitz H (2004) Dynamik der Kraftfahrzeuge. Springer-Verlag, Berlin, Heidelberg, New York
154. Mohandas SU, Sandgren E (1989) Multiobjective optimization dealing with uncertainty. In: Proceedings of the 15th ASME Design Automation Conference, Montreal, Quebec, pp DE Vol. 19–2
155. Naccache PH (1978) Connectedness of the set of nondominated outcomes in multicriteria optimization. Journal of Optimization Theory and Applications 25(3):459–467. https://doi.org/10.1007/BF00932907
156. Narayanan S, Azarm S (1999) On improving multiobjective genetic algorithms for design optimization. Structural Optimization 18:146–155
157. Olhoff N, Du J (2009) On topological design optimization of structures against vibration and noise emission. In: Sandberg G, Ohayon R (eds) Computational Aspects of Structural Acoustics and Vibration, CISM International Centre for Mechanical Sciences, vol 505, pp 217–276
158. Olhoff N, Taylor J (1983) On structural optimization. Journal of Applied Mechanics 50:1134–1151
159. Olhoff N, Lurie K, Cherkaev A, Fedorov A (1981) Sliding regimes of anisotropy in optimal design of vibrating plates. International Journal of Solids and Structures 17:931–948
160. Ostwald M, Rodak M (2013) Multicriteria optimization of cold-formed thin-walled beams with generalized open shape under different loads. Thin-Walled Structures 65:26–33. https://doi.org/10.1016/J.TWS.2013.01.002 https://www.sciencedirect.com/science/article/pii/S0263823113000037
161. Pahl G, Beitz G, Feldhusen W, Grote KH (2007) Engineering Design: A Systematic Approach, third, edit edn. Springer, London
162. Pan F, Zhu P, Zhang Y (2010) Metamodel-based lightweight design of b-pillar with twb structure via support vector regression. Computers and Structures 88:36–44
163. Pantano A, Parks D, Boyce M (2004) Mechanics of deformation of single- and multi-wall carbon nanotubes. Journal of the Mechanics and Physics of Solids 52:789–821
164. Papalambros PY, Wilde DJ (2000) Principles of optimal design. Cambridge Universirity Press, Modeling and computation
165. Pareto V (ed) (1906) Manuale di economia politica. Editrice Libraria, Milano

166. Patel N, Kang BS, Renaud J, Tovar A (2009) Crashworthiness design using topology optimization. Journal of Mechanical Design 131:061013 (12 pages)
167. Pedersen N (2000) Maximization of eigenvalues using topology optimization. Structural and Multidisciplinary Optimization 20:2–11
168. Pedersen P, Pedersen NL (2009) Analytical optimal designs for long and short statically determinate beam structures. Structural and Multidisciplinary Optimization 39(4):343–357. https://doi.org/10.1007/s00158-008-0339-z
169. Petersson J (1999) Some convergence results in perimeter-controlled topology optimization. Computer Methods in Applied Mechanics and Engineering 171:123–140
170. Pilkey W (2004) Formulas for Stress, Strain, and Structural Matrices, 2nd edn. John Wiley & Sons, INC., Hoboken, New Jersey
171. Pingen G, Waidmann M, Evgrafov A, Maute K (2010) A parametric level-set approach for topology optimization of flow domains. Structural and Multidisciplinary Optimization 41:117–131. https://doi.org/10.1007/s00158-009-0405-1
172. Poulsen T (2002) A simple scheme to prevent checkerboard patterns and one-node connected hinges in topology optimization. Structural and Multidisciplinary Optimization 24:396–399. https://doi.org/10.1007/s00158-002-0251-x
173. Prager W, Rozvany G (1977) Optimization of the structural geometry. In: Bednarek A, Cesari L (eds) Dynamical systems (Proceedings of the International Conference, Gainsville, Florida), pp 265–293
174. Qian Z, Ananthasuresh GK (2004) Optimal embedding of rigid objects in the topology design of structures. Mechanics Based Design of Structures and Machines 32(2):165–193. https://doi.org/10.1081/SME-120030555
175. Ragnedda F, Serra M (2005) On optimum thin-walled closed cross section. Structural and Multidisciplinary Optimization 30(3):233–235. https://doi.org/10.1007/s00158-004-0501-1
176. Rajaram L, Sudharsan S (2005) Optimization of caliper housing using fem. SAE International 60: https://doi.org/10.4271/2005-26-060
177. Rao S (1974) A probabilistic approach to the design of gear trains. International Journal of Machine Tool Design and Research 14(3):267–278. https://doi.org/10.1016/0020-7357(74)90035-3 https://www.sciencedirect.com/science/article/pii/0020735774900353
178. Rao SS (1987) Multi-objective optimization of fuzzy structural systems. International Journal for Numerical Methods in Engineering 24(6):1157–1171. https://doi.org/10.1002/nme.1620240608 https://onlinelibrary.wiley.com/doi/abs/10.1002/nme.1620240608, https://onlinelibrary.wiley.com/doi/pdf/10.1002/nme.1620240608
179. Rendall TCS (2006) Static , Buckling and Dynamic Behaviour of Inflatable Beams. Aerospace pp 2006–2006, 10.2514/6.2006-1701
180. Rodriguez J, Seireg A (1988) A geometric rule-based methodology for shape synthesis: 2-d cases. In: Proc 1988 ASME Int Computers in Engineering Conf and Exposition, San Francisco, CA, vol 2, pp 35–41
181. Rondal J, Würker K, Dutta D, Wardenier J, Yeomans N (1992) Structural stability of hollow sections. Verlag TUV Rheinland, Koln
182. Rotter J, Sadowskj A, Chen L (2014) Nonlinear stability of thin elastic cylinders of different length under global bending. International Journal of Solids and Structures 51:2826–2839
183. Rozvany G (1972a) Grillages of maximum stress and maximum stiffness. International Journal of Mechanical Sciences 14:1217–1222
184. Rozvany G (1972b) Optimal load transmission by flexure. Computational Methods for Applied Mechanical Engineering 1:253–263
185. Rozvany G (1976) Optimal Design of Flexural Systems. Pergamon Press, Oxford
186. Rozvany G (1989) Structural design via optimality criteria. Kluwer
187. Rozvany G (1996) Some shortcomings in michell's truss theory. Structural Optimization 12:244–250
188. Rozvany G (1998) Exact analytical solutions for some popular benchmark problems in topology optimization. Structural Optimization 15:42–48

189. Rozvany G (2001a) Aims, scope, methods, history and unified terminology of computer-aided topology optimization in structural mechanics. Structural and Multidisciplinary Optimization 21:90–108
190. Rozvany G (2009) A critical review of estabilished methods of structural topology optimization. Structural and Multidisciplinary Optimization 37:217–237. https://doi.org/10.1007/s00158-007-0217-0 http://link.springer.com/10.1007/s00158-007-0217-0
191. Rozvany G (2011) On symmetry and non-uniqueness in exact topology optimization. Structural and Multidisciplinary Optimization 43:297–317. https://doi.org/10.1007/s00158-010-0564-0
192. Rozvany G, Olhoff N (2000) Topology Optimization of Structures and Composite Continua. Springer, Netherlands
193. Rozvany G, Bendsøe M, Kirsch U, (1995a) Layout optimization of structures. Applied Mechanics Reviews 48:41–119
194. Rozvany GIN (2001b) Stress ratio and compliance based methods in topology optimization - A critical review. Structural and Multidisciplinary Optimization 21(2):109–119. https://doi.org/10.1007/s001580050175
195. Rozvany GIN, Bendsøe MP, Kirsch U (1995b) Layout Optimization of Structures. Applied Mechanics Reviews 48(2):41–119. https://doi.org/10.1115/1.3005097 https://doi.org/10.1115/1.3005097
196. Scheibe F, Smith MC (2009) Analytical solutions for optimal ride comfort and tyre grip for passive vehicle suspensions. International Journal of Vehicle Mechanics and Mobility 47:1229–1252
197. Schmitendorf WE, Leitmann G (1974) A Simple Derivation of Necessary Conditions for Pareto Optimality. IEEE Transactions on Automatic Control 19(5):601–602. https://doi.org/10.1109/TAC.1974.1100673 https://www.scopus.com/inward/record.uri?eid=2-s2.0-0016117373&doi=10.1109%2FTAC.1974.1100673&partnerID=40&md5=43c8b2da2bc2be2ea59a96c7c32f5697
198. Schwerin E (1924) Torsional stability of thin-walled tubes. In: Proc. of the First International Congress for Applied Mechanics at Delft, Delft, Netherlands, pp 255–265
199. Segui WT (2007) Steel Design, fourth edi edn. Thomson, Toronto, Ontario, 10.1080/01436599913901
200. Selyugin SV, Chekhov VV (2001) Multimaterial design of physically nonlinear structures. Structural and Multidisciplinary Optimization 21(3):209–217. https://doi.org/10.1007/s001580050185 https://doi.org/10.1007/s001580050185
201. Sergent N, Tirovic M (2009) Brake caliper design optimisation. Institutions of Mechanical Engineers - Braking pp 131–138
202. Sethian J, Weigmann A (2000) Structural boundary design via level set and immersed interface methods. Journal of Computational Physics 163:489–528
203. Shi Y, Zhu P, Shen L, Lin Z (2007) Lightweight design of automotive front side rails with twb concept. Thin-Walled Structures 45:8–14
204. Shin M, Shin J, Hwang S, Park G (2007) Structural analysis and optimization of a low speed vehicle body. Proceedings of the Institution of Mechanical Engineers, Part D: Journal of Automobile Engineering 221:313–326
205. Shukla A, Misra A, Kumar S (2013) Checkerboard problem in finite element based topology optimization. International Journal of Advances in Engineering and Technology 6:1769–1774
206. Siddal JN (1983) Probabilistic engineering design: principles and applications. MArcel Dekker Inc, New York
207. Siddall JN (1984) A new approach to probability in engineering design and optimization. Journal of Mechanical Design, Transactions of the ASME 106(1):5–10. https://doi.org/10.1115/1.3258562
208. Siddiqui S, Azarm S, Gabriel SA (2012) On improving normal boundary intersection method for generation of Pareto frontier. Structural and Multidisciplinary Optimization 46(6):839–852. https://doi.org/10.1007/s00158-012-0797-1

209. Sigmund O (2001) A 99 line topology optimization code written in matlab. Structural and Multidisciplinary Optimization 21(2):120–127. https://doi.org/10.1007/s001580050176
210. Sigmund O, Petersson J (1998) Numerical instabilities in topology optimization: A survey on procedures dealing with checkerboards, mesh dependence and local minima. Structural Optimization 16:68–75
211. Simon C, Blume L (1994) Mathematics for Economists. W. W, Norton and Company
212. Song C, Lee J (2011) Reliability-based design optimization of knuckle component using conservative method of moving least squares meta-model. Probabilistic Engineering Mechanics 26:364–379
213. Spelsberg-Korspeter G (2010) Structural optimization for the avoidance of self-excited vibrations based on analytical models. Journal of Sound and Vibration 329:4829–4840
214. Stadler W (ed) (1988) Multicriteria Optimization in Engineering and in the Sciences. Springer, US
215. Stanford B, Beran P (2013) Aerothermoelastic topology optimization with flutter and buckling metrics. Structural and Multidisciplinary Optimization 48:149–171. https://doi.org/10.1007/s00158-013-0885-x
216. Stanikov R, Matusov J (1995) Multicriteria Optimization and Engineering. Springer
217. Steuer R, Sun M (1995) The parameter space investigation method of multiple objective nonlinear programming: a computational investigation. Operations Research 43:641–648
218. Stolpe M, Svanberg K (2001) An alternative interpolation scheme for minimum compliance topology optimization. Structural and Multidisciplinary Optimization 22:116–124
219. Stolpe M, Svanberg K (2004) A stress-constrained truss-topology and material-selection problem that can be solved by linear programming. Structural and Multidisciplinary Optimization 27(1–2):126–129. https://doi.org/10.1007/s00158-003-0364-x
220. Strömberg N (2018) Topology optimization of orthotropic elastic design domains with mortar contact conditions. In: Schumacher A, Vietor T, Fiebig S, Bletzinger K-U MK (ed) Advances in Structural and Multidisciplinary Optimization: Proceedings of the 12th World Congress of Structural and Multidisciplinary Optimization, Springer International Publishing, Braunschweig, Germany, pp 1427–1438, https://doi.org/10.1007/978-3-319-67988-4, http://link.springer.com/10.1007/978-3-319-67988-4
221. Suzuki K, Kikuchi N (1991) A homogenization method for shape and topology optimization. Computer Methods in Applied Mechanics and Engineering 93:291–318
222. Takayama A (1993) Analytical Methods in Economics. University of Michigan Press
223. Tang P, Chang K (2001) Integration of topology and shape optimization for design of structural components. Structural and Multidisciplinary Optimization 22:65–82
224. Tavakoli R, Mohseni SM (2014) Alternating active-phase algorithm for multimaterial topology optimization problems: A 115-line MATLAB implementation. Structural and Multidisciplinary Optimization 49(4):621–642. https://doi.org/10.1007/s00158-013-0999-1
225. Thomas JC, Wielgosz C (2004) Deflections of highly inflated fabric tubes. Thin-Walled Structures 42(7):1049–1066. https://doi.org/10.1016/j.tws.2004.03.007
226. Thompson JMT, Hunt CW (1974) DANGERS OF STRUCTURAL OPTIMIZATION. Engineering Optimization 1(2):99–110. https://doi.org/10.1080/03052157408960580 https://doi.org/10.1080/03052157408960580
227. Thota J, OToole B, Trabia M (2011) Optimization of shock response within amilitary vehicle space frame. Structural and Multidisciplinary Optimization44:847–861
228. Timoshenko S (1921) On the correction for shear of the differential equation for transverse vibration of prismatic bars. Phil Mag XLI:744–746
229. Veldman SL (2005) Design and analysis methodologies for inflated beams. DUP Science - Delft University Press, Delft, The Netherlands, http://adsabs.harvard.edu/abs/2005PhDT.........6V
230. Veldman SL (2006) Wrinkling prediction of cylindrical and conical inflated cantilever beams under torsion and bending. Thin-Walled Structures 44(2):211–215. https://doi.org/10.1016/j.tws.2006.01.003

231. Veldman SL, Bergsma OK, Beukers A (2005) Bending of anisotropic inflated cylindrical beams. Thin-Walled Structures 43(3):461–475. https://doi.org/10.1016/j.tws.2004.07.015
232. Vogiatzis P, Chen S, Wang X, Li T, Wang L (2017) Topology optimization of multi-material negative Poisson's ratio metamaterials using a reconciled level set method. Computer-Aided Design 83:15–32. https://doi.org/10.1016/J.CAD.2016.09.009 https://www.sciencedirect.com/science/article/pii/S0010448516301154
233. Wagner A, Spelsberg-Korspeter G, Hagedorn P (2014) Structural optimization of an asymmetric automotive brake disc with cooling channels to avoid squeal. Journal of Sound and Vibration 333:1888–1898
234. Wang C (2013) Optimization of torsion bars with rounded polygonal cross section. Journal of Engineering Mechanics 139:629–634, : 10.1061/ (ASCE)EM.1943-7889.0000447
235. Wang CG, Tan HF, Du XW, He XD (2010) A new model for wrinkling and collapse analysis of membrane inflated beam. Acta Mechanica Sinica 26(4):617–623. https://doi.org/10.1007/s10409-010-0348-1
236. Wang M, Wang X, Guo D (2003) A level set method for structural topology optimization. Computer Methods in Applied Mechanics and Engineering 192:227–246
237. Wang MY, Wang X (2004) "Color" level sets: a multi-phase method for structural topology optimization with multiple materials. Computer Methods in Applied Mechanics and Engineering 193(6–8):469–496. https://doi.org/10.1016/J.CMA.2003.10.008 https://www.sciencedirect.com/science/article/pii/S0045782503005644
238. Wang Y, Luo Z, Zhang X, Kang Z (2014) Topological design of compliant smart structures with embedded movable actuators. Smart Materials and Structures 23(4), 10.1088/0964-1726/23/4/045024
239. Wang Y, Luo Z, Zhang N, Wu T (2016) Topological design for mechanical metamaterials using a multiphase level set method. Structural and Multidisciplinary Optimization 54(4):937–952. https://doi.org/10.1007/s00158-016-1458-6 http://dx.doi.org/10.1007/s00158-016-1458-6
240. Wang Y, Gao J, Luo Z, Brown T, Zhang N (2017) Level-set topology optimization for multimaterial and multifunctional mechanical metamaterials. Engineering Optimization 49(1):22–42. https://doi.org/10.1080/0305215X.2016.1164853 https://doi.org/10.1080/0305215X.2016.1164853
241. Warburton AR (1983) Quasiconcave vector maximization: Connectedness of the sets of Pareto-optimal and weak Pareto-optimal alternatives. Journal of Optimization Theory and Applications 40(4):537–557. https://doi.org/10.1007/BF00933970 https://doi.org/10.1007/BF00933970
242. Webber J (1982) Deflections of inflated cylindrical cantilever beams subjected to bending and torsion. The Aeronautical Journal 86(858):306–312. https://doi.org/10.1017/S000192400001914X
243. Weingarten V, Seide P, Peterson J (1968) Buckling of Thin-walled Circular Cylinders. Tech. Rep. NASA SP-8007, NASA Space Vehicle Design Criteria, 19690013955
244. Wiedemann J (2007) Leichtbau Elemente und Konstruktion. Springer, Berlin
245. Xia L, Zhu J, Zhang W, Breitkopf P (2013) An implicit model for the integrated optimization of component layout and structure topology. Computer Methods in Applied Mechanics and Engineering 257:87–102. https://doi.org/10.1016/J.CMA.2013.01.008 https://www.sciencedirect.com/science/article/pii/S0045782513000194
246. Xia Q, Wang M, Shi T (2014) A level set method for shape and topology optimization of both structure and support of continuum structures. Computer Methods in Applied Mechanics and Engineering 272:340–353
247. Yang BS, Yeun YS, Ruy WS (2002) Managing approximation models in multiobjective optimization. Structural and Multidisciplinary Optimization 24(2):141–156. https://doi.org/10.1007/s00158-002-0224-0
248. Yang XQ, Jeyakumar V (1997) First and second-order optimality conditions for convex composite multiobjective optimization. Journal of Optimization Theory and Applications 95(1):209–224. https://doi.org/10.1023/A:1022695714596

249. Youn B, Choi K, Yang R, Gu L (2004) Reliability-based design optimization for crashworthiness of vehicle side impact. Structural and Multidisciplinary Optimization 26:272–283
250. Young WC, Budynas RG (2002) Roark's Formulas for Stress and Strain. Mc Graw Hill
251. Yulin M, Xiaoming W (2004) A level set method for structural topology optimization with multi-constraints and multi-materials. Acta Mechanica Sinica 20:507–518
252. Zhang Q, Zhang W, Zhu J, Gao T (2012) Layout optimization of multi-component structures under static loads and random excitations. Engineering Structures 43:120–128. https://doi.org/10.1016/J.ENGSTRUCT.2012.05.013 https://www.sciencedirect.com/science/article/pii/S0141029612002490
253. Zhang W, Zhong W, Guo X (2015) Explicit layout control in optimal design of structural systems with multiple embedding components. Computer Methods in Applied Mechanics and Engineering 290:290–313. https://doi.org/10.1016/J.CMA.2015.03.007 https://www.sciencedirect.com/science/article/pii/S0045782515001097
254. Zhang W, Yuan J, Zhang J, Guo X (2016a) A new topology optimization approach based on Moving Morphable Components (MMC) and the ersatz material model. Structural and Multidisciplinary Optimization 53(6):1243–1260. https://doi.org/10.1007/s00158-015-1372-3 http://dx.doi.org/10.1007/s00158-015-1372-3
255. Zhang W, Zhu J, Gao T (2016b) Topology optimization in engineering structure design. Elsevier, Oxford
256. Zhang WH, Yang HC (2002) Efficient gradient calculation of the Pareto optimal curve in multicriteria optimization. Structural and Multidisciplinary Optimization 23(4):311–319. https://doi.org/10.1007/s00158-002-0185-3
257. Zhou M, Rozvany G (1991) The coc algorithm, part ii: topological, geometrical and generalized shape optimization. Computer Methods in Applied Mechanics and Engineering 89:309–336
258. Zhou M, Shyy Y, Thomas H (2001) Checkerboard and minimum member size control in topology optimization. Structural and Multidisciplinary Optimization 21:152–158
259. Zhou S, Li W, Li Q (2010) Level-set based topology optimization for electromagnetic dipole antenna design. Journal of Computational Physics 229:6915–6930
260. Zhu JH, Zhang WH, Xia L (2016) Topology Optimization in Aircraft and Aerospace Structures Design. Archives of Computational Methods in Engineering 23(4):595–622. https://doi.org/10.1007/s11831-015-9151-2 http://link.springer.com/10.1007/s11831-015-9151-2
261. Zhu Z, Hu J, Yuan X, Sun H, Sun H (2010) Research on structural optimization of the aluminum alloy wheel. In: 2010 WASE international Conference on Information Engineering
262. Zhu ZH, Seth RK, Quine BM (2008) Experimental Investigation of Inflatable Cylindrical Cantilevered Beams. JP Journal of Solids and Structures 2(2):95–110, http://www.pphmj.com
263. Zhuang C, Xiong Z, Ding H (2010) Topology optimization of multi-material for the heat conduction problem based on the level set method. Engineering Optimization 42:811–831
264. Ziemian RD (ed) (2010) Guide To Stability Design Criteria for Metal, 6th edn. John Wiley & Sons, INC., Hoboken, New Jersey
265. Zuo Z, Xie Y, Huang X (2011) Reinventing the wheel. Journal of Mechanical Design 133:024502 (4 pages)